WETLAND ARCHAEOLOGY & ENVIRONMENTS
REGIONAL ISSUES, GLOBAL PERSPECTIVES

WETLAND ARCHAEOLOGY & ENVIRONMENTS

REGIONAL ISSUES, GLOBAL PERSPECTIVES

Edited by

Malcolm Lillie and Stephen Ellis

Assistant Editor Helen Fenwick

Oxbow Books

Published by
Oxbow Books, Park End Place, Oxford

© Oxbow Books and the individual authors, 2007

A CIP record of this book is available from the British Library

ISBN 978-1-84217-154-7   1-84217-154-2

This book is available direct from

Oxbow Books, Park End Place, Oxford OX1 1HN
(Phone: 01865-241249; Fax: 01865-794449)

and

The David Brown Book Company
PO Box 511, Oakville, CT 06779, USA
(Phone: 860-945-9329; Fax: 860-945-9468)

or from our website
www.oxbowbooks.com

*Front cover: Flag Fen, © Malcolm Lillie; the Roos Carr model, Holderness,
East Yorkshire, © Hull and East Riding Museum.
Back cover: Drainage ditch through the waterlogged site of the Utti Batue Palace,
Luwu, Sulawesi, Indonesia © David Bulbeck.*

Printed in Great Britain by
Short Run Press Ltd, Exeter

For Lian, Paige and Mya

Contents

PART 1 INTRODUCTION

PART 2 ARCHAEOLOGY

List of Figures

List of Tables

Biographies for Main Authors

Nóra Bermingham has worked in Irish wetlands archaeology since 1992, and with the Irish Archaeological Wetlands unit between 1994–9. She joined the University of Hull as a research assistant in 2001, completing her PhD on the palaeoecology and palaeohydrology of a raised mire system at Kilnagarnagh in the Lemanaghan bog complex in central Ireland in 2005.

Doreen Bowdery received her PhD in Archaeology from the Australian National University in 1996. Her topic investigated the presence of phytoliths in semi arid Australia and their relationship to the archaeology, vegetation and climate change. A large rockshelter, occupied during the Pleistocene and Holocene and the surrounding area were a major study. Bowdery has analysed four 'Origin of Complex Society in South Sulawesi' project sites one of which was Utti Batue. Currently, phytolith analyses of sediments and other samples from different environmental areas are in progress encompassing Northern Territory coastal and arid sites, inland tropical Queensland and a remote Pacific island.

Jeff Blackford is in the Geography Department at Manchester University. His research includes palaeoecology, palaeohazards and climate change, and combining these in studies of prehistoric environmental change. Current projects include environmental change in the Faroe Islands before and during the first human occupation, tephrochronology and volcanic impacts in Alaska, peat bogs and climate change, and the later Mesolithic period in upland Britain.

David Bulbeck received his PhD in Archaeology from the Australian National University (ANU) in 1993. In 1997 he commenced a postdoctoral fellowship at the ANU, funded by the Australian Research Council (ARC), on the "Origin of Complex Society in South Sulawesi" project. This involved joint excavations with Indonesian archaeologists in the shire of Luwu, combined with historical research on the ancient kingdom of Luwu by the English historian Dr Ian Caldwell. In 2002 Bulbeck commenced a second ARC fellowship on the project "The Contribution of South Asia to the Peopling of Australasia" co-ordinated with his senior ANU colleague Professor Colin Groves. This project is investigating the human fossil record from South Asia, Indo-Malaysia and Australia, from about 40,000 years ago to the present, in terms of its evidence for choosing between the "Out of Africa" and "Multi-regional Evolution" theories on the origins of our species.

Jane Bunting completed her PhD in the Sub-Department of Quaternary Research, at Cambridge, and then undertook post-doctoral research in Canada (University of Waterloo, Ontario) and Scotland (University of Stirling, Scotland) before joining the University of Hull in 1997, where she is now a senior lecturer. Between August 2003 and July 2006 she was the director of the Wetland Archaeology and Environments Research Centre at Hull.

Ciara Clarke [BA (mod) hons, MSc, PhD, MIFA] is post excavation divisional manager with AOC Archaeology in Edinburgh. She specialises in palynological (pollen and fungal spores) analyses, and has also conducted several studies in evaluating the archaeological potential of the Scottish peatlands. She has reported and published widely on these subjects.

Mike Corfield is a heritage science and conservation consultant. He has worked in the heritage sector since 1969. After re-joining English Heritage in 1991 he focussed on the preservation of archaeological remains *in situ*, monitoring sites such as the Rose Theatre, amongst others. He initiated two Preserving Archaeological Remains in Situ conferences in 1996 and 2002, and has published numerous papers on the preservation and conservation of archaeological remains.

Anne Crone [BA, PhD, FSA Scot, MIFA] is a post-excavation manager with AOC Archaeology in Edinburgh. She specialises in the analysis of wood from archaeological and historic contexts, and has conducted several fieldwork and research projects on Scottish crannogs. She has published widely on these subjects.

Paul Davies gained a doctorate in environmental archaeology from Cardiff University and subsequently worked on the inaugural year of the Humber Wetlands Project at the University of Hull. He then took up a lectureship at Bath Spa University teaching ecology and palaeoecology. In 1992 he became Head of the Graduate School at Bath Spa. His main research interests are in the application of sub-fossil molluscan analysis to archaeology and land-use histories and in Mesolithic and Neolithic landscape change.

Steven Ellis has research interests in soils and Quaternary environmental change, including wetland responses to landuse change at various timescales. He was director of English Heritage's Humber Wetlands Project at the University of Hull between 1992–2000, and is a member of the university's Wetland Archaeology & Environments Research Centre.

Helen Fenwick is a lecturer in archaeology in the Department of History at the University of Hull, a post she took up on the 1st January 2004. Prior to this she was a Research Assistant on the English Heritage Funded Humber Wetlands Project and Research Fellow at the Wetland Archaeology and Environments Research Centre, until July 2003, both based at the University of Hull. She is currently completing doctoral research on the landscape development and settlement evolution of the Lincolnshire Marsh.

John Field is a Senior Lecturer and Sub-Dean in the School of Resources, Environment and Society in the Faculty of Science, at the Australian National University. He received his PhD from the University of New England in Armidale for his research on the hydro-bio-geography of small rural catchments. He teaches and researches earth sciences, farm silviculture, soil formation and sustainable land management, and regolith and landscape evolution. He was a founding member of the Centre for Australian Regolith Studies, and the Coordinated Research Centre "Landscape Evolution and Mineral Exploration" (a Centre for Excellence funded by the Australian Research Council).

Huw Griffiths was born in Lincolnshire, raised in South Wales and was a zoologist specialising in mammals and freshwater ostracods. His work related primarily to environmental change and conservation. He was involved in active research both in the UK and abroad (notably Turkey, Slovenia, Spain and Cyprus), and had a long term interest in issues related to the study of biodiversity in the Balkans, and worked in Macedonia, Slovenia and Croatia. He published extensively on his main areas of interest including; Late Quaternary Environmental Change, palaeoecology and conservation status of Balkan Lake Dojran, ostracod evolution and extinction, and Mustelids in a modern world, Conservation and management aspects of small carnivore: human interactions. These represent a very small sample of the topics studied by Huw prior to his death.

Peter Halkon is a Lecturer in Archaeology in the Department of History University of Hull. His research interests include landscape archaeology in East Yorkshire's Foulness Valley with a focus on the Iron Age and Roman periods. Heritage Lottery funding has enabled the construction of a web-based virtual landscape 'Valley of the First Iron Masters' (www.ironmasters.hull.ac.uk) based on this research, which was completed to doctoral level in 2006.

Geoff Hope is a professor based in the Department of Archaeology and Natural History, Research School of pacific and Asian Studies, ANU Canberra. He has research interests in swamps, caves and lakes across Asia Pacific and the development of methods to measure environmental change, climate change and high resolution records. His work involves assessing the past impact of people on landscapes by measuring vegetation change (using phytoliths, charcoal and pollen) and geomorphic consequences – erosion, silting and shifts in production. Geoff is also interested in the roles of climate change and fire on human activity. He is currently assessing the long term fire regimes in east Kalimantan, New Guinea, New Caledonia, Myanmar and Fiji in relation to the very different human settlement histories, and is involved in measuring climate change in high altitude sites in Papua and PNG using glacial histories and marine records.

Jim Innes obtained his PhD at Durham in 1989, and he is currently employed as a lecturer in the Department of Geography at Durham University. His current research interests include human palaeoecology, particularly in relation to Mesolithic communities and their impact upon the environment. A recent focus has been on land-use and environment across the Mesolithic-Neolithic Transition, with a particularly emphasis on the palaeoecology of the introduction of cereal cultivation. He is working on the use of fungal spore analyses in palaeoecology and environmental archaeology, microcharcoal analyses, and the applications of very fine resolution palynology.

Sarah Jones completed a PhD in Quaternary Science at Queen Mary. She worked on sea-level changes in northern France and south east England, including analyses of pollen, diatoms and foraminifera. She now works for the Royal Geographical Society/Institute of British Geographers in London.

Abby Kelly completed an MSc in Environmental Modelling and Management at King's College, London and then worked as a palaeoecology research assistant at Queen Mary, University of London, working on web site development and fungal spore projects. She now works for the Groundwork Trust.

Lars Larsson is a Professor at the Department for Archaeology and Ancient History, University of Lund, southern Sweden. He has been doing research in the Mesolithic of Southern Sweden with excavations of bog sites and large coastal sites including cemeteries. His research also includes the Late Palaeolithic of southern Scandinavia, Mesolithic shell middens in southern part of Portugal and the transition between the Middle and Later Stone Age of Northern Zimbabwe.

Malcolm Lillie is a lecturer in the Department of Geography at the University of Hull. He obtained his PhD in 1998 after transferring to part-time research following his appointment to the English Heritage funded Humber Wetlands Survey in 1994. He has been working and researching in wetlands since the late 1980's. Between September 2001 and July 2003 he set-up and Directed the Wetland Archaeology & Environments Research Centre at Hull University. He is currently undertaking projects studying *in situ* preservation in both Scotland and England, and is a researcher on the AHRB funded Vedrovice Project, which is studying the nature of the agricultural transition in the Czech Republic. His research focuses on the study of hunter-gatherers and early farmers alongside research into characterising waterlogged burial environments and the impacts of aggregates extraction on the archaeological resource of wetlands.

Rob Marchant's research interests lie within the themes of vegetation dynamics and ecosystem change. In particular, he uses palaeoecology, vegetation modelling, archaeological, biogeographical and ecological data to determine the role of past events in shaping the present day composition and distribution of tropical vegetation. He is presently co-coordinating the York Institute for Tropical Ecosystem Dynamics (KITE) – a Marie-Curie Excellence Centre located in the Environment Department at the University of York, focusing research on the Eastern Arc Mountains of Kenya and Tanzania.

George Nicholas is Professor of Archaeology, Simon Fraser University, Burnaby, British Columbia, Canada. He was founding Director of Simon Fraser University's Indigenous Archaeology Program in Kamloops (1991–2005). Since moving to British Columbia in 1990 from the United States (he is an American citizen), he has worked closely with the Secwepemc and other First Nations, and has directed a community-based, community supported archaeology program on the Kamloops Indian Reserve for 15 years. His research focuses on the archaeology and human ecology of wetlands, hunter-gatherers past and present, intellectual property rights and archaeology, Indigenous archaeology, and archaeological theory, all of which he has published widely on. He is a member of the Society for American Archaeology's Committee on Curriculum and currently serves as Editor-in-Chief of the Canadian Journal of Archaeology. [www.sfu.ca/archaeology/dept/fac_bio/nicholas/Index.html]

Ian Panter is an archaeological conservator and was employed by English Heritage as the archaeological science advisor for the Yorkshire Region until 2006. He has almost 25 years experience in conservation, having worked for a number of organisations including he Mary Rose Trust and York Archaeological Trust developing expertise in the study, assessment and conservation of waterlogged archaeological wood and leather. Since joining English Heritage, Ian's research areas have focussed on *in situ* preservation issues especially impacts to site hydrology and the development of appropriate in situ monitoring techniques.

Stuart Pasley is a Senior Countryside Officer with the Countryside Agency, now working in North West England. He has a Geography Degree and an MSc in Land Resources Management, and leads "living landscapes" work, encompassing the management of our finest landscapes, sustainable land management and looking at how landscapes might change in the future. He led the Countryside Agency's "Humberhead Levels Land Management Initiative" (which was known locally as "Value in Wetness") whilst working in Yorkshire and the Humber region.

Francis Pryor obtained his PhD at Cambridge in 1984. Francis was the assistant curator of the Royal Ontario Museum, Toronto between 1969–1978, and subsequently became involved in archaeological investigations in the Cambridgeshire Fens from 1978 onwards. From 1987 onwards he has been the Director of the Fenland Archaeological Trust. He has published extensively on his work in the Fens, and has a number of books and research reports published on the archaeology and environments of the Flag Fen Basin. His research interests include, but are not exclusive to, wetland archaeology, prehistoric archaeology and livestock farming, 'ritual landscapes', and the conservation of waterlogged archaeological deposits *in situ*.

Jane Reed obtained her PhD at UCL in 1995. She undertook post-doctoral research at Loughborough, Newcastle and Hull between 1995–2002, with NERC and Leverhulme funding, and joined the Department of geography at Hull University as a lecturer in 2002. Her interests lie in the analysis of diatoms in lake sediment cores as quantitative indicators of past changes in water quality, related to the key environmental themes of climate change, human impact, and the conservation of water quality and biodiversity. Her research combines the quantitative study of modern ecology with that of palaeoecology, as a key to interpreting the past. In general she works closely with teams of international experts, in multi-proxy research efforts which combine a variety of palaeoecological, sedimentological and geochemical techniques for more accurate and comprehensive reconstruction of past changes in aquatic ecosystem status in the Mediterranean. She has worked on lakes in NW Greece, Turkey, the former Yugoslav Republic of Macedonia, SE Spain, Kenya, India and Wales.

Ed Schofield completed a PhD on vegetation succession in the Humber Wetlands at the University of Hull in 2001. Following this he was employed as a Research Fellow with WAERC (2001–2) and, later, as a Post-doctoral RA at Kingston University (2002–4). Dr Schofield is currently a Research Fellow at the University of Aberdeen, working as the palynologist on a project designed to investigate the impact of Viking settlement on the vegetation and landscape of southwest Greenland.

Rob Smith completed a PhD on the preservation and degradation of woody tissues in wetland archaeological and landfill sites at the University of Hull in 2005. He is currently a Research Fellow, also at the University of Hull, working on a project that aims to provide data relating to the effects of 'draw-down', changes in water table dynamics and disruptions to the burial environment concomitant with the creation of areas of aggregates extraction from a floodplain site at Newington, Nottinghamshire, England.

Mickle Zhilin obtained his PhD in 1984 and his Dr. habilis in 1999. He is a leading researcher in the Institute of Archaeology of the Russian Academy of sciences, Moscow. Between 1993 to date, Mickle has been the director of the Upper Volga Expedition, specialising in the study of final Palaeolithic and Mesolithic cultures. He has published widely on the finds from the waterlogged sites of the Volga Basin, with five monographs and in excess of 114 articles published in both Russian and English between 1974 and 2005.

For the past thirty years or so, wetlands have been at the forefront of developments in understanding past cultural activity and associated landscapes. Exceptional finds such as the Ferriby Boats from the foreshore of the Humber Estuary, the trackways from the peatlands of Somerset, England and the raised bogs of Ireland, alongside finds such as the spectacular Windover burial/mortuary pond in Florida and bog bodies such as Tollund Man, Denmark, excite the imagination and provide vivid insights into the past.

The exceptional preservation afforded by waterlogged deposits brought about due to anaerobic conditions, which restrict bacterial decay, are paralleled only by other extremes of preservation such as arid and frozen contexts. As a consequence wetland deposits can provide a wealth of information about the past which is seldom recovered from 'dryland' sites. Waterlogged environments and contexts not only preserve the organic part of the cultural record, but they also provide an archive of the environmental conditions pertaining at the time the deposits form, thereby allowing the detailed reconstruction of their associated environments and landscapes.

The potential of wetland sites to preserve cultural and palaeoenvironmental material has resulted in major funding for wetland research in the UK and elsewhere. For example, English Heritage, the government-funded body responsible for archaeology in England, has recently completed an extensive programme of survey, excavation and environmental assessment in four of England's major wetland areas – the Somerset Levels, the Fens, the wetlands of the Northwest and the Humber wetlands.

The latter project, based at the University of Hull, undertook the systematic investigation of over half a million hectares of land located primarily in the catchment of the Humber Basin, between 1992 and 2000. This project resulted in the publication of a series of six monographs, an on-going series of research papers, the setting-up of the Wetland Archaeology and Environments Research Centre at Hull University, and the development of archaeology teaching and research as an integral part of the teaching curriculum at the University of Hull.

In order to mark the successful completion of the Humber Wetlands Project, the editors invited colleagues from the UK and worldwide (USA, Canada, Australia, Russia, Sweden, The Netherlands, Indonesia and Ireland), many of whom are international authorities, to contribute a series of chapters to this book. The aim was to outline the current state of wetland cultural and palaeoenvironmental knowledge, and to provide multidisciplinary insights into the methodological approaches and theoretical aspects of this important area of study, which is very much at the interface of archaeology, history, geography, ecology and environmental science and management.

The contributors to this volume have all approached the production of their papers from the perspective of a project that is organic in nature, developing over the past four years as our own thought processes and research perspectives have developed. From this standpoint the papers have benefited from a degree of hindsight and reflection which has ensured that the academic process has been a very rewarding and fruitful experience. The editors have enjoyed excellent discussions with colleagues, often via email due to the widely dispersed locations researched, and they extend their thanks to colleagues and friends for their patience during this project.

One final note in this preface is the acknowledgement of the loss we felt at the untimely death of Dr. Huw Griffiths, a friend and colleague here at the University of Hull. He is sorely missed, and we extend our gratitude to his widow, Dr. Jane Reed, for completing the final version of his paper. On a more personal note, the completion of this project was delayed due to the deaths of Malcolm Lillie's mother and sister between August 2004 and April 2005, again as with any personal loss, the effects are immeasurable.

Malcolm Lillie and Steve Ellis
Hull, England
April 2006

Acknowledgements

This volume owes much to the patience of the various authors in seeing it thorough to completion, it has been a project that has been with us through some arduous experiences. We have benefited from the enthusiasm and dedication of colleagues here at Hull, and especially our colleagues past and present who have worked with us in wetland archaeology.

Throughout our time at Hull, colleagues at English Heritage and the Countryside Agency have worked with us at the regional level and constantly inspired us to continue our endeavours at the regional, national, and international levels.

Funding from English Heritage towards the production of this volume is greatly appreciated. At Hull University, John Garner was invaluable in converting and editing a number of the images used in this volume.

PART 1
INTRODUCTION

Zusammenfassung

Feuchtgebiete erhalten eine außergewöhnliche Bandbreite von Zeugnissen vergangener menschlicher Aktivitäten. Es ist weithin bekannt, dass Feuchtgebiete weltweit eine bedrohte Ressource sind, und der potentielle Verlust für unser Verständnis soziokultureller Entwicklungen und der Beziehungen zwischen Mensch und Umwelt ist unermesslich. Ziel dieses Bandes ist es aufzuzeigen, welch reichhaltige Informationsquellen für unser Kulturerbe Feuchtgebiete bieten können, sowie Möglichkeiten für Archäologen vorzulegen, diese Zeugnisse auszugraben und zu interpretieren. Die dreiundzwanzig Kapitel decken eine weite Bandbreite von Fundstellen, methodologischen Ansätzen und geographischen Regionen ab. Gemeinsam unterstreichen diese die Vielfalt und Bedeutung kulturhistorischer Elemente der Befunde aus Feuchtgebieten, ob aus lokaler, regionaler oder globaler Sicht. Die Herausgeber hoffen, dass dieser Band zum wachsenden Bewusstsein um die Bedeutung der Feuchtgebiete und der Archive, die in ihnen enthalten sind, beiträgt und dass die einzelnen Aufsätze Einsicht in die Vielfalt von Forschungsansätzen geben, die notwendig ist, um diese bedeutenden globalen Ressourcen vom Gesichtspunkt des Kulturerbes her zu untersuchen.

(Daniela Hofmann)

Summary

Wetlands preserve an exceptional range of evidence relating to past human activity. The fact that wetlands are a threatened resource globally is well established, and the potential loss in terms of our understanding of past human-landscape interactions and socio-cultural developments is immeasurable. This volume aims to demonstrate the rich heritage resource that wetlands can contain and highlight the ways in which archaeologists excavate and interpret the evidence. The twenty-three chapters cover a wide range of site types, methodological approaches and geographical areas, all of which reinforce the diversity and importance of the cultural-historic elements of the wetland record, whether considered from the local, regional or global perspective. The editors hope that this volume will add to the growing awareness of the significance of wetlands and the archive contained therein, and that the papers provide some insights into the diversity of approaches necessary when investigating the heritage component of this significant global resource.

Résumé

Les zones humides conservent une quantité exceptionnelle de preuves archéologiques des activités humaines passées. Il est bien reconnu que les zones humides sont une ressource menacée mondialement, et la perte potentielle quant à notre compréhension des interactions passées entre les hommes et leur environnement d'une part, et des développements socio-culturels d'autre part est inestimable. Ce livre a pour but de montrer le riche patrimoine que peuvent contenir les zones humides et de présenter les méthodes de fouilles utilisées par les archéologues et leur manière d'interpréter les preuves. Les vingt-trois chapîtres passent en revue une vaste sélection de différents types de sites, d'approches méthodologiques et de régions géographiques, qui soulignent tous la diversité et l'importance des éléments historico-culturels découverts dans les zones humides, qu'ils soient considérés d'un point de vue local, régional ou mondial. Les éditeurs espèrent que ce volume renforcera la prise de conscience croissante de l'importance des zones humides et des archives qu'elles contiennent, et que les différents articles donneront un aperçu de la diversité des approches nécessaires à l'étude de la composante patrimoine contenue dans cette précieuse ressource mondiale.

(Sterenn Girard-Suard)

Chapter 1

Wetland Archaeology and Environments

Malcolm Lillie and Stephen Ellis

> wetland investigations require a multidisciplinary approach or training in a number of fields not routinely studied or combined in university academic programs. (Mitsch and Gosslink 1993, 20)

Whether or not we advocate 'wetland archaeology' as a discipline in its own right (Pryor *this volume*), the above observation of Mitsch and Gosslink (1993), whilst not aimed at archaeologists *per se*, is directly relevant to those of us that investigate the archaeology of wetlands. The significance of waterlogged deposits for the preservation of the organic part of the cultural record cannot be overstressed. However, it is perhaps only relatively recently that the intimate relationship between waterlogged materials and their preserving media – whether a wetland, bog, loch, river, moat ditch or pit, to name but a few – has been seriously recognised and integrated into management strategies for wetlands (*e.g.* Coles and Olivier 2001), and increasingly, sustainability is the goal of conservationists and heritage managers alike.

Integral to the raised awareness of the significance of wetlands globally was the Ramsar Convention on Wetlands (Iran 1971 [http://www.ramsar.org], Davidson 2001); although the species and ecology of wetlands were central to this convention, there was also some mention of the palaeoenvironmental record in Criterion 2: 75(v), and archaeology in Appendix C (19) under social and cultural values.

The impacts of de-watering in relation to the archaeological and palaeoenvironmental records and the global wetland ecosystem are increasingly intertwined, with management strategies now taking account of entire systems, as both natural and human activities impact upon them (*e.g.* Coles 2001a, Crisman *et al.* 2001). As noted by Mitsch and Gosslink (1993: 35–9), while world wetlands cover *c.* 4–6% of the earth's land surface, the rate of wetland loss is not really known, though losses of *c.* 90% have occurred in New Zealand and 60% have occurred in China. Such losses affect not only wetland habitats but also, to varying degrees, the preservation of the cultural and palaeoenvironmental record of the world's wetlands. It is worth remembering that the greatest proportion of peatland is located within the temperate and boreal zones of the planet (Gosselink and Maltby 1993, 297), with 75% of this occurring in the former USSR and Canada. Similarly, in the tropics, in excess of 75% of peat soils occur in a zone bordering the western, southern and eastern margins of the South China Sea, and the threats to these deposits alone is clearly exacerbated by their concentration in distinct regions and their increasing exploitation as fuel (Gosselink and Maltby 1993, 297). It is this exploitation, alongside drainage and other vectors of change at the global level that has led to the identification of the exceptional wealth of wetlands in terms of their culture-historic record.

The nature and state of wetland archaeology in Britain and elsewhere has been outlined in numerous publications. These include those relating to the regional syntheses funded by English Heritage, edited volumes from the Wetland Archaeology Research Project's (WARP) numerous international conferences, and a myriad of research papers (*e.g.* Coles and Coles 1986, 1989, Louwe Kooijmans 1987, Coles 1995, Bernick, 1998, Coles *et al.* 1999, Kenward and Hall 2000, Coles 2001a, Ellis *et al.* 2001, Purdy 2001, Sands and Hale 2001, to name but a few).

From the archaeological perspective, we are now increasingly aware of the fact that excavation, whilst fundamental to an understanding of the nature and state of preservation of the wetland archaeological resource, does in itself introduce an uncontrolled and un-quantified set of processes that will compromise the buried resource (*e.g.* Corfield *this volume*). The precise nature of these processes is not understood, and consequently any efforts at re-establishing equilibrium need to be assessed over the longer term. Limited monitoring of the burial environment has been undertaken at sites in England, such as the Sweet Track, Somerset (Brunning 1999), Flag Fen, Peterborough (Pryor, 1992, Lillie and Cheetham 2002a) and the Knights Hospitaller's Preceptory, a moated site in East Yorkshire (Lillie and Cheetham 2002b). Monitoring and approaches to *in situ* preservation vary, and as we might anticipate, are closely linked to the nature and context of the sites being investigated (Kenward and Hall 2000).

The approaches we adopt when attempting to monitor trends in burial environments are becoming increasingly sophisticated (*e.g.* Corfield *this volume*). However, given that saturation is commonly accepted as a fundamental parameter in wet site preservation (Corfield 1996), the recent research undertaken by James Cheetham at Sutton Common in the Humberhead Levels, England (Van de Noort *et al.* 2001) has provided invaluable insights into water table dynamics and their relationship to the *in situ* preservation of archaeological material. Indeed, Chapman and Cheetham (2002) suggest that by examining a limited range of site-specific variables at a high level of resolution a clearer understanding of the processes affecting *in situ* preservation can be obtained, and consequently the ways in which these may be managed are more easily determined.

Archaeological, environmental, ecological and heritage managers often emphasise the nature of threats to wetlands. As noted above, a primary factor has always been de-watering, whether for landscape reclamation purposes, disease control, peat extraction, or less direct intentions such as water abstraction and the deliberate reduction of water tables to enable greater agricultural viability, and the impact is often catastrophic and irreversible (Mitsch and Gosslink 1993, Crisman *et al.* 2001, Van Heeringen and Theunissen 2001).

Recently, Coles (2001b: 10) has emphasised the fact that preservation in wetlands varies according to context, that the chemistry of survival of wood is not fully explored, and neither are the events that lead to the survival or decay of woody tissue over time. Doctoral research that has recently been undertaken at the Department of Geography, University of Hull has revolved around factors such as raised mire development (Bermingham *this volume*), historic landscape reclamation (Fenwick *this volume*), modelling of water table dynamics (Cheetham 2004), oak woody tissue degradation and microbial studies of waterlogged burial environments (Smith 2005).

As noted above, the study of the degradation of oak wood in differing burial environments was being undertaken by Robert Smith, who has sought to re-create the burial environment in a controlled experiment by using lysimeters to enable different impacts to be assessed over time. Aeration and water tables are artificially controlled, thereby allowing some measure of quantification of the factors that can influence degradation or enhance preservation. Whilst only a limited time depth can be achieved during experimental research of this nature, greater time depth is afforded by the additional study of oak wood from a variety of archaeological contexts. This study marks a significant development in our understanding of the processes acting on the buried cultural resource.

As has been noted by Crisman *et al.* (2001, 254), wetlands are also recognised as major repositories of evidence for past environmental and climate changes. Whilst the above emphasis has been deliberately focussed on the various aspects of wetlands in relation to their archaeological potential, obviously, the significance of archaeology in waterlogged contexts is expanded considerably because of the characteristics of the sediment matrix within which it is contained (*e.g.* Coles and Coles 1986, Hingley *et al.* 1999, Bailey *et al.* 2000). Clearly, in archaeological terms, the diverse range of wetland environments that have the potential to preserve waterlogged archaeology not only produce specific levels of preservation, but also present a varied set of parameters for archaeologists to investigate, as the diversity of regional papers in this volume demonstrate (Figs 1.1 and 1.2). Excavations in the peatlands of the Somerset Levels (Coles and Coles 1986) differ from those encountered in the coastal regions of England and Wales (*e.g.* Bell *et al.* 2000, Abeg and Lewis 2001, Rippon 2001, Davidson 2002), not least because of the fact that intertidal archaeology relies on limited temporal 'windows' for excavations. Perhaps one of the most frustrating aspects of this area of wetland research is the fact that tides have an annoying ability to re-deposit excavated sediments and re-work any archaeology that has been revealed to a point where its surrounding matrix no longer affords protection from tidal inundation.

Despite varying approaches and differing burial environments, the palaeoenvironmental and sedimentological records from the wetland contexts investigated by palaeoenvironmentalists, palaeoecologists and sedimentologists can provide invaluable information for reconstructions of the past. This information can relate to sea level change, the onset of biogenic sedimentation and periodic marine transgression and regression, alongside high resolution environmental data relating to human-landscape interactions and landscape developments (*e.g.* Long *et al.* 1998, Brayshay and Dinnin 1999, Shennan and Andrews 2000, Blackford *et al. this volume*, Bunting and Schofield *this volume*, Hope *et al. this volume*, Marchant *this volume*).

The data generated by studies of landscape development often provide the background against which archaeologists seek to integrate the evidence from excavations, and through which we attempt to understand human responses over time (*e.g.* Akerlund 1996). Environmental studies, when considered in relation to human-landscape interactions, provide us with an important means of understanding and contextualising the archaeological record. Wetlands, their sedimentary archive and the archaeological record they contain, provide a unique dataset for the reconstruction of the past (Coles and Coles 1986).

It was the recognition of this exceptional potential that resulted in the major wetland surveys of the Somerset Levels, the Fens, the Northwest Wetlands and the Humber Wetlands, funded by English Heritage or its predecessors. In the Humber region the work of Ellis and Crowther (1990) led to the funding of the Humber Wetlands Survey, which as noted in the preface, has led to many positive developments, and indeed this volume is an additional and integral aspect of the original research project. The exceptional finds from this region, such as those from Roos Carr (Fig. 1.3) and North Ferriby (Fig. 1.4), depositional contexts containing long palaeoenvironmental records such as 'The

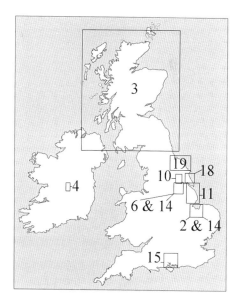

Fig. 1.1 Location of key UK and Irish sites discussed in this volume, numbers are chapter numbers and relate to locations mentioned in the text. Note: for Chapter 14, Beverley is located between map areas 10 and 18.

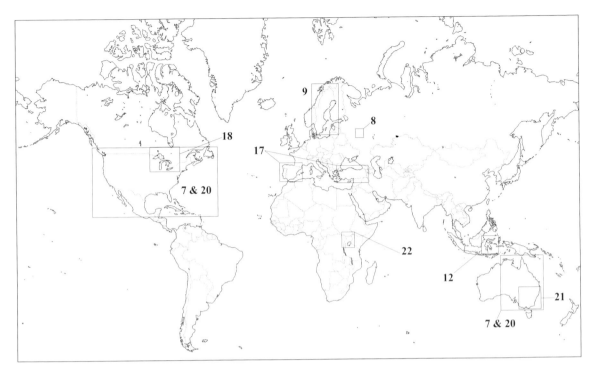

Fig. 1.2 Location of European and World sites discussed in this volume, numbers relate to chapters in which these areas are discussed.

Bog' at Roos and others (*e.g.* Beckett 1981, Tweddle 2001), and the depositional sequences from the Humber and its tributaries (Long *et al.* 1998), all reinforce the significance of this region at the national and international level.

Throughout this introduction, reference has already been made to a number of the papers presented in this volume. In the remaining papers in the current volume the authors outline varying aspects of the perception, use and nature of aquatic ecosystems, and report on a range of archaeological, environmental and methodological approaches to the study of wetlands. The editors feel that these papers serve to reinforce the significant scientific and cultural importance of wetland environments, both nationally and internationally.

Throughout this volume the reader is directed towards the need for integration of all aspects of wetland research in order to facilitate an holistic understanding of the past, as 'if we are to understand wetlands, we need to employ a multidisciplinary approach to their study' (Mitsch and Gosslink 1993, 20).

In order to achieve this goal, the editors felt that the papers could be grouped naturally into broad categories of analysis and research perspective. Consequently the volume commences with a series of contributions that outline a range of issues pertinent to the discussion of wetlands. These include the semantics of wetland archaeology, the development of new initiatives and new approaches to the excavation of the resource, and current approaches to wetland science and management (*e.g.* Pryor, Crone and Clarke, Bermingham, Panter, Pasley). The final paper in this section considers the theoretical issues, economic organisation and resource management strategies of hunter-gatherers in wetlands (Nicholas).

Fig. 1.3 The Iron Age Roos Carr model found in Holderness, England (© Hull and East Riding Museums).

The second section of the volume focuses on archaeological investigations of wetland landscapes and sites in differing environments. This section considers the Mesolithic in the Oka and Volga Basins in central Russia (Zhilin), ritual use of wetlands during the Neolithic period in southern Sweden (Larsson), and prehistoric environments in the Foulness valley, eastern England (Halkon). The final two papers in this section look at the more recent historical period in England and Sulawesi, Indonesia – the first considers landscape development and wetland exploitation in the Lincolnshire Marsh, eastern England during the Medieval period (Fenwick), and the second examines wetland exploitation at the palace centre of Sago City: Utti Batue site, Luwu, dating to the fifteenth and sixteenth centuries AD (Bulbeck *et al.*).

The third section of this volume aims to link a range of methodological approaches to the understanding of the archaeological and environmental records of wetlands. An overview of the current state of wetland archaeological science (Corfield) is followed by an overview of the results of recent fieldwork aimed at characterising burial environments in the UK (Lillie).

Davies then develops the theme of burial environments further in his consideration of molluscan assemblages from wet-ground contexts, which compares the modern and fossil records. Griffiths

Fig. 1.4 A reconstruction of one of the Bronze Age Ferriby Boats, from the Humber at North Ferriby, England.

reviews the use of ostracods as palaeoenvironmental indicators, outlining their past application and potential for future study in environmental reconstructions, and Reed continues the theme of faunal studies by discussing the use of quantitative diatom-based transfer functions as a tool for environmental reconstruction.

Following on from these papers is an assessment of the environmental record from confined wetland contexts (Bunting and Schofield), based on data from a lowland situation in eastern England and similar contexts in Ontario, Canada. In the final paper in this section, Blackford *et al.* focus on detecting and understanding the earliest human impact on the upland vegetation of the North York Moors, England, and advocate the use of peat records as proxy climate indicators that can be used as a robust methodological tool in reconstructing past climate.

In the final section of this volume a series of chapters take the theme of wetland research to its logical culmination in the areas of interpretation and understanding of wetland landscapes as organic environments that are central to human-landscape interactions, and consider exploitation strategies and human activity in relation to the prevailing environment. Nicholas begins this section by considering the ways in which the exploitation of wetland environments can lead to developments in the socio-political organisation of hunter-gatherer groups and shifts in mobility/sedentism patterns associated with such developments. Moving on from the nature of past societies and their exploitation strategies to the understanding of human-landscape interactions via palaeoenvironmental studies, Hope *et al.* consider the coastal swamps of New South Wales, Australia, building a picture of the environment within which the human populations operated, and highlighting the nature of the vegetation that was available for exploitation by these groups. In the final paper in this section Marchant presents a study of the past 2000 years of landscape change around Bwindi-Impenetrable forest, central Africa. In this chapter the author presents a series of palaeoecological maps and considers the mechanisms of change and their concomitant impacts on forest development.

The volume concludes with an editorial discussion of the themes presented, linking these to the present state of wetland historical, cultural and environmental studies, and wetland conservation policy at the regional, national and global levels, and considers the future of research in this important and rapidly expanding area of archaeology.

REFERENCES

Aberg, A. & C. Lewis (eds) 2001. *The Rising Tide: archaeology and coastal landscapes*. Oxford: Oxbow Books.

Akerlund, A. 1996. *Human Responses to Shore Displacement: living by the sea in eastern middle Sweden during the Stone Age*. Stockholm: Riksantikvarieämbetet Arkeologiska undersökningar. Skrifter nr 16.

Bailey, G., R. Charles & N. Winder 2000. *Human Ecodynamics*. Oxford: Oxbow Books.

Beckett, S.C. 1981. Pollen diagrams from Holderness, North Humberside. *Journal of Biogeography* 8, 177–98.

Bell, M., A. Caseldine & H. Neumann 2000. *Prehistoric Intertidal Archaeology in the Welsh Severn Estuary*. York: Council for British Archaeology Research Report 120.

Bernick, K. (ed.) 1998. *Hidden Dimensions: the cultural significance of wetland archaeology*. Vancouver: University of British Columbia Press.

Brayshay, B.A. & M. Dinnin 1999. Integrated palaeoecological evidence for biodiversity at the floodplain-forest margin. *Journal of Biogeography* 26, 115–31.

Brunning, R. 1999. The *in situ* preservation of the Sweet Track, in B. Coles, J. Coles & M. Schou Jørgensen. (eds) *Bog Bodies, Sacred Sites and Wetland Archaeology*, 33–8. Exeter: Warp Occasional Paper 12.

Chapman, H. P. & J. L. Cheetham 2002. Monitoring and modelling saturation as a proxy indicator for in situ preservation in wetlands: a GIS-based approach. *Journal of Archaeological Science* 29, 277–89.

Cheetham, J. L. 2004. An assessment of the Potential for in situ Preservation of Buried Organic Archaeological Remains at Sutton Common, South Yorkshire. University of Hull: Unpublished PhD thesis.

Coles, B. J. 1995. *Wetland Management: a survey for English Heritage*. Exeter: WARP Occasional Paper 9.

Coles, B. J. & J. M. Coles 1986. *Sweet Track to Glastonbury: the Somerset Levels in prehistory*. London: Thames & Hudson.

Coles, B. J. & A. Olivier. (eds) 2001. *The Heritage Management of Wetlands in Europe*. Belgium: Europae Archaeologiae Consilium and WARP.

Coles, B. J., J. M. Coles & M. Schou Jørgensen (eds) 1999. *Bog Bodies, Sacred Sites and Wetland Archaeology.* Exeter: Warp Occasional Paper 12.

Coles, J. M. 2001a. Wetlands, archaeology and conservation at AD 2001, in B. Coles, & A. Olivier (eds), *The Heritage Management of Wetlands in Europe*, 171–84. Belgium: Europae Archaeologiae Consilium and WARP.

Coles, J. M. 2001b. Of water-wings and wellingtons: wetland archaeology and the new journal. *Journal of Wetland Archaeology* 1, 3–13.

Coles, J. M. & B. J. Coles 1989. *The Archaeology of Rural Wetlands in England.* Exeter: WARP & English Heritage.

Corfield, M. 1996. Preventative conservation for archaeological sites, in A. Roy & P. Smith (eds) Archaeological Conservation and Consequences, 32–37. London: International Institute for Conservation of Historic and Artistic Works.

Crisman, T. L., U. A. M. Crisman & J. Prenger 2001. Wetlands and archaeology: the role of ecosystem structure and function, in B. Purdy, (ed.) *Enduring Records: the environmental and cultural heritage of wetlands,* 254–61. Oxford: Oxbow Books.

Davidson, N. 2001. A foreword from the Ramsar Convention on Wetlands, in B. J. Coles & A. Olivier. (eds) *The Heritage Management of Wetlands in Europe,* i. Belgium: Europac Archaeologiae Consilium and WARP.

Ellis, S. & D. R. Crowther (eds) 1990. *Humber Perspectives: a region through the ages.* Hull: Hull University Press.

Ellis, S., H. Fenwick, M. Lillie & R. Van de Noort (eds) 2001. *Wetland Heritage of the Lincolnshire Marsh.* University of Hull: Humber Wetlands Project.

Gosselink, J. G. and E. Maltby 1993. Wetland Losses and Gains, in Williams, M. (ed.) *Wetlands: A Threatened Landscape,* 296–322. Oxford: Blackwell.

Hingley, R., P. Ashmore, C. Clarke & A. Sheridan 1999. Peat, archaeology and palaeoecology in Scotland, in B. Coles, J. Coles & M. Schou Jørgensen (eds) *Bog Bodies, Sacred Sites and Wetland Archaeology,* 105–14. Exeter: Warp Occasional Paper 12.

Kenward, H. & A. Hall 2000. Decay of delicate organic remains in shallow urban deposits: are we at a watershed? *Antiquity* 74, 519–25.

Lillie, M. C. & J. L. Cheetham 2002a. Water Table Monitoring at Flag Fen, Peterborough (TL 227989). Client: Soke Archaeological Services on behalf of English Heritage. Unpublished WAERC Report No. SAS-FF/02-01.

Lillie, M. C. & J. L. Cheetham 2002b. Monitoring of the moat ditch deposits at Knights Hospitaller's Preceptory, Trinity Lane, Beverley (September 2001–August 2002). Client: CgMs Consulting on behalf of Tesco Stores Ltd.. Unpublished WAERC Report No. KHP-BEV/02-02.

Long, A. J., J. B. Innes, J. R. Kirby, J. M. Lloyd, M. M. Rutherford, I. Shennan & M. J. Tooley 1998. Holocene sea-level change and coastal evolution in the Humber Estuary, eastern England: an assessment of rapid coastal change. *The Holocene* 8, 229–47.

Louwe Kooijmans, L. P. 1987. Neolithic settlement and subsistence in the wetlands of the Rhine/Meuse delta of the Netherlands, in J. M. Coles, & A. J. Lawson (eds) *European Wetlands in Prehistory,* 227–51. Oxford: Clarendon Press.

Mitsch, W. J. & J. G. Gosslink 1993. *Wetlands,* 2nd edition. New York: Van Nostrand Reinhold.

Pryor, F. 1992. Current research at Flag Fen, Peterborough. *Antiquity* 66, 439–57.

Purdy, B. (ed.) 2001. *Enduring Records: the environmental and cultural heritage of wetlands.* Oxford: Oxbow Books.

Rippon, S. (ed.) 2001. *Estuarine Archaeology: the Severn and beyond.* Bristol: Severn Estuary Levels Research Committee.

Sands, R. & A. Hale 2001. Evidence from marine crannogs of Later Prehistoric use of the Firth of Clyde. *Journal of Wetland Archaeology* 1, 41–54.

Shennan, I. & J. E. Andrews (eds) 2000. *Holocene Land-Ocean Interaction and Environmental Change around the North Sea.* London: Geological Society Special Publication 166.

Smith, R. J. 2005. The Preservation and Degradation of Wood in Wetland Archaeological and Landfill Sites. University of Hull: Unpublished PhD thesis.

Tweddle, J. C. 2001. Regional vegetation history, in Bateman, M. D., P. C. Buckland, C. D. Frederick & N. J. Whitehouse (eds) *The Quaternary of East Yorkshire and North Lincolnshire,* 35–46. London: Quaternary Research Association.

Van de Noort, R., H. P. Chapman & J. L. Chettham 2001. *In situ* preservation as a dynamic process: the example of Sutton Common, UK. *Antiquity* 75, 94–100.

Van Heeringen, R. & L. Theunissen 2001. Repeated water table lowering in the Dutch delta: a major challenge to the archaeological heritage management of pre- and protohistoric wetlands, in Purdy, B. (ed.) *Enduring Records: the environmental and cultural heritage of wetlands,* 271–6. Oxford: Oxbow Books.

Chapter 2

Beware the Glutinous Ghetto!

Francis Pryor

The publication of the *Current Archaeology* volume devoted to wetland archaeology (No. 172 for February 2001) focussed the archaeological public's attention on the subject. The splendid wealth of illustrations reproduced there showed not just the diversity of sites and landscapes involved, but also the extraordinary results that waterlogging can bring. I must admit that having seen many thousands of examples of prehistoric waterlogged wood, I'm still amazed by every new piece I see. Although explicable in purely scientific terms, it always seems quite astonishing that something so fragile and essentially ephemeral can survive in the ground for so long.

The publication of the Wetland edition of *Current Archaeology* was an important event, but from an entirely selfish point of view it was slightly irritating because *so* many people asked me why it made no mention of the sites I had worked on in the Fens region (Fig. 2.1), such as Flag Fen or Etton (Figs 2.2 and 2.3). Frankly, I don't know why this was, and I can only answer that both sites have already received more than their fair share of attention and the editors doubtless thought it was time to give others a turn in the spotlight. For what it's worth, I did send a short piece on our (then) current work at Flag Fen which didn't find its way into print, for whatever reason. Although it wasn't included I still think that my short note did have something quite important to say and I'd like to take the opportunity presented by the publication of this exciting new collection of papers to give those thoughts another airing.

This is what I wrote, word for word, although I haven't reproduced the final three paragraphs which were mainly about Flag Fen. The title (and maybe that's where I went wrong!) was the one I've used for this paper.

> I well remember the day I told my friend Leendert Louwe-Kooijmans (then Professor of Archaeology at Leiden University) that I considered myself a wetland archaeologist. It was many, many years ago and I had enjoyed enough Dutch beer to talk unadvisedly. He gave me a quizzical look and suggested, in the kindest possible fashion, that I had voluntarily confined myself to an academic ghetto. He explained that in his country nearly all archaeology is either wet, or has wet components, so the term wetland is effectively meaningless – like upland archaeology in Nepal. He also pointed out that, just as there are many words for water in Dutch, there are in turn a multiplicity of wetlands and all of them relate in one way or another to dry land. He advised me to think of myself as a landscape archaeologist. Of course I took his advice.
>
> To my mind an archaeological site, be it wet or dry, gains its importance from context. A superbly preserved bucket of prehistoric butter in the middle of a raised bog will probably mean rather less than (yet another) Iron Age roundhouse in a river valley floodplain. In other words, rarity and good preservation alone

do not confer significance. But before you think I'm being a traitor to the wetland cause, I would say in my defence that I'm not: I'm merely very worried lest so-called wetlands be considered as phenomena that are somehow removed from the rest of archaeology. They're not. In fact many apparently dryland sites, one thinks, for example of Hurst Fen, Mildenhall (the exceedingly important sandy Neolithic site that Grahame Clarke, Eric Higgs and Ian Longworth excavated in the late 50's) are only dry by a fluke of geology (Clark *et al.* 1960). My point is simple: the wet/dry distinction is arbitrary and meaningless. More to the point, did the ancient inhabitants of the Somerset Levels or the Fens consider themselves wetlanders? I very much doubt it – which is why I'm so unhappy about imposing our labels on them.

From time to time I find myself being interviewed about my work in archaeology. Almost without exception, the interviewers have assumed that Flag Fen is the peak of my career, and they're always very disappointed when I tell them that I reckon the causewayed enclosure at Etton was of equal, if not greater importance. Since its publication (by English Heritage), I have been approached by many Neolithic scholars who have been fascinated by our interpretations and the significance they might have for comparable sites elsewhere in Britain and Europe (Pryor 1998). But is Etton a wetland site? Technically I suppose not, as strictly speaking, it falls outside the area of the English Heritage Fenland Survey and is only wet in its lowest-lying bits – but the fact that those bits are highly significant of themselves, and can also be related to large areas of dryland ritual archaeology, is what gives Etton its extraordinary academic potential.

Another area not generally considered a wetland is the Middle Thames Valley. I have always regarded this as one of the key regions of British archaeology, and when I wrote the Fengate Fourth Report (Pryor 1984) I drew heavily on floodplain sites in the Oxford Region, such as Farmoor (Lambrick and Robinson 1979). Today the region is still producing extraordinary material (see, for example, *Current Archaeology*, No. 121). One thinks at once of the Bronze Age 'bridges' at the Eton College Rowing Lake, or that mysterious Middle Bronze Age stone-built causeway dug by Gill Hey's team at Yarnton (Allen and Welsh 1996). These discoveries are important because they can be related to a known and well-studied sequence of (dryland) settlement and land-use. Again, it's their context that is so crucial. These are key sites. They belong in the mainstream, and I would hate to see them confined to the glutinous ghetto.

The post alignment at Flag Fen is undoubtedly significant in its own right (Fig. 2.2), but it gains immeasurably from the fact that it can be linked directly, and at several points, to the less flood-prone landscapes of Fengate (to the west) and Northey (to the east). This is why the major monograph is called *Archaeology of the Flag Fen Basin* (Pryor 2001a). In other words, it's about a gently concave landscape in which the ritual goings-on in the wet can be related to the lives of people who actually lived on and farmed the drier ground of the periphery. I'm not 100% certain, but I would suggest that both areas were of equal importance to each other.

Now I don't want wholly to disregard the term 'Wetland Archaeology', nor indeed more colourful phrases, such as the 'Wetland Revolution'. These can be very useful indeed when one is trying to reach a lay audience, and such communication is urgently needed: almost everywhere it seems wetlands are under a very real threat. Nobody knows that better than I, and we must do everything in our power to raise money and in-kind resources before it's too late. Indeed, in my recent *Seahenge* book I've devoted an entire chapter to the Wetland Revolution and the people who have pioneered wetland research in Britain (Pryor 2001b, 186–208). So I'm far from unsympathetic to the cause, as

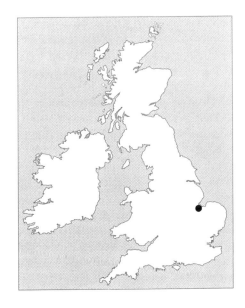

Fig. 2.1 Location of the Fens.

such. It's the idea of Wetland Archaeology as an accepted *academic* subject that worries me profoundly, because if it is given academic credence – and this seems to be happening *de facto* – then from it spring a number of unpalatable corollaries.

If Wetland Archaeology is indeed a *bona fide* subject in its own right, then so is Dryland Archaeology – and to my mind that is patently absurd: what about *Very* Dryland Archaeology (*e.g.* arid zones in the high Andes and Egypt)? Is that also a sub-discipline within the broader field of archaeology? My wife Maisie Taylor was once asked to write a book on Wetland Archaeology that also included chapters on arid regions, where her editor insisted (with an eye to U.S. sales) that preservation could be almost as complete. He was probably right, but so what? The book might sell well, but it would lack all intellectual credibility. It would be a rag-bag, a catch-all book with no narrative thread to hold it together. Sadly, it is the lack of narrative which makes so many routine PPG-16 excavation reports so dry and academically irrelevant.

It seems to me that post-depositional factors, such as comparable conditions of preservation, are not grounds for defining a subject which can then be given an academic ring-fence. As Maisie discovered when she tried to start work on that book-that-never-happened, even the most inventive of authors is pushed to find common cultural themes that link Inca preserved mummies in high mountain caves with Iron Age bog bodies in Lancashire moss peats. It was a futile effort, so she didn't make it – which was a shame, as the publisher was a reputable university press and the extra cash would have been welcome.

I first had qualms about the arbitrary wet/dry distinction in the early days of the Fenland Survey (Hall and Coles 1994). Now I don't for one moment want to criticise that project which has had an enormous impact on our appreciation of the archaeology of eastern England. But it has to be said that what I have observed at Flag Fen – that the wet and dry elements of the landscape are essentially interdependent – can now be seen also to apply elsewhere in the Fens. Areas such as the Welland valley (Pryor and French 1985)

Fig. 2.2 Excavating the post alignment at Flag Fen.

or the higher 'islands' around Ely and Haddenham that were once considered non-Fen (in terms defined by the Survey) are now seen to be of crucial importance (Evans 1988). In fact it is the different and complex pattern of wet/dry interaction through time, that can be so academically stimulating. Hardly a week goes by when I don't discuss their latest findings with Chris Evans and his team at Cambridge University who are working on deep fen and fen-edge sites all over Fenland, from the Ouse valley to within sight of Flag Fen. I don't think anyone in Chris's team gives a moment's thought to whether they're working in wetlands or not. It doesn't matter: what they're doing is regional archaeology – and they're doing it extraordinarily well.

To be fair, I can see one positive advantage in the concept of Wetland Archaeology and it's closely linked to the public perception of threat, which has already been alluded to. In Ireland, for example, the wholesale destruction of huge areas of blanket bog, to fuel power stations and for horticultural use, demanded a rapid response – and hence the formation of the Irish Wetland Unit (see Bermingham *this volume*). It could also be argued that in many parts of the huge peat landscape of the Irish midlands the relationship of wet to dryland was perhaps less crucial (*e.g.* Moloney 1993). Whether or not one accepts that argument (and I'm still not sure that I do), I don't think anyone would deny that the establishment of a specialised wetland unit was both timely and effective, given the resources made available. But can these pragmatic considerations be applied to Britain too? I think not.

For a start, the wetland landscapes of Britain are extraordinarily diverse and I would seek much outside advice before I felt confident enough to cut trenches through Lancashire mosses. They're simply foreign to me. Indeed, even in the Fens, I would hesitate before tackling a site in the Boston silts or the more acid peaty landscapes around Mildenhall, in Suffolk. I've tasted coastal archaeology enough to know that it's a

Fig. 2.3 Excavating at the Etton causewayed enclosure.

special field and one where inland relationships are of overwhelming importance – one looks at the coastal strip in isolation at one's academic peril. I most particularly do not subscribe to the view, which has been expressed from time to time, that good wetland archaeology cannot be done by dryland archaeologists. Laying aside the fact that this view is remarkably patronising, it makes the great mistake of confusing technique and perception. Certain techniques of wetland archaeology can be specialised, I concede, but they are no harder to acquire than any other archaeological procedure, such as, for example, the excavation and removal of cremations – a task often best left to a specialist. But the general techniques of wetland work are not hard to learn. I, for one, made the transition and my brain and body managed to cope with the experience. No, there's something altogether more worrying behind all of this.

I mentioned that technique and perception were confused. Having acquired appropriate wetland techniques, one is not thereby equipped automatically to grapple with the broader issues of perception. Indeed, I would go slightly further than that. In some instances the sheer technical difficulty of some wetland projects has prevented the archaeologists involved from taking a broader view. The site and its immediate natural surroundings become the sole focus of the enquiry. Quite literally, in these cases, the wood is hidden by the trees (or rather, the alder carr), and in this environmentally-deterministic view of the past, the unfortunate folk of antiquity don't stand a chance.

I raised some of these issues at a recent forum, and participants agreed that in principle I was right: the term Wetland Archaeology had no theoretical validity. But, they argued, we live in the real world and wetlands are being drained. Either we hitch our wagon to the conservation/environmental lobby, or we must spend the next few years shouting abuse from the touchlines, while evidence for the historic environment is ignored in favour of other interests such as habitat-creation and so forth. Again, there is much sense in this, and I would certainly not advocate withdrawing from discussions with colleagues in the conservation/environmental fields. I would argue that wet archaeological landscapes have suffered every bit as much as wet natural landscapes and something – anything – has to be done to slow the process down. If that means embracing a concept that is theoretically meaningless, then so be it.

So there are two strong reasons why the Wetland Archaeology concept will be around for some time – wetland conservation and the public perception of threat. Having said that, I would insist that we should be aware that Wetland Archaeology can never be a topic in its own right. It's a pragmatic construction, a fiction, but a (regrettably) necessary one, given the current situation. But we must not use such a flawed concept to structure or justify the way we actually *do* our own research, as archaeologists, away from the realms of SSSIs, Ramsar Sites and the rest.

So my real, gut level, objection to the concept of Wetland Archaeology is that it isn't appropriate to our subject, at a fundamental, a conceptual level. It places the environment and conditions of preservation at its centre. Consequently, people and culture must take second place. In my view you can see this very plainly in many papers devoted specifically to Wetland Archaeology. Sometimes they're vastly more about wetlands than archaeology. Indeed, in the past I have been guilty of wetland environmental determinism myself – and on more than one occasion. But it is not something I or we should be proud of. Given the shortage of research funds, we should not be spending them on stand-alone research into past environments – in the name of 'Wetland Archaeology'.

'That's fine,' I hear a voice say, 'but what new would you put in its place?' My response is that we don't require anything specific to replace it and certainly not something new. In Britain we have a long and distinguished tradition of regional archaeology (and note I have not said Regional Archaeology). Indeed our subject was originally structured around a framework of county and regional archaeological societies. So I'd return to that pre-existing framework and I'd involve the network of local and regional SMRs and the very dedicated people who look after them. It's a network that cries out for closer integration (and the funds needed for that). Apart from anything else, it would be a good way to involve the lay public in the archaeological stories that lie behind their own region's history. It would help to foster a sense of place – as a profession we need to move closer to the public, who have as much right to their past as we do. If we erect further walls – and Wetland Archaeology is yet another jargon term that people interested in the past will have to grapple with – then we do ourselves a long-term disservice.

There is also a danger in creating self-defined ghettoes within the discipline. These ghettoes make for easy targets. For better or for worse, Wetland Archaeology is widely regarded as being very expensive and wet projects will doubtless be chopped when accountants and administrators need to save money. On the other hand, the wet bits within a properly integrated regional research project will not stand out so starkly. So it makes intellectual and practical sense to treat wetland archaeology within a properly considered regional context.

I must finish by adding that I don't like writing negative papers, because it's very hard to end with a stirring clarion call to arms. 'Down with Wetland Archaeology' and 'Arise Regional Archaeology!' are hardly cries that make one stiffen the sinews or summon up the blood. But for the time being they're all I can offer.

REFERENCES

Allen, T. & K. Welsh 1996. Eton Rowing Lake. *Current Archaeology* 148, 124–7.

Clark, J. G. D., E. S. Higgs & I. H. Longworth 1960. Excavations at the Neolithic site at Hurst Fen, Mildenhall, Suffolk. *Proceedings of the Prehistoric Society* 26, 202–45.

Evans, C. 1988. Excavations at Haddenham, Cambridgeshire: a planned enclosure and its regional affinities, in C. Burgess, P. Topping, C. Mordant & M. Maddison (eds) *Enclosures and Defences in the Neolithic of Western Europe*, 127–48. Oxford: British Archaeological Reports International Series 403.

Hall, D. N. & J. M. Coles 1994. *Fenland Survey: an essay in landscape and persistence*, London: English Heritage Archaeological Report 1.

Lambrick, G. H. & M. A. Robinson 1979. *Iron Age and Roman Riverside Settlements at Farmoor Oxfordshire*. London: Council for British Archaeology Research Report 32.

Moloney, A. 1993. *Excavations at Clonfinlough County Offaly*. Dublin: Transactions of the Irish Archaeological Wetland Unit 2.

Pryor, F. M. M. 1984. *Excavation at Fengate, Peterborough, England: the fourth report*. Leicester and Toronto: Northants Archaeological Society Archaeological Monograph 2, and Royal Ontario Museum Archaeological Monograph 7.

Pryor, F. M. M. 1998. *Etton: excavations at a Neolithic causewayed enclosure near Maxey Cambridgeshire, 1982–87*. London: English Heritage Archaeological Report 18.

Pryor, F. M. M. 2001a. *The Flag Fen Basin: archaeology and environment of a Fenland landscape*. London: English Heritage Archaeological Report.

Pryor, F. M. M. 2001b. *Seahenge: new discoveries in prehistoric Britain*. London: HarperCollins.

Pryor, F. M. M. & C. A. I. French 1985. *The Fenland Project Number 1: archaeology and environment in the lower Welland valley*. Cambridge: East Anglian Archaeological Report 27.

Chapter 3

Whither Wetland Archaeology in Scotland in the Twenty-First Century?

Anne Crone and Ciara Clarke

INTRODUCTION

At the WARP (Wetland Archaeology Research Project) conference in Dublin in 1998, John Coles took the Scottish delegates to task for the absence of any strategic programme of wetland archaeology in Scotland. Spurred into action, the delegates established the Scottish Wetland Archaeology Programme (SWAP), an informal group of interested people whose overall aim is to initiate such a programme. This paper briefly summarises what has been achieved in Scotland to date and outlines the SWAP proposal for a strategic programme of works which we hope would see the potential of the archaeological resource of the Scottish wetlands more fully addressed.

We should establish at the outset that SWAP is focusing primarily on freshwater wetlands, because a Scottish forum to develop initiatives in coastal archaeology already exists, *i.e. Shorewatch* (Gilmour 2001). However, it is recognised that there will be a great deal of overlap between respective interest areas.

BACKGROUND

Within Scotland there is a wide range and high concentration of wetland types that combine to give the Scottish landscape its unique character. These include bogs, fens, lochs (lakes), rivers, floodplains, estuaries, coastal marshes and mudflats. However, two particular wetland environments dominate in the Scottish landscape – bogs and lochs. Scottish bogs account for 72% of the British peat resource (Lindsay 1995), and generally comprise widespread but discrete areas of blanket peat with a few isolated areas of raised bog. The Flow country of Caithness and Sutherland (Fig. 3.1), in the extreme north of Scotland, is the largest and most intact area of blanket bog in the world, and is considered to be of global importance due to its unique composition and state of preservation. The largest surviving areas of natural primary raised bog are also to be found in Scotland, predominantly along the Forth valley and on the north Solway shore (Fig. 3.1). With over 30,000 lochs, which comprise approximately 160,000 ha of the total land area, together with associated river systems, the potential of inland freshwater wetland deposits is also substantial.

The character of the Scottish landscape will have influenced the type of settlement and exploitation patterns of its inhabitants and the resulting material remains. As a consequence of the predominant

Anne Crone and Ciara Clarke

Fig. 3.1 Map of Scotland showing location of sites, find spots and wetland areas mentioned in the text.

landforms, wetland studies tend to divide naturally into two research areas – lake settlement, almost exclusively in the form of crannogs (man-made island structures) (Fig. 3.2), and peatland archaeology, which has been characterised by serendipitous finds (Fig. 3.3) and the occasional structure (Fig. 3.4). In Britain, lacustrine archaeology is almost exclusive to Scotland; in England, peatlands and alluviated lowlands are synonymous with 'wetlands', and consequently the English Heritage funded wetland surveys focused on these environmental zones (*e.g.* Coles 1995).

Crannogs are a peculiarly Scottish and Irish phenomenon (Fig. 3.2). Only one crannog is known from the rest of the British Isles, at Llangorse in Wales, and this example is thought to have been built by an Irish prince (Campbell and Lane 1989). Evidence for lake settlement is more extensive elsewhere in Europe, but here this usually takes the form of lakeside settlement

Fig. 3.2 A crannog in Loch Leathan, Argyll, western Scotland (Crown copyright: Royal Commission for the Ancient and Historic Monuments of Scotland).

rather than deliberately created islands. Ireland has both crannogs and lakeside settlements, but in Scotland, despite the probability of its existence, lakeside settlement is currently unknown.

The assumption has always been that the archaeological potential of Scotland's bogs was high, if unrealised. However, 96% of the Scottish peatlands comprise blanket bog, which is found mainly in often inhospitable and inaccessible upland areas, and may therefore never have been intensively exploited in the past. While these upland peats are archaeologically important in that they often seal earlier prehistoric landscapes, they are unlikely to produce organic archaeological remains other than occasional artefacts in pockets of deeper peat. In other European countries, rich organic archaeological remains tend to be found in areas of fen peat and raised bog, which as well as being located in low-lying, more accessible areas were also resource-rich and hence attractive to early populations (*e.g.* Nicholas *this volume*). It therefore seems most likely that the potential for Scotland's peatlands to yield organic archaeological remains is highest in the surviving areas of lowland raised bog and fen.

PREVIOUS DISCOVERIES

The practice of cutting peat for fuel has a long history in Scotland and has often resulted in the accidental recovery of archaeological remains. Increasing antiquarian interest during the nineteenth century meant that these finds began to be recorded, as the acquisitions lists published in the early volumes of the Society of Antiquaries of Scotland testify. The majority of finds from the Scottish bogs have been isolated artefacts, particularly wooden containers such as bog butter kegs and bowls (*e.g.* Earwood 1993; Fig. 3.3). There have been more dramatic finds such as the Deskford carnyx – an Iron Age trumpet found in a moss at Leichestown, Banffshire in 1816 (Alexander Smith 1868, Anderson and Black 1888) – and the famous wooden effigy found when cutting foundations for a wall in North Ballachulish Moss in 1880 (Christison

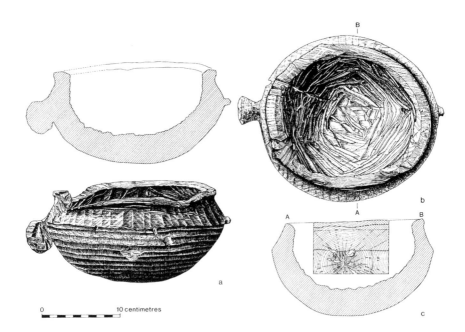

Fig. 3.3 An Iron Age bowl found in a peat bog at Loch a 'Ghlinne Bhig, Bracadale, Isle of Skye. (a) – profile, (b) – interior, (c) – cross section showing alignment of grain.

1881). A trickle of discoveries has continued throughout the twentieth century, despite a reduction in peat-cutting by hand and the mechanisation of activities such as ditching.

The practice of improving the agricultural potential of the land by removing the surface peat, especially in the raised bog complexes of the Forth valley (Cadell 1913), also played a pivotal role in the discovery of archaeological remains. Items such as the tripartite disc wheel from Blairdrummond Moss (Piggott 1959) and the Flanders Moss cauldron (Anderson 1885), as well as numerous trackways (Tait 1794, Sheriff 1796, MacGibbson 1798), were recovered during these operations.

These finds indicate that the Scottish bogs were certainly a focus for human activity in the past, such as for the storage of foodstuffs (for security or perhaps to improve their flavour) and as places for ritual activity. However, there is very little evidence for settlement on, or transport across, the Scottish bogs, a situation which contrasts markedly with the evidence from England and Ireland (*e.g.* Coles and Coles 1996, Moloney 1993). It is possible that this evidence remains to be discovered, but it is equally likely that because blanket bog, the type of bog that predominates in Scotland, was 'resource-poor' in the past it has not been as extensively exploited as raised bog (although in the more recent past blanket peat in particular has been drawn on as a fuel source). It may be more than coincidence that the only records of wooden trackways in the National Monuments Record for Scotland (NMRS), five in all, come from Flanders Moss, the most extensive area of raised bog in the UK. The wheel from Blairdrummond Moss, mentioned above, provides further evidence for transport across this particular raised bog complex.

Evidence for settlement in wetland environments has come almost exclusively from crannog sites. Some 353 crannogs or possible crannogs are recorded in the NMRS, but to date only a handful of lochs have been investigated in any detail. Furthermore, underwater survey has consistently recovered more sites than were

Fig. 3.4 The Neolithic wooden platform at Parks of Garden, Flanders Moss. This is thought to have been used as a base for hunting forays into the Moss (Crown copyright: Historic Scotland).

originally known (*e.g.* McArdle and Morrison 1973, Dixon 1982a), so this figure is probably a gross underestimate. Again, the bulk of the evidence from these site types was gathered during the nineteenth century when they became the focus of antiquarian interest as a result of the revelations of the Swiss lake villages in the middle of that century. In all, 45 crannogs have been excavated to varying degrees, but it is salutary to remember that only eight of these have been investigated since the 1930s to a standard where the excavation report is coherent and can be usefully interrogated. While these investigations serve to demonstrate the wealth of organic and other evidence often preserved on crannogs, they do not provide a data-set of sufficient size to make anything more than broad generalisations about important issues such as chronology, distribution, form and function, among others (Crone 2000).

The second half of the twentieth century has seen little new archaeological evidence being recovered from the Scottish wetlands. This is partly due to the recognition that wetland excavation was time-consuming and costly (despite the obvious returns), while the overwhelming potential of the deposits and lack of knowledge about the location, condition or extent of archaeological remains made prioritisation difficult, a continuing problem to which we will return later. However, recovery of archaeological evidence had also slowed down because the nature of the threats to these environments was changing. What is currently known about wetland archaeology in Scotland was primarily revealed during the hand-cutting of peat and the drainage of bogs and lochs in order to increase and/or improve agricultural land. Hand-cutting of peat for fuel has diminished steadily since the early twentieth century and consequently fewer artefacts have been recovered from this source. Large drainage schemes are no longer countenanced, and consequently fewer crannogs and other site types are exposed and visibly threatened.

THE THREAT?

It is perhaps because wetland resources in Scotland have been perceived as relatively un-threatened that no concerted plan of action has been implemented. Scotland has not suffered the same degree of development pressure that has elsewhere in the British Isles resulted in the exposure and consequent investigation of archaeological deposits. Apart from the Central Lowlands (*i.e.* the Forth and Clyde valleys), Scotland is not heavily populated and therefore has not witnessed the processes of urbanisation, such as road building and housing development, which exposed and threatened many of the prehistoric settlements on the shores of the Swiss lakes (*e.g.* Arnold 1999), and which have contributed to the erosion of the peatlands of northwest England (*e.g.* Hall *et al.* 1995).

Although Scotland does have some commercial peat-harvesting, primarily in the peatlands of the southwest and the Central Lowlands, it is nowhere near the scale of that seen in the Somerset Levels of England or the midland bogs of Ireland. In these areas it was the scale of this visible and imminent threat that led to decisive action to halt the unrecorded destruction of their archaeological heritage. In some ways the situation in Scotland can be characterised as the absence of a sufficiently recognisable and immediate threat to the resource.

While the English fen peats have been subjected to extensive drainage operations in advance of ever-deeper cultivation, most of the blanket peatlands of Scotland are of limited agricultural value, lying over podzolised soils in inhospitable and inaccessible terrain. One of the few areas in Scotland that has seen improvement of peatland for agriculture is the Forth valley raised bog complex. During the nineteenth century, large areas of peat were removed by floating peat blocks into the Firth of Forth, in order to cultivate the underlying mineral soils. By the end of the century this process had ceased, due to its polluting effect on the waters and the consequent intervention of the salmon industries (Cadell 1913). Although many archaeological finds came to light during these operations (see above), it is likely that much important evidence was lost in the large blocks of peat that floated out to sea.

Scotland is also apparently less at risk from the natural processes that threaten wetland deposits elsewhere in the British Isles. In Britain, rising sea-level is greatest along the south coast, with Scotland being least affected (see Coles 1995, 14). Rising sea and river levels have caused erosion in the Severn, Thames and Humber estuaries, which has seen the exposure of significant archaeological remains, prompting targeted archaeological programmes of survey and selected excavation in these areas.

Thus, during a period when other countries were beginning to address the issue of diminishing wetland resources (through the English Heritage wetland surveys and the establishment of the Irish Archaeological Wetland Unit, for instance), Scotland's attitude to the cultural heritage of its wetlands could perhaps be described as complacent. Although there are no clear and obvious threats, Scotland's wetlands are probably as much at risk from the more insidious processes that also threaten wetlands in other parts of the British Isles; acid rain, climate change and water pollution may all be taking their toll on the resource. However, apart from the physical damage to organic deposits recorded on some crannogs and imputed to the use of modern fertilisers and nitrate run-off (*e.g.* Barber and Crone 1993, Henderson 1998a), there are few quantitative data on the impact these factors have on buried archaeological remains, either at the regional or national scale (see Chapman and Cheetham 2002).

In contrast, the impact of more visible processes such as afforestation, mineral extraction and groundwater abstraction can be more easily quantified and appreciated. Until recently, the primary threat to the Scottish upland bogs came from afforestation and associated invasive works, together with the consequent lowering of the water table (Brooks and Stoneman 1997). Following the increased recognition of the nature conservation value of peatlands (*e.g.* Ramsar 1971 and amendments), the last decade has seen a decrease in forestry activities in these environments. Threats from new forestry have now largely ceased, and moves towards bog rehabilitation are underway in some areas (FC 2000). Once the ecosystems have returned to equilibrium, the buried cultural heritage will presumably benefit from the stable waterlogged conditions, but the damage caused to date may be irreversible.

Mineral extraction is identified as a threat, particularly to the raised bogs of central Scotland. These areas are potentially archaeologically rich (see above) but are often located over economically valuable mineral deposits, and as a consequence, decades of open-cast coal mining have altered the integrity of many bogs (Brooks and Stoneman 1997, 232). Evidence of subsidence is widespread, and open-cast mining has resulted in the complete removal of several areas of peat. The repercussions from these alterations may continue to impact in the future.

Whilst Scotland is a region with abundant water resources, the absence of any comprehensive control on water abstraction has resulted in shortages in certain areas, *i.e.* the Spey valley, Dumfries and Fife. The 1999 SEPA (Scottish Environment Protection Agency) report *Improving Scotland's Water Environment* highlights this problem and relates that abstraction in Dumfriesshire has lowered the water table to such an extent that some rivers are drying out. With climate change, the demand for abstraction for agricultural irrigation is likely to rise, with a consequent reduction in groundwater levels.

We must also remember that the burial environment, be it water or sediment, is not passive – even without the perceived threats it is constantly changing and evolving. Neglect has been documented as contributing to the deterioration of wetlands, and many sites continue to degrade due to interventions that may have taken place many years ago and of which there is no obvious visual sign, although the ecological changes continue (Brooks and Stoneman 1997). We simply do not know how much of our wetland heritage will survive without a significant loss of environmental and cultural evidence for future generations to investigate.

RECENT WORK

The last decade of the twentieth century has seen some momentum gathering, partly in recognition of these threats and partly out of a growing realisation that wetland studies in Scotland were stagnating. A condition survey of the crannogs of southwest Scotland, undertaken to investigate the degree to which the resource had diminished since the nineteenth century, revealed substantial losses (Barber and Crone 1993). The location and extent of crannog sites in the Lake of Menteith, Stirlingshire (Henderson 1998a), on the island of Mull, Argyll (Holley 2000), in the Beauly Firth (Hale 1999a) and in Loch Lomond (Baker and Dixon 1998) have been surveyed.

Following on from the southwest Scottish crannog survey, Buiston crannog, Ayrshire, was singled out for extensive excavation which revealed the quality of information existing at these sites (Crone 2000; Fig. 3.5). More limited excavation has taken place on the estuarine crannogs at Dumbuck in the Clyde (Hale 1999b) and at Redcastle in the Beauly Firth (Hale 1999a). The underwater excavation of Oakbank crannog in Loch Tay continues (Dixon 1982b, 1984, Dixon and Andrian 1992), fostering some valuable technical studies (Sands 1997) and engendering the construction of the Loch Tay Crannog Centre, which has been instrumental in raising the profile of this aspect of our wetland heritage. As a consequence of these surveys and excavations there is now an assemblage of 64 radiocarbon dates from crannog deposits (Barber and Crone 1993, Holley and Ralston 1995, Hale 1999a, b, Crone 2000), which has led to some attempts at synthesis (Crone 1993, Henderson 1998b).

The archaeology of the peatlands has also been addressed. The National Museums of Scotland has implemented a programme of radiocarbon dating of those organic artefacts in their collections whose isolated find spots in peat deposits means that there is no associated dating evidence (Sheridan 2002). The find spots of some of these artefacts have also been re-examined to elucidate the circumstances of their

Fig. 3.5 Some of the wooden artefacts found on the seventh century AD crannog at Buiston, Ayrshire (from left to right: churn lid, mallet, fragment of turned bowl, forked tool, bucket lid with handle) (Crown copyright: Historic Scotland).

deposition. For instance, survey and excavation in the area around the find spot of the Deskford carnyx (see above) has located Iron Age activity (Hunter 2001), while at Ballachulish Moss, the find spot of the eponymous wooden effigy, structures and deposits of Late Bronze Age date have been investigated (Clarke *et al.* 1999, Clarke and Stoneman 2001). An evaluation of the archaeological potential of Flanders Moss has been undertaken (Ellis 2001) and this led to the location and excavation of a Neolithic wooden platform at Parks of Garden, on the very edge of the Moss (Ellis *et al.* 2002; Fig. 3.4).

While the work described above has certainly contributed to our knowledge base, it has not been implemented as part of a comprehensive strategy which aims to prioritise and target sites on the basis of informed decisions about aspects such as their age, condition, status or cultural value. Historic Scotland, the government agency responsible for the preservation of the nation's heritage, has for some time recognised the need for a comprehensive policy for the management and preservation of the wetland archaeological resource (Hingley *et al.* 1999). To this end it has funded the establishment of two data-bases – the Scottish Wetland Archaeological Database (SWAD) and the Scottish Palaeoecological Archive Database (SPAD) – both of which are now available on the Internet (http://www.geo.ed.ac.uk/SWAD/ and http://www.geo.ed.ac.uk/SPAD/). SWAD was compiled from desk-based sources and is essentially a site/find spot focused summary of the known evidence for the cultural heritage of the wetlands. It was hoped that the data-base would help to identify those areas of wetland that were of national importance in terms of the condition, nature and extent of the cultural heritage they contained (Hingley *et al.* 1999).

A second phase of work was thus commissioned to test the predictive power of the data-base by interrogation and subsequent field testing (Ellis 1999), and this concluded that there was insufficient data in the data-base to rank known sites in terms of potential. Limited fieldwork indicated that the use of desk-based sources which provide mainly general accounts of past and present landuse, and current and future threats, fails to account for very localised environmental and landuse factors which impact on the status and condition of a site (Ellis 1999). Most importantly perhaps, SWAD only deals with what is already known; it reflects the serendipitous nature of many wetland finds and focuses on those geographic areas where previous researchers chose to work. In its present form it cannot be used to model the potential of other unexplored wetlands.

To summarise, until very recently there has been no systematic, sustained attempt to investigate the wetland archaeological resource. Most investigations have been site-specific, and consequently our knowledge of the resource is currently very patchy.

THE SWAP INITIATIVE

Building a comprehensive picture of current programmes of management and research centring on Scotland's wetlands was considered fundamental to the development of a well-focused archaeological programme. Since many natural heritage agencies aim to preserve tracts of wetland, archaeology could benefit by the preservation of sites and monuments contained within those wetlands, but only if the management regimes proposed take cognisance of the archaeology they contain. Thus, we hoped to prioritise more effectively those areas, both geographic and thematic, where research would facilitate the development of integrated management policies towards the cultural heritage of the wetlands. In this way we hoped to ensure that SWAP's agenda would be in tune with national natural environmental policies so that an integrated appreciation of the cultural landscape would embrace the broader issues of wetland archaeology.

To this end we consulted those key organisations whose activities impact in some way or another on the Scottish wetlands, in order to determine the degree to which the cultural heritage is recognised in their operations. Our consultations have highlighted a number of areas where slight adjustments in the activities and/or attitudes of at least some of the organisations could enhance the survival of the cultural heritage of the Scottish wetlands. It is clear that the 'invisibility' of the resource and the lack of available information on the subject are a hindrance in encouraging organisations to be more aware of the wetland cultural heritage and to be pro-active in its conservation. It is also evident that, in the absence of baseline data, prioritisation of geographic areas and/or thematic topics cannot be implemented as originally hoped. Therefore, it is now important to focus on establishing the nature, extent and condition of archaeological remains extant within the Scottish wetlands, and develop strategies on how to manage and monitor the resource. Thus, the development and implementation of methodologies aimed at locating and monitoring the resource is of paramount concern.

SWAP'S AIMS AND OBJECTIVES

It is against this background that SWAP established its aims and objectives. Our aim is 'The enhancement of our cultural heritage through the exploration of the wetland resource and its full integration into the interpretative frameworks of 'dryland' archaeology'. This integration will be achieved by focusing research within a series of hydrological catchments, rather than concentrating on discrete wetland sites. This will allow the relationships between diverse dryland and wetland archaeological sites and their landscape settings to be more fully investigated. As a result of the EU Water Framework Directive (2000/60/EC), many national environmental agencies will be required to work within catchment units, so by presenting the cultural heritage within the same framework we hope to encourage more active consideration of the archaeological resource.

The development of a coherent research agenda to fulfil the aims outlined above is clearly impeded by the lack of baseline data on the location and extent of the archaeological resource in Scotland's wetlands. A primary objective is, therefore, to establish the location and extent of archaeological deposits within the wetlands.

In keeping with national initiatives on sustainability and the presumption for preservation *in situ* implicit in national planning policy guidelines, conservation of the resource must also be an objective. It is likely that, with the limited resources currently available, it will only be possible to actively conserve the most important sites, and therefore the criteria necessary for ranking wetland sites must be clearly established. To do this, the condition and stability of selected sites must be determined and the nature of the processes impacting on them must be understood. Monitoring the resource is therefore an essential pre-requisite of successful conservation.

SWAP's objectives can be summarised thus: (a) to *establish* the location and extent of the resource, (b) to *monitor* the condition and stability of the resource, and (c) to *conserve* the resource with effective management. These are comparable to the objectives of the large and successful wetland projects undertaken by English Heritage in recent years, and we hope that within the Scottish context they will help to focus what could seem like an overwhelming project into a series of discrete and achievable tasks.

Obviously, it would be a Herculean task to establish the location and extent of the wetland resource throughout the length and breadth of Scotland. Instead, specific catchments will be selected and predictive models that can be used to determine those locations with the greatest potential for surviving archaeological deposits will be developed. Single artefact finds, which comprise 17.7% of the entries in SWAD, will by their very nature always be serendipitous and their location unpredictable. However, the location of structures relating to settlement, movement and economic activities will be predicated by variables such as underlying topography and geomorphology, while their survival will be determined by factors such as the local hydrology, depth of peat and alluvium, and the nature of existing threats within the catchment. Geophysical advances such as the application of Ground Penetrating Radar (GPR) to wetland environments may provide information on anomalies within wetland deposits that could signify archaeological remains (Clarke *et al.* 1999). By modelling these and other variables it may be possible to predict where in the wetland landscape we might expect to find archaeological deposits. Models may simply take the form of GIS databases collating these layers of information. A pilot study on the suitability of GPR to establish peat depths and the location and extent of the archaeological resource in Moine Mhor, Argyll is planned and follows on from an earlier desk based exercise to predict areas of archaeological potential. All of this information is stored on a GIS database.

Earlier survey work on crannogs in south-west Scotland had highlighted their vulnerability to changing agricultural practices (Barber and Crone 1993) and, consequently this region has been targeted as the locus for a long-term monitoring programme. Recent fieldwork in the region has identified prospective candidates within a number of catchments on the basis of accessibility and evidence of recent degradation and/or erosion (Henderson *et al.* 2003) and plans are afoot to implement monitoring on a number of these sites. The results of the monitoring programme will eventually feed into strategies for conservation.

CONCLUSION

The global issues that threaten wetlands worldwide apply equally to Scotland, where, from an archaeological perspective, the wetlands can be considered as either lacustrine or peatland. In the same way that the peatlands of Scotland have achieved international significance for their ecological properties and condition, the crannogs, as a resource found only in Scotland and Ireland, should likewise be seen as being of international archaeological importance.

Scotland has lagged behind in wetland studies in comparison with its British and European neighbours. One benefit of this is that SWAP will be able to harness the technological and methodological developments that have taken place elsewhere and apply them to the Scottish situation. In particular, we recognise that working closely with the natural environmental agencies from the outset will enhance the chances of implementing effective management strategies.

SWAP's work to date has demonstrated that the absence of much baseline data is a significant impediment in the formulation of strategies for the management and conservation of Scotland's wetland archaeological resource. Acquisition of those data must therefore be a major priority. It is also a major impediment to the formulation of research strategies, and we must never lose sight of the fact that the aim, in conserving the resource for future generations, is ultimately the understanding of our past. Disseminating knowledge about our wetland cultural heritage is the most effective means of ensuring the conservation and sustainability of the resource, and will be central to any programme of work that SWAP undertakes.

In the last pages of *Enlarging the Past*, Coles and Coles (1996, 157–8) presented a 'shopping list' of actions that they considered necessary to galvanise wetland archaeology in Scotland. These include the implementation of research projects into particular environments or monuments, fostering relationships with other natural environmental bodies, establishing the condition of sites and raising the profile of Scottish wetland archaeology. With this latest initiative we are hopeful that at least some of these actions will soon be implemented.

ACKNOWLEDGEMENTS

Much of this paper is based on a document summarising Phase 1 of the SWAP programme (Crone and Clarke 2001) and the authors are grateful for the support of Historic Scotland, the Society of Antiquaries of Scotland and the Scottish Trust for Archaeological Research in their endeavours. The authors would like to thank the other members of SWAP – John Barber, Mike Cressey, Alex Hale, Jon Henderson, Rupert Housley, Rob Sands and Alison Sheridan – for reading and commenting on this paper.

REFERENCES

Alexander Smith, J. 1868. On a Bronze Age ornament found in Banffshire. *Proceedings of the Society of Antiquaries of Scotland* 7, 342–7.

Anderson, J. 1885. Notice of a bronze cauldron with several kegs of butter in a moss near Kyleakin, in Skye; with notes of other cauldrons of bronze found in Scotland. *Proceedings of the Society of Antiquaries of Scotland* 19, 313.

Anderson, J. & F. Black. 1888. Reports on local museums in Scotland, obtained through Dr RH Gunnings jubilee gift to the society. *Proceedings of the Society of Antiquaries of Scotland* 22, 370.

Arnold, B. 1999. Archaeology on the shores of Lake Neuchatel: past and present, in B. Coles, J. Coles & M. Schou Joergensen (eds) *Bog Bodies, Sacred Sites and Wetland Archaeology*, 11–16. Exeter: Wetland Archaeology Research Project.

Baker, F. & N. Dixon 1998. Loch Lomond Islands Survey. *Discovery & Excavation Scotland 1998*: 23–96.

Barber, J. W. & B. A. Crone 1993. Crannogs; a diminishing resource? A survey of the crannogs of South West Scotland and excavations at Buiston Crannog. *Antiquity* 67, 520–33.

Brooks, S. & R. Stoneman 1997. *Conserving Bogs: the management handbook*. Edinburgh: The Stationery Office.

Cadell, H. M. M. 1913. *The Story of the Forth*. Glasgow: James Maclehose & Sons.

Campbell, E. & A. Lane 1989. Llangorse: a tenth century royal crannog in Wales. *Antiquity* 63, 675–81.

Chapman, H. P. & J. L. Cheetham 2002. Monitoring and modelling saturation as a proxy indicator for *in situ* preservation in wetlands: a GIS-based approach. *Journal of Archaeological Science* 29, 277–89.

Christison, R. 1881. On an ancient wooden image, found in November last at Ballachulish Peat Moss. *Proceedings of the Society of Antiquaries of Scotland* 15, 158–78.

Clarke, C. M., E. Utsi & V. Utsi 1999. Ground penetrating radar. Investigations at North Ballachulish Moss, Highland, Scotland. *Archaeological Prospection* 6, 107–21.

Clarke, C. & R. Stoneman 2001. Archaeological and palaeoenvironmental investigations at North Ballachulish Moss, Highland, Scotland, in B. Raftery & J. Hickey (eds) *Recent Developments in Wetland Research*: 201–14. Dublin: Seandálaíocht Monograph 2 & WARP Occasional Paper 14.

Coles, B. 1995. *Wetland Management: a survey for English Heritage*. Exeter: WARP Occasional Paper 9.

Coles, J. & B. Coles 1996. *Enlarging the Past*. Edinburgh: Society of Antiquaries of Scotland Monograph Series 11.

Crone, B. A. 1993. Crannogs and chronologies. *Proceedings of the Society of Antiquaries of Scotland* 123, 245–54.

Crone, B. A. 2000. *The History of a Scottish Lowland Crannog: excavations at Buiston, Ayrshire 1989–90*. Edinburgh: STAR Monograph Series 4.

Crone, B.A. & C. Clarke 2001. Scottish wetland archaeology programme; Phase 1. Unpublished STAR report.

Dixon, T. N. 1982a. A survey of crannogs in Loch Tay. *Proceedings of the Society of Antiquaries of Scotland* 112, 17–38.

Dixon, T. N. 1982b. Excavation at Oakbank crannog, Loch Tay; An interim report. *International Journal of Nautical Archaeology* 11, 125–32.

Dixon, T. N. 1984. Scottish Crannogs: underwater excavation of artificial islands with special reference to Oakbank Crannog, Loch Tay. Unpublished PhD thesis, Edinburgh University.

Dixon, T. N. & B. L. Andrian 1992. Oakbank Crannog, Loch Tay. *Edinburgh University Department of Archaeology 38th Annual Report*, 27–9.

Earwood, C. 1993. *Domestic Wooden Artefacts in Britain and Ireland from Neolithic to Viking Times*. Exeter: Exeter University Press.

Ellis, C. 1999. An Archaeological Assessment of the Scottish Wetlands. Unpublished Report for Historic Scotland.

Ellis, C. 2001. Realising the archaeological potential of the Scottish peatlands; recent work in the Carse of Stirling, Scotland, in B. A. Purdy (ed.) *Enduring records. The environmental and cultural heritage of wetlands*, 172–82. Oxford: Oxbow.

Ellis, C., A. Crone, E. Reilly & P. Hughes 2002. Excavation of a Neolithic wooden platform, Stirlingshire. *Proceedings of the Prehistoric Society* 68, 247–56.

FC 2000. *Forests and Peatland Habitats*. Forestry Commission Guideline Note.

Gilmour, S. 2001. Shorewatch: monitoring Scotland's coastal archaeology, *Pilot Phase 3*, Unpublished Report for the Council for Scottish Archaeology.

Hale, A. 1999a. Marine Crannogs: the archaeological and palaeoenvironmental potential. Unpublished PhD thesis, Edinburgh University.

Hale, A. 1999b. Past and present research on the Dumbuck marine crannog. *Glasgow Archaeological Society Bulletin* 42, 6–11.

Hall, D., C. E. Wells & E. Huckerby 1995. *The Wetlands of Greater Manchester*. Lancaster: Lancaster University Archaeological Unit.

Henderson, J. C. 1998a. A survey of crannogs in the Lake of Menteith, Stirlingshire, Scotland. *Proceedings of the Society of Antiquaries of Scotland* 128, 273–92.

Henderson, J. C. 1998b. Islets through time: the definition, dating and distribution of Scottish crannogs. *Oxford Journal of Archaeology* 17, 227–44.

Henderson J. C., Crone, B. A. & M. G. Cavers 2003. A condition survey of selected crannogs in south-west Scotland. *Transactions Dumfriesshire and Galloway Natural History and Antiquarian Society* 77, 79–102.

Hingley, R., P. Ashmore, C. Clarke & J. A. Sheridan 1999. Peat, archaeology and palaeoecology in Scotland, in B Coles, J. Coles & M. Schou Joergensen (eds) *Bog Bodies, Sacred Sites and Wetland Archaeology*, 105–14. Exeter: Wetland Archaeology Research Project.

Holley, M. 2000. *The Artificial Islets/Crannogs of the Central Inner Hebrides*. Oxford: British Archaeological Reports, British Series 303.

Holley, M. & I. B. M. Ralston 1995. Radiocarbon dates for two crannogs on the Isle of Mull, Strathclyde Region, Scotland. *Antiquity* 69, 595–6.

Hunter, F. 2001. The carnyx in Iron Age Europe. *Antiquaries Journal* 81, 77–108.

Lindsay, R. 1995. *Bogs: the ecology, classification and conservation of ombrotrophic mires*. Edinburgh: Scottish Natural Heritage.

McArdle, A. & I. Morrison 1973. Scottish lake-dwellings. Survey, archaeology and geomorphology in Loch Awe, Argyllshire. *International Journal of Nautical Archaeology* 2, 281–2.

MacGibbson, A. 1798. Parish of Kilmadock or Doune. No. 3, in J. Sinclair (ed.) *The Statistical Account of Scotland, Drawn Up from the Communications of the Ministers of the Different Parishes,* 20, 40–91. Edinburgh: William Creech.

Moloney, A. 1993. *Survey of the Raised Bogs of County Longford*. Dublin: Irish Archaeological Wetland Unit.

Piggott, S. 1959. A tripartite disc wheel from Blairdrummond, Perthshire. *Proceedings of the Society of Antiquaries of Scotland* 90, 238–40.

Ramsar 1971. *Convention on Wetlands of International Importance especially as Waterfowl Habitat*. Paris: UNESCO.

Sands, R. 1997. *Prehistoric Woodworking: the analysis and interpretation of Bronze and Iron Age toolmarks*. London: University College London, Institute of Archaeology.

Sheridan, A. 2002. The radiocarbon dating programmes of the National Museums of Scotland. *Antiquity* 76, 794–6.

Sheriff, J. 1796. Parish of St. Ninians, in J. Sinclair (ed.) *The Statistical Account of Scotland, Drawn Up from the Communications of the Ministers of the Different Parishes,* 18, 385–410. Edinburgh: William Creech.

Tait, C. 1794. An account of the peat mosses of Kincardine and Flanders in Perthshire. *Transactions of the Royal Society of Edinburgh* 3, 266–78.

Chapter 4

The Tumbeagh Bog Body and a Consideration of Raised Bog Archaeology in Ireland

Nóra Bermingham

INTRODUCTION

Raised bogs are undeniably rich repositories of the residue of thousands of years of human activity. This residue can take many forms, including, for example, wooden structures, artefacts and palaeoenvironmental evidence. Less common but generally better known are the finds of human remains or bog bodies. Raised bogs are also considerable peat fuel reserves, and it is the greatest irony that the industrial exploitation of bogs in the Irish midlands has allowed for a heightened recognition of the wetlands of the central plain as threatened reservoirs of archaeological heritage. For the past decade or so state agencies involved in peat extraction and conservation of the archaeological resource have grappled with this irony. Over recent years progress has begun to be made, largely in the form of mitigation of select sites, in dealing with the persistent discovery of archaeological sites and the continuous milling of peat, and one of the first major finds which had to be dealt with under a developing programme of archaeological mitigation was that of a bog body. In September 1998 the partial remains of a body were discovered by archaeologists in a cut-over bog in Tumbeagh, Co. Offaly, Ireland. The discovery represents the first archaeological excavation of a bog body in Ireland and the first find since 1978 (Delaney and Ó Floinn 1995).

Before describing the find and its excavation a brief background of the different agencies with responsibility for archaeological sites in industrial raised bogs is presented. All are involved in the excavation and post excavation of the Tumbeagh bog body, as well as being responsible for the treatment of raised bogs as an archaeological resource. The two main agencies are the Department of Arts, Heritage, Gaeltacht and the Islands (DAHGI), incorporating *Dúchas* (the Heritage Service) (Note: as of 2002 *Dúchas* became part of the Department of the Environment. The DAHGI is no longer in existence.), the National Museum of Ireland, and the Department of Public Enterprise to which *Bord na Móna*, the state's largest peat extraction company, reports. Within the DAHGI the protection and monitoring of archaeological sites falls to *Dúchas*. Legislation has been in place for many years which gives protection to known sites and sets out a series of procedures to which all developers, regardless of the scale of the development, should adhere if they threaten a known site or a site of archaeological potential. Notice of intent to develop within the immediate vicinity of a recorded monument must be submitted to *Dúchas*. As the custodian of Ireland's artefactual heritage, all finds, which are automatically owned by the state, must be reported and submitted to the National Museum of Ireland. This applies to objects found by members of the public as well as to those retrieved during the course of excavations (Deevy 1998).

In the mid-1940s, peatlands, by then regarded as the most significant indigenous source of fuel, became a fundamental part of national development plans concerned with self-sufficiency, employment and the provision of electricity for the industrial and domestic sectors. The earliest peat-fired stations at Portarlington, Co. Laois and Ferbane, Co. Offaly were built in the 1950s, followed in the 1960s by others elsewhere in Offaly, Longford and Mayo. Construction of a new station near Edenderry, Co. Offaly began in 1999 which is to be supplied directly by *Bord na Móna* from its 'fuel peat reserves' for the coming 30 years (www.bnm.ie 2001). Today around 55,000 hectares of peat throughout Ireland are under production, this includes raised and blanket bogs.

Archaeological discoveries from Irish bogs have been made since the nineteenth century but it was with the industrial exploitation of raised bogs that the discovery of archaeological structures became more common. In the 1960s occasional short exploratory excavations were conducted by the National Museum on roadways, most of which were already known locally but were increasingly exposed as a bog became reduced (Rynne 1961–3, 1964–5, 1965). The late 1970s saw the Museum excavate the pre-bog Mesolithic lakeside settlement of Lough Boora in Co. Offaly, exposed as a result of drainage and peat extraction (Ryan 1984). It was not, however, until the 1980s that raised bogs were unequivocally identified as zones of high archaeological potential through the survey and excavations of bogs in Co. Longford (Raftery 1996). Here several hundred sites were identified, ranging in date from the early Neolithic to the early Medieval period, and consequently the Irish Archaeological Wetland Unit (IAWU) was established. Funded by *Dúchas*, the IAWU have, since 1991, conducted surveys of *Bord na Móna* raised bogs in nine counties – Longford, Tipperary, Galway, Roscommon, Westmeath, Meath, Offaly, Mayo and Kildare. More than 2500 sites are now known, although it cannot be said that they all still survive. In general the majority of sites are incorporated into the Record of Monuments and Places, and as such are subjected to the same legislatory protection as other known sites.

The last ten years have not only seen an increasing number of archaeological sites identified but have also been a time of periodic discussions between *Bord na Móna*, *Dúchas* and the DAHGI. Clearly *Bord na Móna*, *Dúchas* and their respective government departments hold differing views on the industrial development of raised bogs and its impact on archaeological sites and finds within the bogs. To address this difference, *Bord na Móna*, the Department of Public Enterprise and the DAHGI have established *Agreed principles for the protection of wetland archaeology in* Bord na Móna *bogs*. First drawn up in 1998, the agreement was launched in the Autumn of 2000 (Anon. 1999, 2000). In it both parties acknowledge the other's position regarding industrial exploitation of bogs and its impact on the archaeological heritage. In addition, a formal bilateral committee serves to address the conflict of interest between peat extraction and safeguarding the archaeological resource.

Arising out of this, *Bord na Móna* have hired a consultant archaeological company which conducts re-assessments of bogs previously surveyed and limited excavations of targeted archaeological sites within these areas, the extent of excavation being agreed by committee. Prior to this development the first re-assessment of a previously surveyed area was carried out in 1998 by the IAWU on behalf of *Bord na Móna*, and it was arising from this re-assessment that the discovery of the human remains in Tumbeagh Bog, and discussed below, was made.

THE TUMBEAGH BOG BODY

Tumbeagh forms part of a larger bog complex known as Lemanaghan, and is situated in the western part of Co. Offaly (Fig. 4.1). Lemanaghan is around 1300 ha and was brought into production, at different

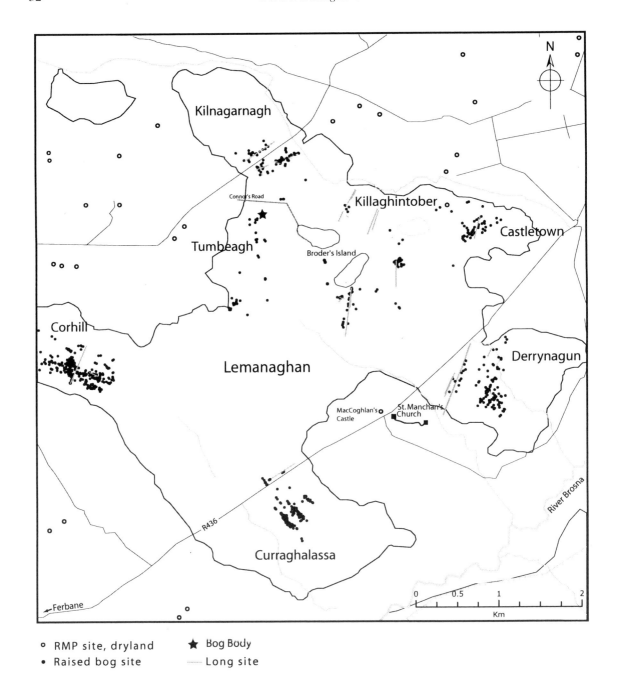

Fig. 4.1 The Lemanaghan complex situated in northwest Co. Offaly showing all known raised bog archaeological sites and those on the surrounding dryland (site locations derived from IAWU and RMP).

stages, since the 1950s. The first archaeological surveys took place in 1993 with
further follow-up seasons in 1994, 1996 and
1997. As a result of these surveys around
600 new archaeological sites were identified
(IAWU 1997).

In 1998, selected areas of Lemanaghan
were chosen for re-assessment by *Bord na
Móna*. One of these was Tumbeagh bog,
which had been surveyed by the IAWU the
previous summer. However, the area was
subject to further milling between 1997 and
September 1998. The distribution and number of archaeological sites known from
Tumbeagh had not changed greatly during
this time, although new structures were
identified and some known sites no longer
existed or had been heavily damaged
(Bermingham 1998). In a part of the bog
where no sites had previously been identified, the well preserved, although damaged,
remains of the bog body were found.

The archaeological survey of Tumbeagh
involved examination of the field surfaces
and drain faces (Fig. 4.2). Every 13 m,
drains 1 m wide are cut to allow unimpeded
drainage of the bog surface prior to milling.
As the bog reduces drains are re-cut.
Remarkably, during the course of the survey
a small, brown flap of skin was identified on
the surface, close to the centre of a field (Fig.
4.3). Lying next to this was a single broken
birch roundwood and a loose bone, subsequently identified as the navicular of the
left foot. Limited clearance of loose peat
from around the skin flap revealed some
blackened bone and more tanned skin. Small
white crumbs of body fat speckled the milled
peat lying on top of and around the
remains.

Fig. 4.2 Aerial view of Tumbeagh showing location of find in the bog and in relation to a nearby bog island. Dryland is just off to the left of the picture (photo: Dúchas *the Heritage Service).*

Fig. 4.3 Flap of skin as exposed on the field surface (photo: IAWU).

The positive identification of the find as human and the securing of it, particularly in terms of bringing
an end to milling on the field in which it lay, culminated in an excavation project conducted under the
direction of the author on behalf of *Bord na Móna* and *Dúchas*. The excavation was carried out in close co-
operation with Dr Máire Delaney of Trinity College Dublin and Mr Rolly Read, Head of Conservation in
the National Museum of Ireland.

Field excavation

As an archaeological excavation of a bog body had never before taken place in Ireland, a strategy was devised which would allow the recovery of the *in situ* remains and associated finds. Alongside this the recovery of disturbed or re-deposited material, the taking of peat samples for pollen and beetle analysis, the completion of a peat strati-graphic survey and the partial recording of the excavation by a film crew was under-taken. All of this had to be done in a 10-day period in early October just as the weather and conditions in the bog were degenerating.

The excavation was concerned with the recovery of the surviving human remains in a block of peat that would undergo excava-tion in the laboratory of the National Museum. This approach was known to be both productive and sensitive as excavations such as Lindow, England (Stead *et al.* 1986) and those in Denmark (Glob 1969), have shown.

From the find's location, at the highest point in the field, it was clear that milling had taken its toll, as parts of the body had already been removed. Consequently the excavation began by defining the extent of the remains through careful hand-excavation of a series of test pits in the immediate vicinity of the find spot, as well as a thorough cleaning, using trowels, of the wider field surface (Fig. 4.4). Testing showed the *in situ* material to be confined to a 1 m square area, a result in line with original predictions. At this stage it was likely that either the lower legs or the upper and lower parts of one leg had survived. The foot or feet lay to the south-

Fig. 4.4 Hand-dug test pits were put in place around the metre square in which the remains would eventually be isolated (photo: N. Bermingham).

Fig. 4.5 The heavily damaged hazel withy found on the field surface just to the south of the feet (photo: M. Moraghan).

southeast and the knee or thigh area lay to the north-northwest.

Following the removal of surface finds, which included a horizontal, worked, piece of roundwood, some smaller pieces of worked hazel and a withy or plaited wooden rope also of hazel (Fig. 4.5), a hand-dug moat was excavated around the metre square which contained the remains (Fig. 4.6). This central pedestal measuring 1 cubic metre was eventually reduced in size, following the excavation of further test pits, to a block measuring 0.80 × 0.50 × 0.30 m. Prior to reducing the block, but subsequent to testing, a monolith for pollen analysis was taken from one corner.

In order to remove the block from the bog safely, it was wrapped in cling-film and framed by a wooden box into which expanding polyurethane foam was poured. The foam hardened to form a protective shell, maintaining the block's integrity and also inhibiting aerobic activity and preventing loss of moisture either from the peat or the surviving human remains. Once prepared, the block was removed to cold storage at the National Museum, in advance of the laboratory excavation.

In the field, the presence of the original skin flap on the bog surface indicated that the human remains were damaged and more than likely had been subjected to continuous milling for at least five years. Given the manner in which bogs are milled, there was every chance that parts of the body could be recovered from elsewhere in the bog. Milled peat is transferred into stockpiles for subsequent collection by custom-made bog trains. During the process, anything already mixed with the peat is also automatically transferred.

To the west of the find spot was a stockpile formed by the re-deposition of partially dried, dusty peat including peat from the field on which the bog body find was made. The potential for finding bog body parts incorporated in the stockpile was considered to be high and it was decided to hand search the stockpile for traces of the body. This investigation began roughly in line with the find spot and then moved in a northerly direction for 36 m. Eighteen 2 m wide transects were established across the stockpile and each was hand searched (Fig. 4.7).

Fig. 4.6 The hand-dug moat nears completion (photo: N. Bermingham).

Fig. 4.7 The crew hand-searches the peat stockpile which was 4 m wide at the base and around 1.5 m high in the centre (photo: D. Jennings).

Despite the dust whipped up by the hand sifting of peat and the October winds the search proved to be fruitful with finds including some skin, a patella, a proximal tibial epiphyses and a single vertebra. In addition, from the field between the stockpile and the find spot, part of a human rib was found. This field had a heavy covering of loose peat and was raked but did not produce any additional finds.

Laboratory excavation

The laboratory excavation of the peat block took place in a number of stages, beginning with exposure of the upper surface of the block. At all times sterile gloves were worn and the remains were kept moist using de-ionised water. The peat that was removed from above the remains was used to cover them once the excavation was completed, and the temperature never exceeded 11–12°C. The same excavation procedure was repeated for the underside of the remains. This took place once the upper side was recorded, packed with damp peat, secured in a foam shell and inverted. This allowed excavation of the underside without having to lift or move the remains out of the position in which they lay.

The excavation took place quite quickly, with most time being spent on the descriptive, photographic and drawn records. Interestingly the remains proved more durable than anticipated, allowing longer periods of working time than originally estimated. All peat that was in immediate contact with the remains from the upper and lower surfaces was retained for beetle analysis.

Excavation revealed the lower legs of an individual, extending from just above the knee to the feet. Skin, bone, tendon, body fat, hyaline cartilage and toenails had survived (Fig. 4.8). Overall preservation was excellent, with the underside only slightly better preserved than the upper side. The feet had skeletalised and some of the phalanges were displaced as a result. Others had been pulled out of position by recent machine action. There was no visible pathology in evidence on the surviving limbs.

The excavation also located three birch stakes in the area of the knees. Initially it appeared that the stakes lay under the legs and it was thought that they had formed part of a light wooden structure on top of which the body may have been placed (Bermingham and Delaney 2000). As the excavation progressed, however, the stakes were found to have been damaged by milling machinery. They appear to have been turned more or less in their original location but their original position in relation to the legs was now uncertain. The stakes could as easily have lain over the legs as beneath them.

Fig. 4.8 The upper side of the surviving remains revealed in the first stage of the laboratory excavation (photo: National Museum of Ireland).

PRELIMINARY RESULTS AND CONSIDERATION OF THE BODY'S ORIGINS

Examination by Dr Delaney has shown that the remains are those of an individual between 16 and 18 years of age and of undetermined gender, an interpretation unlikely to change unless there are improvements in DNA analyses of human samples from raised bogs. Both male and female bog bodies are known from Irish bogs, and while the majority of these are male, the sex of a significant number of other individuals remains indeterminate (Ó Floinn 1995a).

Apart from the wooden finds, displaced skeletal elements and skin fragments, no other associated archaeological finds were made. The body of this young adult was isolated within the bog. There was no trace of clothing found in the area covered by the investigation and the legs and feet were bare. The body may have been naked or perhaps only the upper body was once clothed. However, leather or wool, materials known to preserve well in raised bogs, were absent. If clothing had been removed by milling machines then we perhaps could have expected to recover fragments on the milled surface or in the stockpile. Other fabrics which do not survive well in raised bogs, such as linen, could have been worn, and the degradation of a linen shirt or shroud would thus present a false picture of nakedness.

When the Tumbeagh remains were first discovered there was much speculation that the find might belong to the Iron Age European tradition of depositing bodies in bogs, where the bodies belonging to persons apparently executed in an elaborate and ritualistic manner occur (*e.g.* Glob 1969, Turner and Scaife 1995). At Tumbeagh the apparent nudity and the presence of stakes could perhaps support such an interpretation. However, dating has shown the Tumbeagh remains to be Medieval. Samples of skin, wood and peat submitted to the Centrum voor Isotopen Onderzoek (CIO) of Groningen University have shown the Tumbeagh youth to have lived in the 15th or 16th century AD (van der Sanden and van der Plicht forthcoming).

Most Irish bodies from raised bogs are medieval in date (Ó Floinn 1995b, 142). Historically it is known that bodies were deliberately buried in bogs, with such irregular burial perhaps being the result of a murder, a suicide, or a contagious illness. The bog may equally have become the final resting place following an accident; sometimes people just lost their way and became trapped (Ó Floinn 1995b). A key to understanding how or why a body should come to rest in a bog is the cause of death. Are there signs of fatal wounds or other ante-mortem injuries which might suggest premature death perhaps inflicted by another party? If not did they die of natural causes such as old age or sickness or exposure? Such evidence would help establish if the person was pushed or if they fell (Briggs 1995). However even in cases where bodies are complete the cause of death cannot always be established (*cf.* Meenybraddan woman – Delaney and Ó Floinn 1995). At Tumbeagh where the body only partially survives we have no way of knowing how this person died and must rely on a combination of the associated evidence and knowledge of other finds in order to suggest the possible cause/s of death.

For example if this individual had wanted to cross the bog they chose their route badly. A stratigraphic survey of peat faces in the vicinity, conducted by Dr Wil Casparie, found the location in which the body lay to have been very unstable and very wet. Safer places across the bog lay within 100 m, both to the north and the south. The physical environment might suggest that this person lost their way, that they became stuck in an area surrounded by rush fields and were unable to make their way to less difficult terrain. The position of the legs suggests that the most probable position in which the body lay was on its left side with its hips and knees slightly bent. A person could have been placed in this way but a stranded person could also fall or lie in this manner (Delaney pers. comm.).

However, other evidence suggests that the encapsulation of this body in the peat was not entirely independent of other people. The birch stakes driven into the peat by the knees could have served to pin the body down, thus implying that another party (or parties) was involved in the 'burial'. The reason as to why it was thought necessary to bury this youth in the bog is not known but there is a range of possibilities, some perhaps more or less sinister than others.

POSTSCRIPT

At the time of writing, the study of the human remains from Tumbeagh is far from complete. Some investigations, such as trace metal analysis, have been undertaken but others remain to be completed. Imaging which will include X-ray and MRI will take place in advance of conservation. In terms of archaeo-environmental aspects of the project, analysis of pollen and beetle samples taken during the course of the excavations has been funded by the Heritage Council of Ireland. At present the human remains are still packed in peat and stored in a refrigerator, and do not show any visible signs of decay or distortion.

The Tumbeagh find was made as part of an increasing archaeological presence in *Bord na Móna* bogs. This is likely to result in more bog body finds, perhaps more complete finds. Despite the fact that the Tumbeagh bog body only partly survives, it has allowed us the opportunity to develop, in Ireland, strategies and expertise that will permit better and more efficient treatment of future finds, even where these finds are not initially made by archaeologists.

ACKNOWLEDGEMENTS

I would like to thank the following people for their contributions to this paper. The samples as dated formed part of a research project on the reliability of radiocarbon dates of bog bodies being conducted by Wijnand van der Sanden, Drents Plateau, Assen, the Netherlands and Jan van der Plicht and Centrum voor Isotopen Onderzoek (CIO) of Groningen University.

REFERENCES

Anonymous 1999. A new structured approach to archaeology at Bord na Móna. *Sceal na Móna* 13, 14.

Anonymous 2000. Bord na Móna-NRA codes of Practice. *Archaeology Ireland* 14, 6.

Bermingham, N. 1998. Preliminary report on the excavation of human remains in Tumbeagh bog, Lemanaghan, Co. Offaly (98E452). Report prepared on behalf of *Dúchas* the Heritage Service and *Bord na Móna*.

Bermingham, N. & M. Delaney 2000. The Tumbeagh Bog Body, in J. Moore (ed.) *The Yearbook of the Institute of Field Archaeologists,* 40–42. Reading: IFA.

Briggs, C. S. 1995. Did they fall or were they pushed? Some unresolved questions about bog bodies, in R. Turner & R. Scaife (eds) *Bog Bodies: new discoveries and new perspective*s, 168–82. London: British Museum Press.

Deevy, M. (ed.) 1998. *The Irish Heritage and Environment Directory 1999.* Wicklow: Archaeology Ireland.

Delaney, M. & R. Ó Floinn 1995. A bog body from Meenybraddan, Co. Donegal, in R. Turner & R. Scaife (eds) *Bog Bodies: new discoveries and new perspectives,* 123–32. London: British Museum Press.

Glob, P. V. 1969. *The Bog People: Iron Age man preserved.* London: Faber & Faber.

Irish Archaeological Wetland Unit 1997. Filling in the blanks: an archaeological survey of the Lemanaghan bogs, Co. Offaly. *Archaeology Ireland* 11 (2), 22–5.

Ó Floinn, R. 1995a. Ireland: a gazetteer of bog bodies in the British Isles, in R. Turner & R. Scaife (eds) *Bog Bodies: new discoveries and new perspectives,* 221–34. London: British Museum Press.

Ó Floinn, R. 1995b. Recent research into Irish bog bodies, in R. Turner and R. Scaife (eds) *Bog Bodies: new discoveries and new perspectives,* 137–45. London: British Museum Press.

Raftery, B. 1996. *Trackway Excavations in the Mountdillon Bogs, Co. Longford 1985–1991.* Dublin: Transactions of the Irish Archaeological Wetland Unit 4.

Ryan, M. 1984 Archaeological excavations at Lough Boora, Broughal townland, Co. Offaly, 1977, in J. Cooke (ed.) *Proceedings Seventh International Peat Congress, Dublin, 1*, 407–413. Dublin: The Irish National Peat Committee.

Rynne, E. 1961–3. The Danes' road, a togher near Monasterevin. *Journal of the Kildare Archaeological Society* 13, 449–56.

Rynne, E. 1964–5. A togher and a bog road in Lullymore Bog. *Journal of the Kildare Archaeological Society* 14, 34–40.

Rynne, E. 1965. Toghers in Littleton Bog, Co. Tipperary. *North Munster Antiquity Journal* 9, 138–44.

Stead, I., J. B. Bourke & D. Brothwell 1986. *Lindow Man: the body in the bog.* London: British Museum Publications.

Turner, R. C. & R. G. Scaife (eds) 1995. *Bog Bodies: new discoveries and new perspectives.* London: British Museum Press.

van der Sanden, W. & J. van der Plicht, forthcoming. The radiocarbon dates in N. Bermingham & M. Delaney (eds) *The Tumbeagh Bog Body Project.*

www.bnm.ie 2001. *The Peatlands of Ireland.*

Chapter 5

The English Heritage Strategy for the Conservation and Management of Monuments at Risk in England's Wetlands

Ian Panter

World Wetlands Day 2002 saw the publication of the English Heritage strategy for wetlands. This strategy set out the approach that has been adopted to ensure more effective conservation, protection and management of the cultural heritage resource of wetlands (Olivier and Van de Noort, 2002). This paper looks briefly at the rationale behind the development of the strategy, and how the strategy has been implemented.

BACKGROUND

In 2001English Heritage commissioned the University of Exeter to collect and collate data on the degree of survival of wetland monuments and identify the threats that compromised long-term preservation (Van de Noort *et al*, 2002). This study, commonly referred to as the MAREW study (Monuments at Risk in England's Wetlands), formed the basis of the English Heritage wetlands strategy.

The archaeological and palaeoenvironmental potential of wetlands has long been recognised following programmes of work focussing on the four principle lowland wetland regions, namely the Somerset Levels, the Fenlands of eastern England, the North West region and latterly the Humber Wetlands region. However, the threats posed to this vulnerable resource are all too apparent and the data collected by the team at Exeter suggested that an estimated 10,450 wetland monuments had been destroyed or damaged during the last fifty years. Furthermore, at least 50% of the original extent of lowland peatland had been lost during the same period. The causes for these losses included increased drainage for agriculture, water abstraction, arable reversion, peat wastage and erosion, peat extraction and urban and industrial development.

The wholesale physical destruction of many peatlands has been brought about by both urban expansion as well as the demand for peat in horticultural and related activities. However, the recent discovery of a late Neolithic trackway and platform on Hatfield Moor (South Yorkshire) demonstrates the potential for "spectacular" discoveries even in those areas ravaged by peat abstraction. Hatfield Moor and the adjacent Thorne Moor are now undergoing active restoration by English Nature – a very positive step forward in preserving these wetland ecosystems indeed. The nationally driven policy for "brownfield" development may ease development pressure on extant wetland areas too.

Nevertheless, wetlands are still at risk from activities that impact upon the hydrological processes operating at both the "site" level and more importantly on the "landscape" scale (e.g. Lillie *this volume*).

Organic archaeological and palaeoenvironmental remains are preserved best where oxygen is excluded from the burial environment. Such anoxic conditions are established and maintained by saturation of the deposits through waterlogging which inhibits the activity of anaerobic microorganisms, which are the principle agents of decay of organic material. Any activity that results in a lowering of the water table will compromise preservation of the archaeological and palaeoenvironmental remains, as well as the wetlands themselves.

Saturation is only part of the equation; the quality of the ground and surface waters entering the wetland ecosystem is also critical. An influx of water rich in chemicals such as nitrates and phosphates can act to alter the equilibrium established in the burial environment. Decay processes can then be initiated even in the absence of oxygen (see Corfield, *this volume*). Regulations that have been introduced to limit levels of nitrates in water run off should go some way towards alleviating this problem

Fundamental to understanding the dynamics of a wetland ecosystem, and therefore understanding how preservation can be brought about, is more detailed knowledge concerning the hydrological processes functioning at both the "site" level as well as the "landscape" scale. Wetlands cannot and should not be considered in isolation – they exist as a unit operating within a much wider hydrogeomorphic context. Recent monitoring exercises at several rural wetland sites including Sutton Common in South Yorkshire (Van de Noort *et al*, 2001, Lillie *this volume*) have demonstrated this all too well. Preservation therefore must be focussed on entire wetlands rather than what has been described previously as "monument islands" (Darvill and Fulton, 1998).

The English Heritage strategy has been developed with this concept at its core – the strategy aims to protect, conserve and manage both archaeological sites within the landscape as well as the wetlands themselves.

How the strategy works

The strategy can be described as a high level framework that permits the development of specific policies and programmes within four broad themes, which are:

> **Management** – promoting practical mechanisms to conserve and protect the cultural heritage by developing guidance and best practice for the integration of cultural heritage and nature conservation in wetland management;

> **Outreach** – promoting and disseminating understanding and appreciation of the cultural heritage of wetlands by making the results of wetland research easily accessible to the general public, to landowners and managers, and to professional interests;

> **Policy** – promoting the cultural heritage interests of wetlands in the work of local authorities, national, international and intergovernmental agencies;

> **Research** – continuing with programmes of survey and excavation as an essential pre-condition for the development of successful management practices and promoting applied research to underpin good management of wetlands and to inform future policy development.

A number of programmes have already been developed and completed since the launch, and these serve to illustrate how the strategy has been instrumental in beginning to address the issues with wetland preservation.

For example, following the compilation of an inventory of nationally important wetland monuments and landscapes, outline management plans have been drawn up for each as part of the heritage management project initiated by the University of Exeter and commissioned by English Heritage (Van de Noort, 2002). A total of 25 plans have been prepared and published on the web, and whilst the majority of the plans refer to rural wetland sites, including Fiskerton in the Witham Valley and Star Carr in the Vale of Pickering, it is worthwhile noting that plans have also been produced for York, Carlisle and the City of London. Whilst the common perception is that wetlands are a rural issue, we must not neglect the nationally significant waterlogged remains that lie buried beneath many of our urban centres, often referred to as "urban wetlands" for want of a better term.

Neither must we neglect the fact that we are not the only agency concerned with wetland preservation, and underpinning the strategy is the desire to develop partnerships between archaeologists and non-archaeologists alike. As alluded to above, effective preservation will be dependent upon managing the wetland landscape in its entirety – this is perhaps the only effective way of having some degree of influence over the key hydrological processes, short of purchasing the land and isolating it from its hydrological context through the construction of clay bunds or similar engineering schemes. Partnership is crucial and we have begun to develop further conservation management strategies with organisations such as the Environment Agency and English Nature to ensure that such plans provide a "wide-range of cross-disciplinary benefits including *in situ* preservation and archaeological research, nature conservation, recreation access and local co-operation and ownership".

One of the "headline" statistics from the MAREW study was that 72% of local authorities have no policies on the identification, assessment, preservation or management of wetland archaeology. This is not surprising given the fact that there are very few wetland-specific policies embedded in the planning process, and as wetlands are under threat from continued urban expansion and other demands then this is a situation where further assistance is urgently required. To partly rectify this situation a wetland GIS–based resource has been developed in order to provide base-line data on the extent of the various types of wetland landscape (lowland and upland peat, alluvial etc.) within each planning authority. The system provides an assessment of the wetland archaeological potential based on existing data from previous research throughout England, and will hopefully assist in the formulation of appropriate mitigation strategies relevant to existing planning legislation.

Outreach and education is also a core tenet of the strategy. To this end, English Heritage will continue to support and participate in World Wetland Day activities to promote cultural heritage issues and will continue to contribute to other awareness raising initiatives.

FUTURE DIRECTIONS?

The strategy makes reference to a tranche of programmes and initiatives that are either ongoing or will be developed over the ensuing years. Many of these will be subject to prioritisation and others will be developed in partnership with other organisations. It is not possible to list all the projects that are proposed but a few are described below to demonstrate the directions in which the strategy is moving:

Continue with the programme of identifying monuments at risk in England's wetlands

The focus to date has been on the potential of lowland rural wetlands and associated threats and it is recognised that similar studies to MAREW need to be developed for the upland peat zones, urban waterlogged deposits and inter-tidal deposits.

The development of appropriate techniques capable of adequately assessing the archaeological potential of wetlands

The desire for wetland habitat re-creation is likely to continue, and whilst such moves are welcome, these schemes are often centred around existing wetlands where the potential for cultural heritage remains are going to be high. Evaluating this potential is problematic though as "conventional" methods have their drawbacks. Geophysical techniques tend not to respond well in saturated environments, although the potential of ground penetrating radar would appear to be improving and would perhaps benefit from further research. Augering and test pitting are useful but often unreliable at locating linear features such as trackways unless large-scale extensive surveys are employed. It is unlikely that a single means of assessment will work alone; instead a suite of techniques will be required. A corollary to this will be the production of good practice guidance for assessing wetland potential that would be of benefit to all working in this area.

In situ *preservation*

There are a number of specific issues that require further study, including the effect of chemical contaminants (for example nitrates and phosphates) on wetlands under pasture, and the effects of re-wetting artefacts and ecofacts in partially desiccated wetlands.

DISCUSSION

In addition to the above, one key issue that warrants further study is that of the effects of climate change on long-term preservation of wetlands. The response of wetland ecosystems to changes in rainfall and temperature is currently not well understood and therefore further study into the dynamics of the hydrological cycle and how this is likely to alter with changing rainfall patterns is necessary.

However, this is clearly an issue that extends beyond the boundaries of cultural heritage and wetland preservation. Wetlands are known to be one of the largest repositories of terrestrial carbon, which will be released into the atmosphere as CO_2 in ever greater quantities if wetlands continue to be ploughed and drained (thereby contributing to the cycle of global warming and climate change). Therefore it is in everyone's interest to ensure long term preservation of wetlands.

To conclude, there is no doubt that the publication of the English Heritage strategy has helped raised awareness of the cultural heritage issues of wetlands throughout the archaeological and non-archaeological communities alike. However whilst the strategy has been in existence for only three years there is still much to do in order to preserve these unique ecosystems and the cultural heritage resource that they contain.

REFERENCES

Darvill, T. & A. K. Fulton. 1998. MARS: the Monuments at Risk Survey of England, 1995, Main Report. London: Bournemouth: School of Conservation Sciences and English Heritage.

Olivier, A., & R. Van de Noort. 2002. A Strategy for the Conservation and Management of Monuments at Risk in England's Wetlands. London: English Heritage and Exeter University.

Van de Noort, R., Fletcher, W., Thomas, G., Carstairs, I. & D. Patrick. 2001, Monuments at Risk in England's Wetlands, Exeter. Available as a pdf download from: http://www.projects.ex.ac.uk/marew/

Van de Noort, R., Chapman, H.P., & Cheetham, J. L. 2001 *In situ preservation as a dynamic process: the example of Sutton Common, UK.* Antiquity 75 No. 287, 94–100.

The outline management plans for the 25 wetland sites and landscapes produced by the University of Exeter can be downloaded at: http://www.sogaer.ex.ac.uk/wetlandsresearch/hmew/plans.htm

The strategy document for Conservation and Management of Monuments at Risk in England's Wetlands is available as a pdf download from: http://www.english-heritage.org.uk/upload/pdf/wetlands_strategy.pdfChapter 5

Chapter 6

"Value in Wetness": The Humberhead Levels Land Management Initiative

Stuart Pasley

INTRODUCTION

The "Humberhead Levels" of Yorkshire and North Lincolnshire, England – the large, flat, predominantly arable agricultural area surrounding the top end of the Humber Estuary – is an area with a fascinating history, much of it closely related to water and land management (Fig. 6.1). The Internal Drainage Boards (IDBs) have reclaimed the wetlands for the benefit of agriculture and the people who live there. Once drained, the silted-up lake bed that we see today provides one of the most fertile agricultural areas in England, but water management is still critical for many of the interests within the area:

– Flood defence is critical for settlement

– Water abstraction is critical for domestic, agricultural and industrial users

– Drainage is critical for agriculture

– Water availability is critical for wetland heritage

Water management issues are therefore critical for many interest groups. The degrees and perspectives vary. The primary objectives vary – and may even conflict – but they share a deep interest in the same basic issues and therefore share the need to address the same basic problems. By working together, we believe that the various interests can find shared solutions to these shared problems.

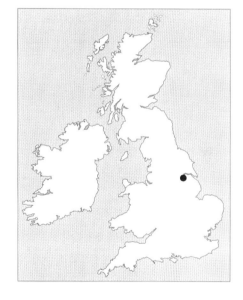

Fig. 6.1 Location map for the Humberhead Levels.

EFFECTIVE WATER MANAGEMENT

Trying to balance all of the different demands for water management outlined above is difficult and to date policy has not yet 'cracked it'. That's why we are seeking to look at water management through the "Value in Wetness" initiative. It is about learning lessons. Hopefully some real things are happening on some real sites – such as at Sutton Common, South Yorkshire, England – but the main activity is about learning lessons so that we can seek to influence policy development at the regional, national and European levels. At these levels approaches such as the England Rural Development Programme and the Common Agricultural Policy have a big impact on the way water and land are being managed.

In particular, we seek the following *Key Outcomes:*

1. Integrated economic, social and environmental objectives for sustainable water and land management in arable areas researched and defined;

2. Proof that alternative management of water could bring environmental, social and economic benefits of high value;

3. Practical methods of achieving the defined integrated objectives (including small site demonstrations);

4. Practical ways of implementing the EU Water Frameworks Directive demonstrated. Recommendations concerning the changes required to other policies to assist and add value to implementation of the Directive;

5. Analysis of the potential for the Internal Drainage Boards to have a broader and more positive role in water management.

Our starting point has been recognition that concerns and problems associated with water management in the Humberhead Levels are not confined to any single interest group. We believe there is scope to find common solutions to shared problems and we therefore aim to integrate social, economic and environmental demands upon water. It must be emphasised that an experimental, voluntary and incentive-led approach underpins our thinking.

To achieve these aims, we have embarked upon a five-year action programme to:

–   Involve all appropriate stakeholders in an open and experimental process.

–   Agree – through research and consultation – the integrated economic, social and environmental objectives that Government policies should be seeking to achieve within the Humberhead Levels.

–   Run a series of land use experiments in order to gain a better understanding of the interrelationships between water management, economically viable land uses, social benefits and environmental conservation.

–   Co-ordinate a monitoring network to build up a picture of water levels and qualities across the area.

–   Map the water network and associated wetland sites into a computer based Geographic Information System to allow better analysis – and perhaps modelling – of the linkages and interdependencies between them.

–   Identify – through research and modeling – optimal policy mechanisms by which Regional, National and European Governments can seek to deliver the agreed objectives.

–   Promote findings and lessons to all levels of Government in order to influence policy development to the benefit of the Humberhead Levels.

The Initiative is being developed in partnership with the Environment Agency, Grantham Brundell and Farran (Consulting Engineers who act on behalf of many Internal Drainage Boards), English Heritage and English Nature. The National Farmers Union and the Country Land and Business Association have welcomed the work.

INTEGRATING OBJECTIVES

The Initiative's activity can be summarised by three questions. Firstly, why change water management? What are the forces that are pushing us or should be pushing us towards changing the way we are currently doing things? Secondly, what changes will bring about benefits? If we are working on an assumption that there are benefits to be gained, then what changes are needed to attain them? and thirdly, how do we bring about that change?

We are looking at nine pilot sites as part of this learning process. "Value in Wetness" has an interest in the detailed archaeology, but it is not our main reason for being involved. English Heritage and the Universities [such as Hull and Exeter] provide the expertise. We have an interest in the biodiversity value, but again, it is not our main reason for being involved, that is English Nature's area of expertise. What we are interested in is what we can learn in terms of how the sites are being managed, and the implications for the way that other sites of this sort should be managed – and, therefore, how policies need to be shaped for the future.

DISCUSSION

We believe there is scope to influence four policy targets to benefit water and land management within the study region:

– The Government is in the process of reviewing the England Rural Development Plan, the way in which a lot of European money is delivered to rural areas within England. What can we learn in terms of how that needs to change for the future?

– The Curry Commission was established to address the problems facing British farming following the Foot and Mouth Disease outbreak in 2001. One of the recommendations was that more should be done to "farm water", directly linking agriculture to water management. So we have a concept, but what does it actually mean in practice? Are there things we can learn from our experiences that can help turn it into a profitable activity for farmers? The Government's response to the Curry Commission's report was the Strategy for Sustainable Farming and Food (SSFF). This is being implemented in Yorkshire and The Humber region through the "Framework for Change", which has identified the Humberhead Levels as a priority area for action, building upon the work of the Value in Wetness Initiative.

– The Common Agricultural Policy is the source of large amounts of money for the agricultural community. It has recently been reformed, but is expected to be reformed again 2005/2006. Again, things that we are learning through the Value in Wetness Initiative will be feeding into the discussions.

– Finally, also coming from Europe, there is the European Union Water Framework Directive. This establishes some fundamental rules about the way water will be managed in the future, but they have to be made practical. We hope that the things that we are learning from the Value in Wetness Initiative will help turn these European rules into practical procedures for use in England.

Chapter 7

Prehistoric Hunter-Gatherers in Wetland Environments: Theoretical Issues, Economic Organization and Resource Management Strategies

George P. Nicholas

INTRODUCTION

Wetlands, in their many different manifestations, have long been a significant feature in the global landscape. The term itself refers to a wide array of seasonally inundated and/or semi-terrestrial lands, including swamps, marshes, bogs, fens, estuaries and wadis that span subarctic to tropical to arid settings (Finlayson and Moser 1991, Mitsch and Gosselink 2000). The unique ecological values that wetlands possess are closely linked to their transitional state between terrestrial and aquatic systems. They thus tend to have high primary productivity levels and a dependable water supply, two of many factors that contribute to supporting an impressive variety of floral and faunal communities.

Although their number, density and extent have diminished significantly in the past several centuries, especially in recent decades, in North America and elsewhere, wetlands remain a prominent feature in most locations. Today, they are valued for their contributions to water purification, hydrologic regime stabilisation, plant and animal habitat, and environmental aesthetics. However, despite the fact that the importance of wetlands is now more widely recognised and understood than ever before, they remain a politically contentious issue, as illustrated in the United States by recent changes to federal conservation and protection policies, and by continuing challenges to protect them elsewhere.

Wetlands have also had a significant role in human affairs for a very long time, perhaps extending to the origins of the *Homo* lineage. Although this is best documented in the historical and archaeological records of the last few millennia, as represented by trackways, inundated sites and even bodies (*e.g.* Coles and Coles 1986, 1989), the association between people and wetlands spans both the Pleistocene and Holocene periods (*e.g.* Coles and Lawson 1987, Mellars and Dark 1998, Nicholas 1998a). This is evidenced, for example, by the presence of hominid sites near what would have been marshy, lakeside settings in the Plio/Pleistocene landscapes of Africa (Walker and Leakey 1993), and by site distribution patterns and other data for the Middle Palaeolithic in the Sahara (Wendorf 1993). In addition, the Upper Palaeolithic in the Mediterranean (Higgs 1961), the late Pleistocene in South America (Dillehay 1988), the early Holocene in North America (Kuehn 1998), the Mesolithic in Europe (Bonsall 1989) and the late Holocene in Australia (Meehan *et al.* 1985), are but a few examples of such associations.

In many parts of the world, wetlands appear to have been frequently utilised by those small-scale societies generally referred to as hunter-gatherers. This association is strongly supported by both ethnographic and archaeological data. In the Great Basin region of the western United States, for example, this is clearly illustrated by ethnographic groups, such as the *Toidikado* (Cattail-eater Northern Paiute) associated with Stillwater Marsh, Nevada (Wheat 1967, Fowler 1982), and their prehistoric counterparts (Janetski and Madsen 1990, Hemphill and Larsen 1999, Kelly 2001). This is also the case in the swamps of the Kakadu region of northern Australia, where our knowledge of wetland utilisation also extends from the present to the distant past (*e.g.* Jones 1985, Isaacs 1987, Meehan 1991). Figs 7.1 and 7.2 identify the locations of these and other sites mentioned in the text.

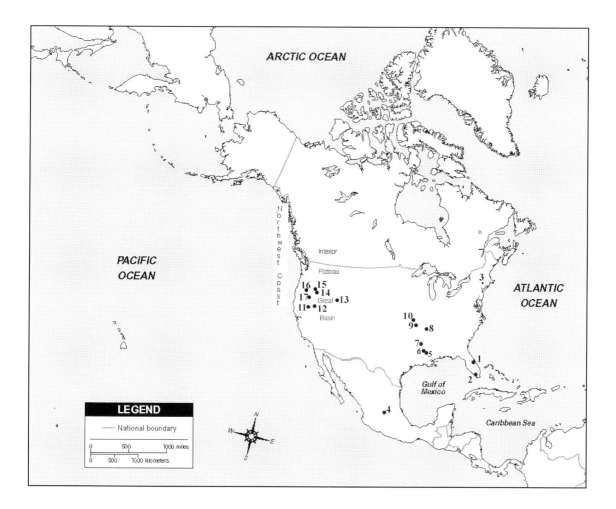

Fig. 7.1 Prominent North American sites and locations mentioned in the text.
Key: 1. Windover, 2. Calusa, 3. Robbins Swamp, 4. Teotihuacan, 5. Watson Brake, 6. Poverty Point, 7. Sloan, 8. Carrier Mills, 9. Cahokia, 10. Koster, 11. Stillwater Marsh, 12. Spirit Cave, 13. Great Salt Lake, 14. Diamond Swamp, 15. Malheur Lake, 16. Lake Abert/Chewaucan Marsh, 17. Nightfire Island.

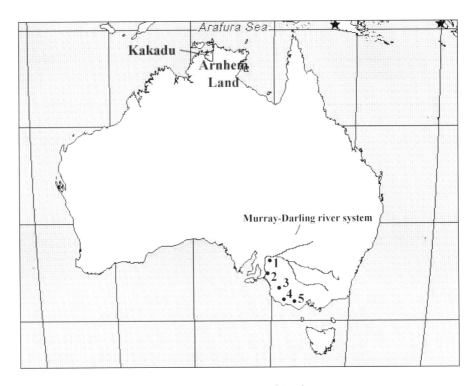

Fig. 7.2 Prominent Australian sites and locations mentioned in the text.
Key: 1. Lake Victoria, 2. Roonka Flat, 3. Toolondo, 4. Caramut, 5. Mt. Eccles, Lake Condah, MacArthur
Cree area (all locations are approximate).

The value of wetlands to human societies, particularly small-scale groups, is closely linked to a variety of ecological factors that include high values for resource productivity, reliability and diversity, and also the contribution of wetlands to landscape heterogeneity. In the Pacific Northwest Coast, resources such as wapato (*Sagittaria latifolia*) and cattail (*Typha latifolia*) were widely and intensively utilised as staple foods, as well as for more utilitarian uses. The economy of the Klamath of Oregon was based on fish and *wokas*, pond lily seeds (*Nuphar polysepalum*) (Oetting 1999, 212). While it is often difficult to demonstrate resource use in more distant times, especially of plants, wetland flora and fauna are often found in archaeological sites (*e.g.* Thomas 1985, Legge and Rowley-Conwy 1986, Perry 1999). In addition, even when such materials are absent due to preservation biases, the location of sites adjacent to, or sometimes even within, wetlands may provide indirect evidence of this resource exploitation (although there are clearly many non-economic factors affecting site location).

In a recent article, Nicholas (1998a) reviewed the long-term and global extent of the association between hunter-gatherers and wetlands, primarily in terms of different scales of economic relationships and landuse patterns (Fig. 7.3). Aspects of this are represented in the archaeological record of different regions in terms of site distribution patterns (*e.g.* McBride 1992), presence of wetland-associated flora and fauna (*e.g.* Schmidt and Sharp 1990), evidence of human modification of wetland environments (*e.g.* Lourandos 1987, Builth 1999) and other indicators. Additional research will flesh out (perhaps literally) both the evidence for, and our understanding of, this relationship. Of course, this is not to imply that all wetlands were

TIME

		Short-Term	Long-Term
S **P** **A** **C** **E**	Local	Individual Sites/Wetlands and Annual Land-Use Range	Culture History and Process of Adaptation
	Regional	Cultural Diversity and the Cultural Landscape	Landscape Ecology and the Regional Archaeological Record

Fig. 7.3 Four scalar aspects of human landuse (Nicholas 1998c, 3).

productive or attractive, as there is considerable variability between them and, in addition, systematic archaeological surveys in their vicinity may reveal no evidence of human presence (*e.g.* Nicholas 1994, 31), but then again, not all activities leave an archaeological signature.

CHALLENGING THE *STATUS QUO*

Despite considerable variation in wetland composition, productivity and related factors, the general notion that wetlands have had an important role in historic and ancient human ecosystems is widely supported. To date, we have learned much from the study of wetlands and hunter-gatherers for the reasons noted above. Continued study of the archaeology and human ecology of such settings offers many opportunities to explore short-term patterns of human behaviour, as well as longer-term responses to environmental change during the Holocene – issues that the present author has explored elsewhere (Nicholas 1988, 1998a, in prep.). However, in contrast this paper moves in a different direction, by asking 'how might knowledge of the prehistoric use of wetland environments contribute to increasing our understanding of small-scale societies?'

To date, wetlands-based research has generally focused on identifying site distribution and subsistence patterns, or describing preserved technologies and human remains. Our knowledge of the types and extent of means by which temperate swamps or northern bogs, for example, were utilised has increased substantially as a result. However, no matter how interesting this research is, it has remained peripheral to some of the 'big issues' in archaeology, such as the trend towards sedentism, the development of cultural complexity and the origins of food domestication. There are a few exceptions to this statement, such as Kelly's (2001) study of sedentism/mobility at Stillwater Marsh, but most wetlands-based research continues to address wetlands-focused questions. Nonetheless, new approaches to the archaeology and human ecology of wetlands may yield important insights into the foraging spectrum.

The focus here is to examine some of the ways that wetlands provide an avenue to re-evaluating and illuminating hunter-gatherers. To do so, two complementary questions are posed. First, what does the relationship between hunter-gatherers and wetlands reveal about how small-scale societies function and change? Of interest here is the degree to which wetlands may have influenced, controlled or channelled shifts in population density, the development of hierarchical socio-political systems and other features atypical of hunter-gatherers. The second question is how does this research contribute to evaluating ideas about the representativeness of ideas about past hunter-gatherers and the nature of their archaeological record? Identifying the broadest evidence for past human behaviour, landuse patterns and related topics is a pre-requisite not only for making sense of the archaeological record and the human lifeways it represents, but also it has significant implications for cultural resource management.

The discussion that follows has a strong processual orientation, and builds upon an extensive knowledge base derived from archaeology, ethnography and environmental studies. The overall goal of this paper is to de-marginalise wetland archaeology. It is not intended, however, to suggest that wetlands are the key to everything, but that they are nonetheless an important setting that has often been overlooked by archaeologists and others. For reasons of convenience and focus, the case studies cited here are drawn primarily, but not exclusively, from North America and Australia. A much more inclusive approach is adopted in another work now in progress (Nicholas in prep.).

THEORETICAL PROSPECTS

The peoples that we refer to as hunter-gatherers have long been considered as a benchmark in the story of humanity. Regardless of how they have been defined, our perceptions of hunting and gathering peoples (and indeed of the concept itself) have changed significantly over time. Once positioned on the lower (indeed lowest) rungs of the cultural evolutionary ladder constructed by Edward Tylor (1871) and subsequently Lewis Henry Morgan (1877), in recent years they have taken the helm as ecologically savvy peoples occupying a special niche (Bettinger 1991, Lee and Daly 2000). Hunter-gatherers have also attracted our attention owing to their once ubiquitous tenure, to their frequent positioning as analogues for prehistoric societies and to what many have viewed as the practical advantages of working with small-scale social constructs.

Many of the now-classic notions of hunter-gatherers that developed in the 1960s took form at the 1966 *Man the Hunter* conference (Lee and Devore 1968), and these have left a lasting impression. Our understanding of this general way of life was, and continues to be, associated with a series of frequently cited characteristics that include: a high degree of mobility, small-sized groups, a low population density, limited food storage, a generally egalitarian political system, limited material culture, absence of territoriality and an overall light footprint on the landscape – all components of what has come to be known as the generalised foraging model. The small number of non-agricultural societies that were clearly different, notably the Kwakiutl, Nu-Chal-Nuth and other ranked or stratified societies of the Northwest Coast of North America (see Suttles 1990), were usually considered hunter-gatherers in farmers' clothing, so to speak. By the 1970s, the traits noted above were firmly entrenched in the literature; frequently viewed as living examples of 'Stone Age peoples', the !Kung, the Mardudjara and other groups came to be the most widely-cited, and studied, representatives of this vanishing way of life.

Hunter-gatherers have had a particularly important role in archaeology in the last half-century in North America, Australia and other locations where much of the archaeological record is seen as a product of this way of life. As Kelly (1995, 338) has observed, 'archaeologists are perhaps even more susceptible than social anthropologists to the urge to create a hunter-gatherer stereotype. Given the usually impoverished nature of the archaeological remains of hunter-gatherer societies, especially those of the Pleistocene, archaeologists understandably are tempted to look elsewhere for ways to reconstruct the past.' Binford (1980), for example, used ethnographic G/wi San and Nunamiut data to develop his foragers/collectors model as a means to illuminate prehistoric settlement patterns, while Gould (1990) turned to the Ngatatjara to interpret ecological relationships and rock art. In fact, two of the most important recent studies of hunter-gatherers were written by archaeologists (Bettinger 1991, Kelly 1995). There is also widespread recognition today by both archaeologists and ethnographers that the term 'hunter-gatherers' is, at the least, deceptive because it obscures the diversity represented by what is more accurately described as gatherer-hunters or hunter-fisher-gatherers and other combinations.

Beyond semantics, what is potentially wrong with our reliance upon the 'classic' notions of hunter-gatherers? There are two major limitations, both of which constrain significantly or channel our understanding of past cultural systems. The first is that there is substantial variation between hunter-gatherer societies in terms of subsistence strategies and economic organisation, type of socio-political system, group size, degree of mobility and many other key factors, as Kelly (1995) and others have documented. The second is that most of what we know about these small-scale, non-agricultural societies is based upon groups living in relatively marginal environments (*i.e.* the Arctic, the Great Basin, the Kalihari, the Western Desert of Australia and tropical rainforests), and thus these will not be representative of their counterparts living in more temperate environments. In other words, before they were displaced by horticulturalists thousands of years ago or impacted by colonialism and disease more recently, hunter-gatherers occupied the most attractive places in the world, not just the marginal locations they were inhabiting in the last century.

If we wish to learn what hunter-gatherers may have looked like in 'attractive settings', one logical place to look is in wetland environments. After all, temperate swamps, marshes and estuaries rank among the most productive of all ecosystems (Whittaker 1975, 83), and possess other characteristics that make them ecologically distinct places in the landscape. These are also locations largely overlooked by archaeologists in the past. Depending upon their type, size and composition, wetlands could have made substantial contributions to the resource base available to hunter-gatherers (Nicholas 1998a) and otherwise have supported lifestyles outside the range of contemporary hunter-gatherers.

To illustrate the potential contributions of wetlands-based research to illuminating hunter-gatherer lifeways, the following sections review two productive and important topics – first, wetland resources and economic organisation; and second, resource management strategies and landscape modification. These are complemented by a discussion of two additional topics – landuse patterns and the question of mobility versus sedentism, and models of socio-political organisation – that appear later in this volume (see Chapter 20).

WETLAND RESOURCES AND ECONOMIC ORGANISATION

People have been attracted to wetland environments for many reasons – for the food and other resources found there; for the security they offer, in terms of both places of refuge (*e.g.* McBride 1992) and places known to contain relatively secure, reliable resources that can be returned to; and for social or ceremonial reasons, such as ties to important Dreamtime locations in Australia, *e.g.* 'Nerrima has many swamps where they had their big dancing grounds; it is also a sacred place' (David Mowaljarli, cited in Isaacs 1991,188).

The primary association that hunter-gatherers have had with wetland environments is most clearly linked with hunting, harvesting or resource management practices. The attractiveness of wetlands in this context can be correlated to five variables that define the nature of the resource base, namely type, diversity, productivity, reliability and seasonal availability of the flora and fauna sought. We can also look at the positioning of wetlands as part of the larger landscape, and consider the degree of contrast between wetlands and other environmental settings, as well as degrees of environmental heterogeneity (or patchiness) and other variables. Wetlands tend to have high values for all of these elements, making them often (but not always) economically important settings.

A wide array of plant resources are found in wetland settings, ranging from grasses and sedges, to tubers, to trees. Some of these were, and still are, economically important to hunting and gathering people as primary, secondary or emergency foods, and for systematically or fortuitously harvested materials for utilitarian purposes (see Finlayson and Moser 1991, Nicholas 1998a). For example, water lilies (*Nuphar* sp., *Nympheae* sp.) were collected only as food, bullrush (*Scirpus acutus*) was utilised both as food and building material, and reed (*Phragmites* spp.) was used only for mats, arrow shafts and similar purposes. Many other plants were not utilised by people but figured prominently in the diet and habitat of wetland-associated fauna such as moose (*Alces alces*), lechwe (*Kocus leche*) and ducks (*Anas* spp.).

Some wetland plants were particularly important as staple foods, often growing in profusion and being easily harvested. Cattail, which is found almost everywhere, was perhaps the most widely utilised of these. A multi-purpose plant, it is known ethnographically to be a nutritious food, and a versatile material for mats, baskets, sandals and a host of other applications (Fig. 7.4). Although its calorific return value, at 172 kcal/hr, is lower than many other gathered foods, the edible pollen has a return rate of up to 9,000 calories/hour (Simms 1987, 133). Cattail is occasionally found at archaeological sites. At Hidden Cave in Nevada, it is represented in mat, basketry and sandal fragments, and by quids (Goodman 1985), with significant quantities of its pollen present in coprolites (Wigand and Mehringer 1985, 118). There is also some evidence of cattail pollen being present in early European sites such as Coombe Grenal in France (Binford, cited in Wendorf 1993, 355). One emerging field of study that is shedding additional light on prehistoric dietary patterns and preferences is isotopic studies (*e.g.* Larsen 2000, 37–41, Lillie and Richards 2000).

*Fig. 7.4 Cattail (*Typha *spp.), one of the most useful of all wetland plants (photo: G. Nicholas).*

Wapato is an herbaceous perennial that was highly valued by aboriginal peoples across Canada and northern North America, and for Northwest Coast groups the tuber was an important source of carbohydrates (Kuhnlein and Turner 1991, 70–71). In many parts of the world, such as in Australia, where their harvesting and overall importance is well documented, water lilies (*Nymphaea* sp.) were a staple, highly valued food. For example, in 1901 Baldwin Spencer observed that 'men, more especially the women, wade in, often breast high, pulling up stalks and roots. The former they eat, the latter they bake in hot ashes. They collect the hard seeds in great numbers, pound them up and knead them into loaves that are baked, and when cold are almost as hard as a rock. Members of the Kotandji tribe... acknowledged such a close relationship between themselves and waterlilies that they buried the remains of their dead in the banks of the lagoon to keep them cool and to assist in the growth of the lilies' (Mulvaney 1987, 108).

What these three plants have in common is that they were widely utilised by hunter-gatherers, were easily gathered, had relatively high caloric values and could be processed by drying for storage. Huge amounts of cattail, wapato and various water lilies are known to have been harvested, and often traded, by North American groups (Erichsen-Brown 1979, 209–214, Kuhnlein and Turner 1991, 72). There is also evidence of ownership of productive gathering areas for some of these plants. For example, patches of wapato and another wetland plant, spike rush (*Eleocharis dulcis*), were owned by families (Kuhnlein and Turner 1991, 72, Isaacs 1987, 96).

Fig. 7.5 Beaver lodge and associated pond/wetland in British Columbia, Canada (photo: G. Nicholas).

The other side of the economic equation is the fauna. Many different species of animals, birds, reptiles and fish are found within, or are associated with, wetlands (Tiner 1998, Weller 1999). Two that may have been particularly important to hunter-gatherers are beaver (*Castor canadensis*) and waterfowl (*e.g.* magpie goose (*Anseranas semipalmata*)). Both are widely reported in the ethnographic literature, and remain highly sought-after today both for their meat and eggs (*e.g.* Brocknell *et al.* 1995), and both species are also well represented in the archaeological record. In northeastern North America, beaver remains are found in Paleoindian sites (*e.g.* Curran 1987, 262, Spiess *et al.* 1985, 147), and even earlier in the Middle East (Legge and Rowley-Conwy 1986). Dincauze (2000, 36) has proposed that flyways may have been a particularly important factor in landuse patterns and colonising trends in late Pleistocene America. Both of these species clearly represented productive and reliable types of resources. In fact, beaver and moose are very closely tied to swamps and similar habitats (Fig. 7.5); although waterfowl obviously have a much greater range, wetlands are nonetheless regular in their travel stops and thus they end up as frequently expected dinner guests.

Moving beyond what people are eating or harvesting, there are other topics related to economic organisation that bear mentioning here. Closely tied to the resource base is the carrying capacity of the environment. There is extensive evidence, particularly in temperate and tropical regions, that swamps, marshes, estuaries and other such settings were very productive. On this basis we can predict that population density may be higher in some wetland-rich areas than elsewhere. This is the case in the Murray-Darling river area of South Australia, which had one of highest population densities not only on the continent at the time of contact (Butlin 1993, Lourandos 1997), but perhaps elsewhere as well; the recent discovery of an estimated 6000 to 16,000 burials in the Lake Victoria area (O'Neill 1994) attests to this. Of course population figures are notoriously difficult to obtain, especially for comparative research; nonetheless, in most areas where there is a rich and productive resource base associated with wetlands, including both the

American Southeast and Northwest Coast, population numbers appear to have been relatively high.

It is important to note, however, that wetlands were seldom the sole economic focus, even seasonally, but were part of the larger environmental context in which people are situated. Catchment areas centred on wetlands are known ethnographically and, presumably, archaeologically. The Anbarra of Arnhem Land, northern Australia (Meehan 1991), occupy a landscape that is dominated by freshwater and mangrove swamps, but that also has forest, riverine and coastal elements. Such mosaics, comprising such ecologically diverse elements, tend to be exceptionally productive. The archaeological record also demonstrates that human presence in such areas was both widespread and long-term, as reflected by habitation sites, resource harvesting or processing loci, rock shelters and earth mounds (Schrire 1984, Jones 1985).

There are several other aspects of economic organisation that need be mentioned. The availability of many wetland resources was seasonal, meaning that their exploitation and harvesting had to be scheduled. When key resources were not available, lower ranked foods might have been sought, or other locations exploited. Certain resources, such as magpie geese or eels, arrived in substantial numbers in the swamps at particular times of the year. Any concerted harvesting of these would likely require advance planning and the organisation of a labour force, especially if large nets had to be made and employed (*e.g.* Napton 1969), or eel traps constructed, maintained or operated. Lourandos (1980, 381) has calculated that the Toolondo wetlands drainage system in Victoria required a total of 13,000 hours of labour to develop. The division and organisation of labour is clearly an important but little understood element in hunter-gatherer research because the opportunity for studying such features (other than for fish weirs) is so uncommon (see Lourandos 1988).

Another aspect of labour organisation is the sexual division of activities. At Malheur Lake, the high degree of sexual dimorphism present in skeletal robusticity and osteoarthritis (Fig. 7.6) suggests that men and women had different activity patterns (Bettinger 1999, Hemphill 1999). It appears that women were engaged primarily in gathering and other activities around the marsh, while men were often on extended forays, possibly for hunting. Determining the degree to which these skeletal indices are represented in other wetland-associated populations (*e.g.* Windover, Florida; Lake Victoria, Australia) would provide important points of comparison in terms of division of labour and degree of mobility.

RESOURCE MANAGEMENT STRATEGIES AND LANDSCAPE MODIFICATION

There is growing evidence that resource management was a regular and widespread practice of hunting and gathering peoples in many parts of the world. Recent recognition of the scope and rationale of aboriginal resource management strategies, usually under the auspices of traditional ecological knowledge, suggests that we need to re-think the nature of hunting and gathering, at least as presented in the generalised foraging model. Some harvesting strategies employed by ethnographic hunter-gatherers, such as leaving a portion of harvested tubers, may qualify as proto-horticultural, and provide valuable insights into the process of domestication and the development or adoption of farming.

The most impressive and best documented evidence for managing the landscape relates to the use of fire to clear land, encourage new growth to attract game, rejuvenate specific resource patches and otherwise renew the land (*e.g.* Lewis 1982, Head 1994). In aboriginal Australia, it was so widely practiced as to be termed 'fire-stick farming'. Although ethnographic and historic observations are available for many other regions (*e.g.* Boyd 1999), the systematic use of fire as a resource management strategy undoubtedly has a history that extends well into the past, although the evidence is often sketchy and frequently restricted to charcoal influx values in pollen cores (*e.g.* Patterson and Sassaman 1988). There is some evidence that fire was used to manipulate wetland resources. For example, the Spokane of northwestern North America regularly burned tule and cattail patches after harvesting, a practice that improved both the yield and quality of future growth (Ross 1999, 283). The same is also true in the Kakadu region of northern Australia (Russell-Smith 1995). Archaeologically recovered charcoal from the Mesolithic site of Star Carr in England indicates frequent, if not regular, burning of the reed (*Phragmites australis*) beds there (Hather 1998, 196).

Hunter-gatherers have also influenced their environment in other ways. One way is by intentionally shifting or controlling water flow and drainage by digging channels and constructing stonework in and around wetlands. The primary evidence for this is found at Toolondo (Fig. 7.7), Lake Condah, and other locations in southeastern Australia (Lourandos 1997, Builth 1999), which have been linked to eeling. The age of most of these constructions is unknown, but is probably late Holocene. However, wetland canals related to horticulture have also been found at Kuk Swamp in New Guinea (Bayliss-Smith and Golson 1992, 20), with reported dates of about 9000 years BP (uncalibrated). Although comparable evidence has not been found in the Americas, the widespread practice of chinampas farming in Mesoamerica and raised fields in parts of Central and South America likely have their antecedents in the wetland-associated activities of earlier hunting and gathering peoples.

There is also evidence of artificial mounds within and adjacent to swamps in both northern and especially southeastern Australia. The construction of these low mounds may have been based upon naturally developed ones, and these served as dryland access during periods of high water. Meehan (1991, 205) has noted the dimensions of one as being 20 m long, 10 m wide and 2 m high. In Arnhem Land these are widely scattered in and amongst swamps and riverine features (Meehan *et al.* 1985), while in the southwestern Victoria area they are not only more numerous but frequently appear in clusters, the largest cluster containing 28 within a 4 hectare area (Lourandos 1997, 216). These constructions are further discussed in Nicholas (this volume).

Hunter-gatherers may have had a significant impact on the landscape as a result of beaver hunting practices. In North America, Europe and elsewhere, beaver not only created new wetlands due to the

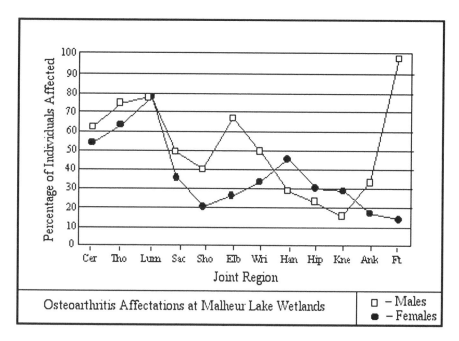

Fig. 7.6 Evidence of osteoarthritis in Malheur Lake skeletal population (afterHemphill 1999, 262).

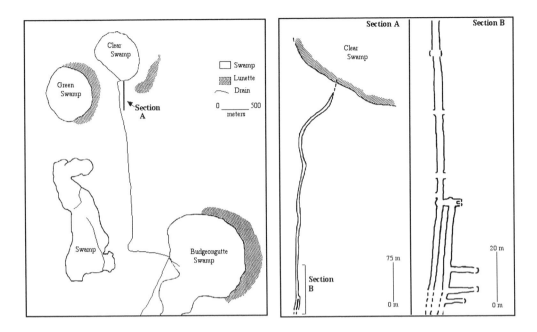

Fig. 7.7 Canal system in the Toolondo wetlands, southeastern Australia (after Lourandos 1980).

impoundments that formed behind their dams, but also expanded existing ones. Increased hunting would not only decrease the size of the beaver population in an area, but also lead to a reduction in the number or size of wetlands, which, in turn, could influence local and regional hydrologic regimes, and floral and faunal communities (see Nicholas 1999, also Krech 1999).

Finally, it is likely that hunter-gatherers had other impacts on the landscape, especially at the local level, such as by occasional over-hunting or the intentional transplanting of favoured plants (*e.g.* sweet flag, *Acorus americans* (Marles *et al.* 2000, 271)). Given that some wetland plants such as tule and cattail were very intensively harvested as materials for lodge coverings and other uses, we need to explore how resilient such species are to selective harvesting, or even small-scale clear-cutting. A mat lodge (Fig. 7.8), such as found in the Interior Plateau, may require as many as 1000 tule or cattail stems for one structure (N. Turner pers. comm. 2001); a camp of four such structures would thus require 4000 stems, which would likely have to be renewed every other year. Based on estimated density of modern cattail (*e.g.* 20 plants/m^2), approximately 200 m^2 would be required for each lodge every year or two. Controlled burns may have had a positive effect on density, while digging up the rhizomes of cattail, scirpus and other plants for food may have had a comparable positive influence on their growth, as has been documented for such terrestrial plants as Balsamroot (*Balsamorhiza sagittata*) (Peacock and Turner 1999).

EXPANDING THE HORIZONS OF WETLAND ARCHAEOLOGY

Wetland archaeology has steadily gained in prominence in the years following the publication of John Coles' seminal work *The Archaeology of Wetlands* (1984). What has followed is a steady stream of publications that identify, inventory and sometimes explain the relationship between prehistoric humans and wetland environments, including a series of edited volumes produced in association with the Wetland

Archaeological Research Group (WARP) (*e.g.* Purdy 1988, 2001, Coles and Coles 1989, Bernick 1998, Coles *et al.* 1999).

Interestingly, many of the publications that fall within the rubric of wetland archaeology are more accurately described as 'wet site' archaeology. The difference between them is significant: 'Wetland sites are defined here by a *relationship* between people and the particular types of ecological settings represented by wetlands, and the archaeological record it has produced...Wet sites, on the other hand, are defined by the *association* between artifacts and the context of preservation' (Nicholas 2001).

Wet site archaeology is and will continue to be an important dimension of archaeology, and provides sometimes startling insights into past lifeways as the result of the remarkable preservation found in water-saturated contexts. However, when we begin to look at the overall role that swamps, marshes and comparable settings had on past cultural landscapes, we begin to learn some very important things about past landuse patterns of hunter-gatherers, and potentially even aspects of their social and political organisation.

Human responses to dynamic environments

Wetland environments are highly variable, highly dynamic entities. In both temperate and tropical regions, the number, size and composition of marshes have responded to climatic shifts, to changes in hydrologic regime and to the influences of both human and non-human (*e.g.* beaver) agencies. Post-glacial changes in sea-level and the land-

Fig. 7.8 Modern example of a traditional Interior Plateau tule mat lodge, Secwepemc Heritage Park, Kamloops, B.C. (photo: G. Nicholas).

sea gradient influenced not only coastal wetlands, but interior drainage patterns as well. Subsequent climatic events, such as the warm/dry Hypsithermal that occurred in North America during the early to middle Holocene, would have resulted in a possible reduction in the extent of wetlands, but an even more pronounced reduction in other water sources elsewhere in the landscape, perhaps leading to more intensive utilisation of the former (Nicholas 1998b).

In Arnhem Land, large areas of coastal inundation occurred around 6000 years ago, as the result of sea-level rise, transforming the nature of the already extensive wetland systems in the region (Hope *et al.* 1985). These factors indicate that wetlands offer considerable potential to study environmental change, and especially human responses to those changes. In addition, it is important to note that wetlands that were active thousands of years ago may today be dry land. It may be necessary to incorporate geological studies of past landforms, and hydric soils to locate former wetlands in order to ensure adequate representation in survey coverage.

In considering these factors within the context of wetland archaeology, we can not only identify the processes of environmental change (the record of which is often very well preserved in such settings), but also illuminate the means by which people responded to such change. In other words, wetlands are an ideal setting in which to explore important elements of past human ecosystems both at particular points in time and over very long periods of time. The result may be a deeper understanding of hunter-gatherer landuse, and new insights into more general human responses to environmental and climatic change.

Wetlands as a heritage management concern

One of the greatest challenges facing wetlands-orientated archaeologists is to insure that knowledge of the rich and diverse wetland heritage is incorporated into resource management policies. If one of the mandates of cultural resource management is to identify and manage significant resources representative of past lifeways (Nicholas 1992, 1994), then this must be done. Worldwide, the number, condition and extent of wetlands continues to diminish at an alarming rate. In the United States alone, of the estimated 215 million acres of wetlands that existed at the time of colonisation, less than 46% remained by the mid-1970s (Tiner 1984, 25–26, see also Williams 1990).

Although conservation efforts and legislation have slowed appreciably the rate of loss, the impact of wetland destruction or drainage on the archaeological resource is that sites are frequently found on wetland margins, or sometimes even within them, as was the case of the Windover cemetery site that was found through dredging. Drainage may well have destroyed hitherto preserved organic materials within wetlands (*e.g.* Purdy 1991). Some wetland conservation strategies seek to maintain the current number of swamps and marshes by constructing replacements; while this may restore valuable habitats for plants and animals, it cannot replace the archaeological components of wetlands lost to development.

The impetus is on those of us who are knowledgeable of the archaeological heritage of wetlands to increase the awareness of others. This requires us not only to inform our colleagues involved in the realm of cultural resource management, but also to share our knowledge and understanding of wetlands, and why they are important, with the public at large. The potential of wetlands to contribute to important questions about past human behaviour is considerable.

CONCLUSIONS

In a review of the wetland archaeological record in the Great Basin, Bettinger (1999, 327) posed the question, 'Do marshes bring happiness, health, and mobility?' I agree with his answer – 'No!' – in a qualified way. However, as much as we promote the notion that wetlands were an important component of past human ecosystems, they were not a utopian environment, nor were hunter-gatherer lifestyles necessarily freer of the types of social, demographic or health problems associated with farming communities. There is, for example, evidence in skeletal remains from both the Great Basin wetlands and the Murray River area of frequent hypoplasias and other indicators of childhood or adult dietary stress (Webb 1995, Larsen and Hutchinson 1999), in addition to the evidence (at the former) of personal injuries, including those inflicted by others.

It is not this author's intention to imply that wetlands were necessarily critical to the emergence of complex cultural systems (but see Nicholas, Chapter 20); the presence of extensive swamps and marshes adjacent to the Mississipian city of Cahokia or the Aztec capital of Tenochtitlan should not lead us to conclude 'Aha, so that's the reason those societies developed!' While the examples provided here focus on those often very productive wetland systems that were utilised by historic or prehistoric hunter-gatherers, not all wetlands are equal. Short or long lived, low to high species diversity, low to high primary productivity – all combinations of these and other factors are possible. Certainly we have to be careful about the assumptions we make about wetlands; I have been involved in systematic, intensive surveys of what I considered attractive wetland settings only to find a virtual absence of any archaeological indicators of use (Nicholas 1994, 31).

On the other hand, there is ample evidence worldwide that wetlands were a major component of many regions (*e.g.* Coles and Lawson 1987), frequently utilised by many different types of prehistoric peoples including complex hunter-gatherers and their progeny. Based on site distribution patterns, and floral and faunal remains, some peoples appear to have exploited wetlands on a very limited level, and perhaps only fortuitously. For others, however, wetlands were regularly utilised on a seasonal basis. In some regions, it appears that wetlands provided both the economic foundation and geographic focus of hunter-gatherer lifeways. As Kelly (1995) and others have noted, there is substantial variation within the hunter-gatherer spectrum. To increase our understanding of this diversity, we must either carefully integrate or carefully discount wetlands in our investigations of hunter-gatherers, but we cannot ignore them.

ACKNOWLEDGEMENTS

I thank Malcolm Lillie for his invitation to participate in this project, and also Robert Kelly, Dena Dincauze, Heather Builth, Nancy Turner and Catherine Fowler for providing information and helpful insights into my on-going research on wetlands. In Australia, I am particularly grateful to Heather Builth, with whom I spent several days touring wetland sites in the Mt. Condah/Mt. Eccles area of western Victoria, and to Donald Pate for organising a visit to Rooka Flat.

REFERENCES

Bayliss-Smith, T. & J. Golson 1992. Wetland agriculture in New Guinea Highlands prehistory, in B. Coles (ed.) *The Wetland Revolution in Prehistory*, 15–28. Exeter: WARP Occasional Paper 6.

Bernick, K. (ed.) 1998. *Hidden Dimensions: the cultural significance of wetland archaeology*. Vancouver: UBC Press.

Bettinger, R. L. 1991. *Hunter-Gatherers: archaeological and evolutionary theory*. New York: Plenum.

Bettinger, R. L. 1999. Faces in prehistory: Great Basin wetlands skeletal populations, in B. E. Hemphill & C. S. Larsen (eds) *Prehistoric Lifeways in the Great Basin Wetlands: bioarchaeological reconstruction and interpretation*, 321–32. Salt Lake City: University of Utah Press.

Binford, L.R. 1980. Willow smoke and dogs' tales: hunter-gatherer settlement systems and archaeological site formation. *American Antiquity* 45, 4–20.

Bonsall, C. (ed.) 1989. *The Mesolithic in Europe*. Edinburgh: John Donald Publishers.

Boyd, R. (ed.) 1999. *Indians, Fire and the Land in the Pacific Northwest*. Corvallis: Oregon State University Press.

Brockwell, S., R. Levitus, J. Russell-Smith & P. Forrest 1995. Aboriginal heritage, in T. Press, D. Lawrence, T. Press, D. Lea, A. Webb & A. Graham (eds) *Kakadu: natural and cultural heritage and management*, 15–63. Darwin: Australian Nature Conservation Agency.

Builth, H. 1999. The connection between the Gunditjmara people and their environment: the case for complex hunter-gatherers in Australia. Unpublished ms. in author's possession.

Butlin, N. G. 1993. *Economics of the Dreamtime: a hypothetical history*. Cambridge: Cambridge University Press.

Coles, B. & J. M. Coles 1986. *Sweet Track to Glastonbury*. New York: Thames & Hudson.

Coles, B. & J. M. Coles 1989. *People of the Wetlands: bogs, bodies and lake-dwellers*. New York: Thames & Hudson.

Coles, B., J. M. Coles & M. S. Jorgensen (eds) 1999. *Bog Bodies, Sacred Sites and Wetland Archaeology*. Exeter: WARP Occasional Paper 12.

Coles, J. M. 1984. *The Archaeology of Wetlands*. Edinburgh: Edinburgh University Press.

Coles, J. M. & A. J. Lawson (eds) 1987. *European Wetlands in Prehistory*. Oxford: Clarendon Press.

Curran, M. L. 1987. The Spatial Organization of Paleoindian Populations in the Late Pleistocene of the Northeast. Unpublished PhD thesis, University of Massachusetts, Amherst.

Dillehay, T. 1988. *Monte Verde: a Pleistocene settlement in Chile. Vol. 1. Paleoenvironment and Site Context*. Washington, D.C.: Smithsonian Institution Press.

Dincauze, D. F. 2000. The Earliest Americans: the Northeast. *Common Ground: Archaeology and Ethnography in the Public Interest*, Spring/Summer, 34–43.

Erichsen-Brown, C. 1979. *Medicinal and Other Uses of North American Plants: a historical survey with special reference to the eastern Indian tribes*. New York: Dover Publications.

Finlayson, M. & M. Moser (eds) 1991. *Wetlands*. New York: Facts on File.

Fowler, C. S. 1982. Food-named groups among Northern Paiute in North America's Great Basin: an ecological interpretation, in N. M. Williams, & E. S. Hunn (eds) *Resource Managers: North American and Australian hunter-gatherers*, 113–29. Boulder: Westview Press. American Association for the Advancement of Science, Selected Symposium 67.

Goodman, S. 1985. Material culture: basketry and fiber artifacts, in D. H. Thomas (ed.) *The Archaeology of Hidden Cave, Nevada*, 262–98. New York: Anthropological Papers 61, American Museum of Natural History.

Gould, R. A. 1990. *Recovering the Past*. New Mexico: University of New Mexico Press.

Hather, J. G. 1998. Identification of macroscopic charcoal assemblages, in P. Mellars, & P. Dark. (eds) *Star Carr in Context: new archaeological and palaeoecological investigations at the Early Mesolithic site of Star Carr, North Yorkshire*, 183–96. Cambridge: McDonald Institute Monographs.

Head, L. 1994. Landscapes socialized by fire: post-contact changes in aboriginal fire use in northern Australia, and implications for prehistory. *Archaeology in Oceania* 29, 172–81.

Hemphill, B. E. 1999. Wear and tear: osteoarthritis as an indicator of mobility among Great Basin hunter-gatherers, in Hemphill, B. E. & C. S. Larsen (eds) *Prehistoric Lifeways in the Great Basin Wetlands: bioarchaeological reconstruction and interpretation,* 241–89. Salt Lake City: University of Utah Press.

Hemphill, B. E. & C. S. Larsen (eds) 1999. *Prehistoric Lifeways in the Great Basin Wetlands: bioarchaeological reconstruction and interpretation*. Salt Lake City: University of Utah Press.

Higgs, E. S. 1961. Some Pleistocene faunas of the Mediterranean coastal areas. *Proceedings of the Prehistoric Society* 27, 144–54.

Hope, G., P. J. Hughes & J. Russell-Smith 1985. Geomorphological fieldwork and the evolution of the landscape of Kakadu National Park, in R. Jones, (ed.) *Archaeological Research in Kakadu National Park*, 229–40. Canberra: Australian National Parks and Wildlife Service and the Department of Prehistory, Australian National University.

Isaacs, J. 1987. *Bush Food:aboriginal food and herbal medicine*. Sydney: Weldon.

Isaacs, J. 1991. *Australian Dreaming: 40,000 years of aboriginal history*. Sydney: Ure Smith.

Janetski, J. C. & D. B. Madsen (eds) 1990. *Wetland Adaptations in the Great Basin*. Provo, Utah: Brigham Young University Museum of Peoples and Cultures Occasional Papers 1.

Jones, R. (ed.) 1985. *Archaeological Research in Kakadu National Park*. Canberra: Australian National Parks and Wildlife Service and the Department of Prehistory, Australian National University.

Kelly, R. 1995. *The Foraging Spectrum: diversity in hunter-gatherer lifeways*. Washington, D.C.: Smithsonian Institution Press.

Kelly, R. 2001. *Prehistory of the Carson Desert and Stillwater Mountains: environment, mobility, and subsistence in a Great Basin wetland*. Salt Lake City, Utah: University of Utah Anthropological Paper 123.

Krech, S. III 1999. *The Ecological Indian: myth and history*. New York: W.W. Norton.

Kuehn, S. R. 1998. New evidence for Late Paleoindian-Early Archaic subsistence behavior in the Western Great Lakes. *American Antiquity* 63, 457–76.

Kuhnlein, H. V. & N. J. Turner 1991. *Traditional Plant Foods of Canadian Indigenous Peoples: nutrition, botany and use*. Philadelphia: Gordon and Breach.

Larsen, C. S. 2000. *Skeletons in Our Closet: revealing our past through bioarchaeology*. Princeton, NJ: Princeton University Press.

Larsen, C. S. & D. L. Hutchinson 1999. Osteopathology of Carson Desert Foragers: reconstructing prehistoric lifeways in the Western Great Basin, in B. E. Hemphill, & C. S. Larsen (eds) *Prehistoric Lifeways in the Great Basin Wetlands: bioarchaeological reconstruction and interpretation,* 184–202. Salt Lake City: University of Utah Press.

Lee, R. B. & R. Daly (eds) 2000. *The Cambridge Encyclopedia of Hunters and Gatherers*. Cambridge: Cambridge University Press.

Lee, R. B. & I. Devore (eds) 1968. *Man the Hunter*. Chicago: Aldine.

Legge, A. J. & P. Rowley-Conwy 1986. The beaver (*Castor fiber* L.) in the Tigris-Euphrates basin. *Journal of Archaeological Science* 13, 469–76.

Lewis, H. T. 1982. Fire technology and resource management in aboriginal North America and Australia, in N. M. Williams & E. S. Hunn (eds) *Resource Managers: North American and Australian hunter-gatherers*, 45–68. Boulder: Westview Press.

Lillie, M. C. & M. P. Richards 2000. Stable isotope analysis and dental evidence of diet at the Mesolithic-Neolithic transition in Ukraine. *Journal of Archaeological Science* 27, 965–72.

Lourandos, H. 1980. Forces of Change: aboriginal technology and population in southwestern Victoria. Unpublished PhD thesis, Department of Anthropology, University of Sydney.

Lourandos, H. 1987. Swamp managers of southwestern Victoria, in D. J. Mulvaney, & J.P. White (eds) *Australians to 1788*, 292–307. Sydney: Fairfax, Syme & Weldon.

Lourandos, H. 1988. Palaeopolitics: resource intensification in aboriginal Australia and Papua New Guinea, in T. Ingold, D. Riches, & J. Woodburn (eds) *Hunters and Gatherers 1: history, evolution and social change,* 148–60. New York: Berg.

Lourandos, H. 1997. *Continent of Hunter-Gatherers.* Cambridge: Cambridge University Press.

McBride, K. A. 1992. Prehistoric and historic patterns of wetland use in eastern Connecticut. *Man in the Northeast* 43, 10–24.

Marles, R. J., C. Clavelle, L. Monteleone, N. Tays & D. Burns (eds) 2000. *Aboriginal Plant Use in Canada's Northwest Boreal Forest.* Vancouver: UBC Press.

Meehan, B. 1991. Wetland hunters: some reflections, in C. D. Haynes, M. D. Ridpath & M. A. J. Williams (eds) *Monsoonal Australia: landscape, ecology, and man in the Northern Lowlands*, 197–206. Rotterdam: A.A. Balkema.

Meehan, B., S. Brockwell, J. Allen & R. Jones 1985. The wetlands sites, in R. Jones (ed.) *Archaeological Research in Kakadu National Park*, 103–54. Canberra City: Australian National Parks and Wildlife Service, Special Publication 13.

Mellars, P. & P. Dark 1998. *Star Carr in Context: new archaeological and palaeoecological investigations at the Early Mesolithic site of Star Carr, North Yorkshire.* Cambridge: McDonald Institute Monographs.

Mitsch, W. J. & J. G. Gosselink 2000. *Wetlands* (3rd ed.). New York: John Wiley & Sons.

Morgan, L. H. 1877. *Ancient Society.* New York: Holt.

Mulvaney, J. 1987. *The Aboriginal Photographs of Baldwin Spencer.* Victoria: Viking O'Neil.

Napton, L. K. 1969. *Archaeological and Paleobiological Investigations in Lovelock Cave, Nevada.* Berkeley: Kroeber Anthropological Society Papers, Special Publication 2.

Nicholas, G. P. 1988. Ecological leveling: the archaeology and environmental dynamics of early postglacial land-use, in. G. P. Nicholas, (ed.) *Holocene Human Ecology in Northeastern North America,* 257–96. New York: Plenum Press.

Nicholas, G. P. 1992. Directions in wetlands research. *Man in the Northeast* 43, 1–9.

Nicholas, G. P. 1994. Prehistoric human ecology as cultural resource management, in J. Kerber, (ed.) *Cultural Resource Management: archaeological research, preservation planning, and public education in the northeastern United States*, 17–50. Greenwich: CT: Bergin & Garvey.

Nicholas, G. P. 1998a. Wetlands and hunter-gatherers: a global perspective. *Current Anthropology* 39, 720–733.

Nicholas, G. P. 1998b. Assessing climatic influences on human affairs: wetlands and the maximum Holocene warming in the Northeast. *Journal of Middle Atlantic Archaeology* 14, 147–60.

Nicholas, G. P. 1998c. Wetlands and hunter-gatherer land use in North America, in K. Bernick (ed.) *Hidden Dimensions: the cultural significance of wetland archaeology*, 31–46. Vancouver: UBC Press.

Nicholas, G. P. 1999. A light but lasting footprint: human influences on the Northeastern landscape, in M. L. Levine, M. S.. Nassaney & K.E. Sassaman (eds) *The Archaeological Northeast*, 25–38. Greenwich, CT: Bergin and Garvey.

Nicholas, G. P. 2001. Wet sites, wetland sites, and cultural resource management strategies, in B. A. Purdy, (ed.) *Enduring Records: the environmental and cultural heritage of wetlands*, 262–270. Oxford: Oxbow Books.

Nicholas, G. P. in prep. *Wetland Ecology and Hunter-Gatherer Archaeology.*

Oetting, A. C. 1999. An examination of wetland adaptive strategies in Harney Basin: comparing ethnographic paradigms and the archaeological record, in B. E. Hemphill, & C. S. Larsen (eds) *Prehistoric Lifeways in the Great Basin Wetlands: bioarchaeological reconstruction and interpretation*, 203–18. Salt Lake City: University of Utah Press.

O'Neill, G. 1994. Cemetery reveals complex aboriginal society. *Science* 264, 1403.

Patterson, W. A. III & K. E. Sassaman 1988. Indian fires in the prehistory of New England, in G. P. Nicholas, (ed.) *Holocene Human Ecology in Northeastern North America*, 107–36. New York: Plenum Press.

Peacock, S. & N.J. Turner 1999. 'Just Like a Garden': traditional plant resource management and biodiversity conservation on the British Columbia Plateau, in P. Minnis & W. Elisens (eds) *Biodiversity and Native North America*, 133–79. Norman: University of Oklahoma Press.

Perry, D. 1999. Vegetative tissues from Mesolithic sites in the northern Netherlands. *Current Anthropology* 40, 231–7.

Purdy, B. A. (ed.) 1988. *Wet Site Archaeology*. Caldwell, New Jersey: Telford Press.

Purdy, B. A. 1991. *The Art and Archaeology of Florida's Wetlands*. Boca Raton, FL: CRC Press.

Purdy, B. A. (ed.) 2001. *Enduring Records: the environmental and cultural heritage of wetlands*. Oxford: Oxbow Books.

Ross, J. 1999. Proto-historical and historical Spokan prescribed burning and stewardship of resource areas, in R. Boyd, (ed.) *Indians, Fire and the Land in the Pacific Northwest*, 277–91. Corvallis: Oregon State University Press.

Russell-Smith, J. 1995. Fire management, in T. Press, D. Lea, A. Webb & A. Graham (eds) *Kakadu: natural and cultural heritage and management*, 217–37. Darwin: Australian Nature Conservation Agency and Australian National University.

Schmidt, D. D. & N. D. Sharp 1990. Mammals in the marsh: zooarchaeological analysis of six sites in the Stillwater Wildlife Refuge, Western Nevada, in Janetski, J. C. & D. B. Madsen (eds) *Wetland Adaptations in the Great Basin*, 75–96. Provo Utah: Brigham Young University Museum of Peoples and Cultures, Occasional Papers 1.

Schrire, C. 1984. Interpretations of past and present in Arnhem Land, North Australia, in S. Schrire (ed.) *Past and Present in Hunter Gatherer Studies*, 67–94. San Diego: Academic Press.

Simms, S. R. 1987. *Behavioral Ecology and Hunter-Gatherer Foraging: an example from the Great Basin*. Oxford: British Archaeological Reports International Series 381.

Spiess, A. E., M. L. Curran & J. R. Grimes 1985. Caribou (*Rangifer tanandus* L.) bone from New England Paleoindian sites. *North American Archaeologist* 6, 145–9.

Suttles, W. (ed.) 1990. *Northwest Coast. Handbook of North American Indians, Vol. 7*. Washington, D.C: Smithsonian Institution Press.

Thomas, D. H. (ed.) 1985. *The Archaeology of Hidden Cave, Nevada*. New York: Anthropological Papers 61, American Museum of Natural History, New York.

Tiner, R. W. 1984. *Wetlands of the United States: current status and recent trends*. Newton Corner, MA: US Fish and Wildlife Service.

Tiner, R. W. 1998 *In Search of Swampland: a wetland sourcebook and field guide*. New Brunswick, N.J.: Rutgers University Press.

Tylor, E. B. 1871. *Primitive Culture*. London: John Murray.

Walker, A. & R. Leakey (eds) 1993. *The Nariokotome* Homo erectus *Skeleton*. Cambridge: Harvard University Press.

Webb, S. G. 1995. *Palaeopathology of Aboriginal Australians*. Cambridge: Cambridge University Press.

Weller, M. W. 1999. *Wetland Birds: habitat resources and conservation implications*. Cambridge: Cambridge University Press.

Wendorf, F. 1993. E-87-5: Occupations dating to the grey phases, in F. Wendorf, R. Schild & A. E. Close (eds) *Egypt During the Last Interglacial: the Middle Paleolithic of Bir Tarfawi and Bir Sahara East*, 345–55. New York: Plenum Press.

Wheat, M. M. 1967. *Survival Arts of the Primitive Paiutes*. Reno: University of Nevada Press.

Whittaker, R. H. 1975. *Communities and Ecosystems*. New York: Macmillan.

Wigand, P. E. & P. J. Mehringer Jr. 1985. Pollen and seed analyses, in D. H. Thomas (ed.) *The Archaeology of Hidden Cave, Nevada*, 108–24. New York: Anthropological Papers 61, American Museum of Natural History.

Williams, M. (ed.) 1990. *Wetlands: a threatened landscape*. Oxford: Blackwell.

PART 2
ARCHAEOLOGY

Chapter 8

Mesolithic Wetland Sites in Central Russia

Mickle G. Zhilin

INTRODUCTION

Extensive field surveys and excavations undertaken in recent decades in Central Russia have brought to light about 300 new Mesolithic sites, more than 70 of which have been excavated. The sites are attributed to three main archaeological cultures – Butovo, Ienevo and Resseta (Zhilin 1996, Koltsov and Zhilin 1999). Almost all of these sites are situated on mineral soils where organic materials are not preserved, the stratigraphy is often not reliable, and radiocarbon and pollen data are either scarce or absent. However, field surveys and excavations carried out by the author in 1988–2000 revealed about 60 sites in wetland locations, which exhibited good preservation of organic remains, reliable stratigraphy and ideal opportunities for scientific analyses. Twelve of these were excavated, producing a good sequence of cultural layers dating from the beginning to the end of the Mesolithic. Most of them also have upper layers with good preservation of organic materials, covering the whole of the Neolithic and sometimes also the Bronze Age and later periods. The focus of the current paper is the Mesolithic stage of human activity at these sites.

Almost all of the sites are situated in large peat bogs which occupy glacial depressions. Layers of clay and gyttja beneath the peat deposits attest the presence of large lake systems during the Stone Age. These lakes were connected by small rivers to the main rivers of Central Russia – the Oka and Volga (Fig. 8.1). Climatic changes during the early Holocene caused periodic lake transgressions and regressions, resulting in shoreline displacement. Sites inhabited during regressions were subsequently submerged, and their cultural layers were sealed by gyttja and peat, thereby producing excellent conditions for the preservation of bone, antler, wood, bark, seeds, plant fibres, resin, coprolites, insects and other organic remains. During subsequent regression episodes, certain sites were re-occupied, and again subsequently submerged. This process led to the formation of sites with several Mesolithic cultural layers, divided by sterile horizons, or cultural horizons that were otherwise easily separated from each other stratigraphically. In some cases settlements occupied dryland areas near a lake shore, and the associated cultural remains, including various organic materials, are found at these sites in both lake and bog sediments near the contemporary shore. The cultural layers in such deposits are usually also divided by sterile horizons. This paper provides a brief description of the most important wetland sites with Mesolithic cultural layers in the Volga-Oka area, and pays particular attention to the well preserved artefacts made from a variety of organic materials.

THE OZERKI PEAT BOG

This wetland is situated *c.* 20 km south of the town of Tver, and is connected to the Upper Volga by the rivers Inyuha and Shosha. About 20 sites were discovered in the western part of this peat bog, with three of them containing Mesolithic layers with good preservation of organic materials (Zhilin 1996, 1998a, 1999). The site of Ozerki 17 was situated at the outlet of a small river, originating from a large lake measuring around 5 × 3 km. The lower middle Mesolithic cultural layer (IV) of the site is radiocarbon dated to 8830±40 BP (GIN-6655, pine fishtrap fragments) and 8840±50 BP (GIN-7474, birch stake), and also to the first third of the Boreal period (8800–8600 BP) by palynological analysis (all radiocarbon determinations reported in this paper are conventional and uncalibrated). Sixty-seven square metres of the site were excavated.

Finds include bone arrowheads, which comprised needle-shaped forms with a bi-conical head, asymmetrical one-winged forms, and one blunt-headed example, which were probably used for fur hunting. In addition, fragments of lance heads, elk scapula knives, awls, an antler scraper, 'ice-pick', elk and beaver incisor pendants and a small perforated disk were recovered (Zhilin 1996, 281). Of particular interest is an intact fishing hook with the remains of a line, which was attached to its head by a knot. The line is a twisted cord 1 mm in diameter, made from an indeterminate plant fibre. A small fragment of a fishing net made from a similar line, with a cell area of 20 × 20 mm was recovered, and this was accompanied by half of an elongated rhomboidal pine bark float, which was perforated in the middle.

A series of net sinkers, including large pebbles, which were bound in the middle with a strip of lime bark or some bog grass, were also recovered. Some sinkers had shallow hollows made by direct percussion of their sides for the fixing of binding. The latter is either preserved, or else indicated by transverse dark stripes on the surface of the pebble. More usually, the binding is preserved only under the sinker, while its upper part was torn off during attempts by their users to extract the sinkers, which were deeply embedded in the gyttja layer at the bottom of the lake. One sinker, made of a limestone slab, had a natural perforation with a fragment of a twisted rope, 5 mm in diameter and made of lime bark, still preserved in it. Numerous long straight pine splinters with rectangular cross-section were also found, and it appears that these were used for making conical fishtraps, which were usually woven

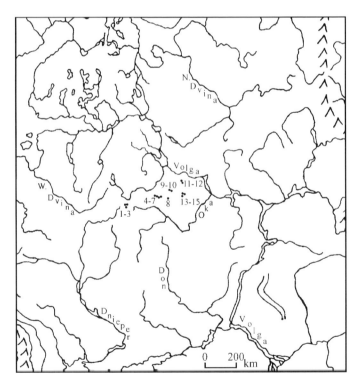

Fig. 8.1 Mesolithic wetland sites in the Volga Basin. 1–3: Ozerki 5, 16, 17; 4–7: Nushpoli 11, Okajomovo 4, 5, 18a; 8: Chernetskoje 8; 9–10: Ivanovskoje 3, 7; 11–12: Stanovoje 1, 4; 13–15: Sahtysh 2a, 9, 14.

transversely with lime bark or plant fibres. Two long splinters were found near a pine stake 60–70 mm in diameter, which was evidently sharpened with a flint adze. The ends of such stakes, stuck in the lake bottom, are numerous in the shallow water deposits near both Mesolithic and Neolithic sites. Most of these objects are associated with ancient fishing activities and served for fixing nets and fish traps.

The site of Ozerki 16 is situated *c*. 150 m from Ozerki 17, and here the lower cultural layer (II) is radiocarbon dated to 8870±40 BP (GIN-6654, pine fishtrap fragments) and by pollen analysis is dated to the first third of the Boreal period (*c*. 8800–8600 BP). Twelve square metres of Ozerki 16 were excavated and the finds included a perforated pine bark float, pebble sinkers, one with hollows at the sides and with the lime bark binding preserved, a fragment of a lance head and two long bone arrowheads (Zhilin 1999, 296). The first of these has a conical head, very carefully carved, and the surface is smoothed in a manner reminiscent of lathe turning. The second example has a double bi-conical head and a very long tang.

Ozerki 5 is located only 50 m from Ozerki 17, on the opposite side of the ancient river at its outlet from the lake. The lower, terminal Mesolithic cultural layer of the site (IV) has a number of radiocarbon dates – 7410±90 (GIN-6659, charcoal), 7310±120 BP (GIN-7218, worked log), 7190±180 BP (GIN-6660, charcoal), 7120±50 BP (GIN-7217, worked log), 6970±120 BP (GIN-6662, wooden splinters) and 6930±70 BP (GIN-7216, wooden splinters) – and is dated to the beginning of the Atlantic period by palynology. About 220 square metres of Ozerki 5 were excavated and the remains of a hearth and three pits were found, with various activity areas being traced throughout the excavated area. The finds include more than a thousand lithics and about two thousand bone and antler artefacts, fragments of wooden tools, rope and cord fragments, and worked birch bark (Zhilin 1999, 298–302). All find groups and most types of late Mesolithic bone and antler artefacts are encountered at this site. Arrowheads include, amongst others, needle-shaped forms, short arrowheads with irregular bi-conical shape, those with a wide flat head, asymmetric one-winged types, and blunt massive forms. Other finds include barbed points and harpoons, daggers, intact fishing hooks, various knives, scrapers, awls, chisels, antler axes and adzes, a range of chisels, knives and scrapers made from beaver mandibles (Zhilin 1998b), and antler punches, pressure flakers and numerous pendants made of mammal teeth.

THE DUBNA PEAT BOG

This particular bog is situated about 100 km to the north of Moscow, in the middle reaches of the river Dubna, a tributary of the Upper Volga. About 40 sites were discovered at this location, ten of which have Mesolithic layers with good preservation of organic materials (Zhilin 1999). The most interesting site is Okajomovo 5, which is situated on a small islet near a palaeochannel of the Dubna, which connected several large ancient lakes. Its lower late Mesolithic layer (III) is radiocarbon dated to 7910±80 BP (GIN-6191, peat from lower part of cultural layer) and 7730±60 BP (GIN-6192, peat from upper part of cultural layer). Pollen analysis dates it to the late Boreal at *c*. 8000–7800 BP. A total of fifty-two square metres of the site were excavated, with the finds including about 140 bone and antler artefacts (Fig. 8.2). Among these, arrowheads comprising short needle-shaped forms with an irregular bi-conical head, symmetric two-winged arrowheads with barbs, and asymmetric one and two-winged examples (some of the latter with grooves for flint insets), were recovered.

In addition, barbed points, massive intact and flat composite daggers with slots for flint insets, a series of intact and composite fishing hooks, broad elk scapula knives and narrow fish scaling knives made of split ribs were recovered. Other finds include rib side scrapers, awls, fragments of antler axes and adzes, a series of beaver mandible tools, 'ice-picks', an antler punch and a number of elk and beaver incisor pendants. Wooden artefacts are represented by various fragmentary remains, among which a paddle blade is of particular interest. This item is long, rather wide near the shaft and very gently narrowing towards the tip,

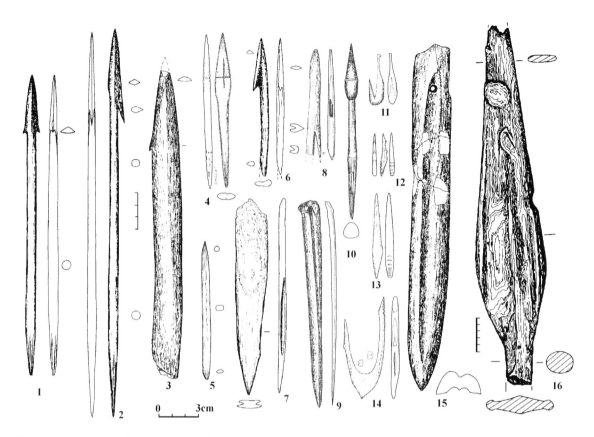

Fig. 8.2 Okajomovo 5, lower layer. 1–15: bone artefacts; 1–2, 4–6, 8, 10: arrowheads; 3: spearhead; 7, 9, 15: daggers; 11: fishing hook; 12–14: details of composite fishing hooks; 16: wooden paddle blade.

with three longitudinal ribs in the centre of the blade and along its sides in the wider part (Fig. 8.2, 16). This type of paddle is perfectly suited for manoeuvring a dugout canoe in boggy surroundings and attests the degree of boat handling skill of the Mesolithic fishers in this region.

The site of Nushpoli 11 is situated about 15 km downstream from the previously described group of sites, at the western border of the Dubna peat bog. Eighty-five square metres of the site were excavated. The terminal Mesolithic layer (III) is radiocarbon dated to 7310±40 BP (GIN-6657, worked stake) and dated by pollen analysis to the beginning of the Atlantic period at *c.* 7600–7200 BP.

Bone and antler artefacts include arrowheads of needle shape, asymmetric two-winged examples, those with fine barbs near the point and a long example with several sparse barbs along the stem and a hollow for a flint point at the tip. A fragment of a barbed point, a small harpoon, the point of a composite fishing hook, awls, knives, beaver mandible tools, and elk and beaver teeth pendants were also recovered. The lower middle Mesolithic cultural layer (IV) at Nushpoli 11 is dated by pollen analysis to the first third of the Boreal period at *c.* 8800–8600 BP.

Amongst the various bone tools recovered from layer IV are arrowheads of needle shape, also with massive bi-conical head, and one blunt example, a fragment of a barbed point, a wide elk scapula knife, and elk and beaver pendants.

THE IVANOVSKOYE PEAT BOG

Ivanovskoye is situated about 150 km to the northeast of Moscow, between Moscow and Yaroslavl in the middle reaches of the River Nerl, which ran through a large lake during the Stone Age, connecting it with the River Klyazma, a tributary of the Oka. Ten sites were discovered here, with Ivanovskoye 7 being the most interesting in archaeological terms. 106 square metres of this site were excavated by D. A. Krainov in 1974–75, and 332 square metres were excavated by the present author in 1992–96. The site has three Mesolithic and two Neolithic cultural layers.

The Mesolithic settlement occupied a low promontory during lake regressions, which was submerged during subsequent transgressions. The lower, early Mesolithic layer (IV) is radiocarbon dated to 9650±110 BP (GIN-9520, wood gnawed by beaver) and 9640±60 BP (GIN-9516, worked bone). It is dated by palynology to the first half of the Pre-Boreal period, before its climatic optimum. During the middle Pre-Boreal lake transgression it was submerged, and the base of the gyttja layer, overlapping the lower cultural layer, is radiocarbon dated to 9690±120BP (GIN-9367), 9500±110 BP (GIN-9517) and 9500±100 BP (GIN-9385).

About 300 bone and antler artefacts were found, amongst which long needle-shaped arrowheads are the most numerous, with some having a slot for insets. One of the arrowheads had an area near the tang which was treated in a lathe-like manner, and an area of relief was also developed on the shaft (Fig. 8.3). Other artefact types include long examples of arrowheads with regular bi-conical ornamented head (Fig. 8.3, 6), and narrow tanged slotted forms with micro-blades preserved in slots at both sides, fixed by glue made from coniferous resin, beeswax and charcoal dust (Fig. 8.3, 3). Other forms include an asymmetric two-winged arrowhead with a slot for insets opposing the wing, and a small barbed type, possibly for shooting pike. In addition, unilateral barbed points with sparse or dense teeth, massive harpoons, massive lance heads, fragments of slotted spearhead, and intact and slotted daggers were found, together with intact fishing hooks (Fig. 8.3, 7–9) and a short double-pointed rod, which served as a hook for catching large pike and pike-perch, were recovered. Of particular interest is a spearhead made of obliquely cut tubular reindeer bone, richly ornamented with geometric designs over the whole surface.

Other tools include hollow end scrapers, side scrapers made of tubular elk bones, wide elk scapula knives and fish scaling knives made of split ribs, awls, needles and a needle case, beaver mandible tools, antler axes, adzes, chisels and gouges, perforated antler sleeves for mounting axe and adze blades, punches and pressure flakers, and animal teeth pendants. Several pebble sinkers with traces of plant binding were also recovered, one of which had concave sides.

The next phase of occupation, the middle Mesolithic cultural layer (III), is dated by palynology to the second quarter of the Boreal period, while the radiocarbon dates on peat and gyttja for this layer are 8780±120 BP (GIN-9383), 8550±100 BP (GIN-9366), 8530±50 BP (GIN-9373 II) and 8290±160 BP (GIN-9372). At about 8500 BP, on the basis of the radiocarbon dates obtained from the overlying gyttja layer, this settlement phase was again submerged. Bone artefacts include needle-shaped arrowheads with a bi-conical head, narrow flat and long forms with a barb near the point, a short unilateral flat finely barbed point and fragments of two other points, with sparse and dense teeth. Other finds include fragments of daggers, one with a slot for insets, two waste pieces from fishing hook production, a wide elk scapula knife, a fragment of a hide polisher, awls, beaver mandible tools, a side scraper made from a beaver upper incisor, an antler axe blade, a fragment of an 'ice-pick', a pressure flaker made of bear canine, and a wedge.

Decorative ornaments are represented by elk incisor and wolf canine pendants and several flat rectangular perforated pendants made of split ribs. Of special interest is a small bone figurine, representing a merganser head (identified by A. A. Karhu) with a long beak, produced in a very life-like manner (Fig. 8.4, 16). Another figurine made of wood has the head broken, and it probably represents a swimming swan, judging

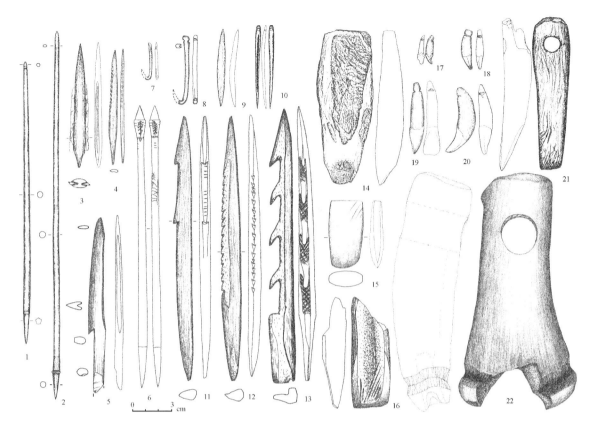

Fig. 8.3 Ivanovskoje 7, lower Mesolithic layer (IV), bone and antler artefacts. 1–6: arrowheads; 7–9: fishing hooks; 10: needle; 11–12: barbed points; 13: harpoon head; 14: antler adze blade; 15: antler axe blade; 16: hollow edge endscraper; 17–20: mammal teeth pendants; 21: shaft straightener; 22: antler adze sleeve.

by its long neck and short massive body (Fig. 8.4, 17). Such images of waterfowl are also evident in Neolithic petroglifs and pottery. Several pebble sinkers with preserved bog plant binding were also found in this layer, accompanied by fragments of pine wood from fishtraps.

The terminal Mesolithic settlement (layer IIa) occurred at this site during the next regression stage in the early Atlantic period, as defined by pollen analysis. Peat, with associated cultural remains, is radiocarbon dated to 7530±150 BP (GIN-9361 I), 7520±60 BP (GIN-9361 II), 7490±120 BP (LE-1260), 7375±170 BP (LE-1261) and 7320±190 BP (GIN-9369 I). Bone and antler artefacts include short and long needle-shaped arrowheads (Fig. 8.4, 1–5), one with a short slot filled with glue and with an imprint of a micro-blade, a long leaf-shaped blade with a barb at one side and a slot with glue on the other, a symmetrical two-winged type with a hollow for a flint point at the end, and two massive blunt examples which were probably used for fur hunting. Several unilateral points with sparse barbs, a massive lance head, fragments of flat narrow slotted daggers and a dagger with oblique blade, and intact fishing hooks (Fig. 8.4, 13–14) with slender or thick stems were also encountered in this layer. Other artefacts include narrow knives made of split ribs and various knives of bone splinters, rib side scrapers, awls, beaver mandible tools, 'ice-picks', a chisel, antler adzes and fragments of axes, wedges, fragments of a punch and a pressure flaker, an antler spoon, and pendants made

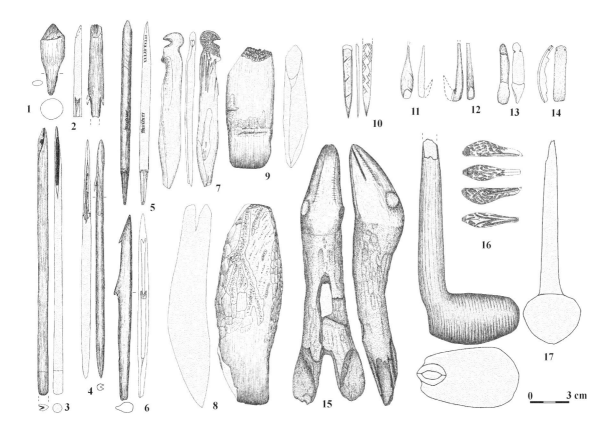

Fig. 8.4 Ivanovskoje 7. 1–15: upper Mesolithic layer (II a), bone and antler artefacts; 1–5: arrowheads; 6: barbed point; 7: knife; 8–9: antler adze blades; 10: awl; 11–12: fishing hooks; 13–14: teeth pendants; 15: antler figurine (staff top); 16–17: middle Mesolithic layer; 16: bone merganser head; 17: wooden swimming swan figurine.

of elk and beaver incisors. Of special interest is a sculpture of a fantastic creature (Fig. 8.4, 15), the head of which is made of elk skull bone, while the body and tail are carved from the antler growing from the skull. The face, with large protruding eyes and firmly shut jaws, combines elements of various animals, but no specific type can be determined. The elongated body, seemingly too small for such a head, gradually changes into a flat fork-like tail. An oval hole is carefully cut in the central part of the figurine, apparently for hafting it to a handle. Use-wear analysis revealed no traces of use, and it is assumed that this object most probably served as the decorative top of a ceremonial staff, of a type known from petroglifs and ethnographic records.

THE PODOZERSKOYE PEAT BOG

Podozerskoye is situated between the towns of Ivanovo and Yaroslavl, *c.* 50 km to the southeast of the latter. The River Lahost, a tributary of the River Kotorosl, itself a tributary of the Upper Volga, originates at this bog. Stanovoye 4 is the most interesting archaeological site in this area. It is situated on a promontory located between a shallow inlet of a large ancient lake and a small lake, which is connected to the larger

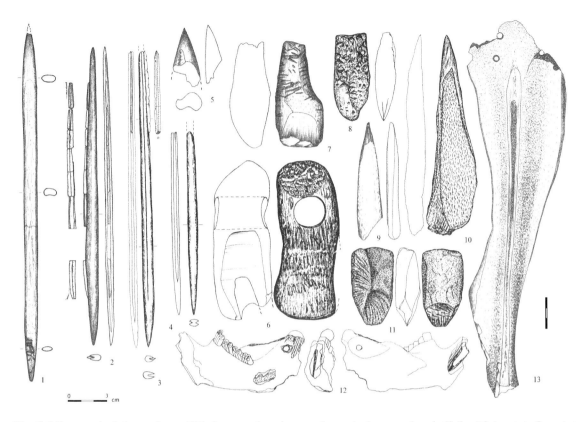

Fig. 8.5 Stanovoje 4, lower layer (IV), bone and antler artefacts. 1–4: arrowheads (2,3 with insets); 5: point of a lance head; 6: antler adze sleeve; 7, 11: antler adze blades; 8: antler axe blade; 9–10: awls; 12: beaver mandible burin-scraper; 13: elk scapula knife.

lake via a small outlet. A total of 385 square metres of the site were excavated in 1992–2000. Cut 1 (16 square metres in area) was on the dryland part of the site, with cut 2 (139 square metres) located in the bog, close to the dry land, and cut 3 (230 square meters) located further into the bog.

The site has four Mesolithic cultural layers. The lower, earliest layer (IV) was investigated in cut 3 and was also traced in cut 2. Pollen data indicate that the formation of this layer took place during the transition from the Younger Dryas to the Pre-Boreal. Radiocarbon dates on the peat incorporating cultural remains in the central, deepest part of cut 3 are 10,060±120 BP (GIN-10127 I), 10,040±90 BP (GIN-10027 II), 9970±50 BP (GIN-10026 I) and 9940±50 BP (GIN-10026 II). A sample of peat, covering the upper areas with artefacts deposited on the gravel subsurface at a distance of *c.* 7 m to the northwest, was dated to 9850±60 BP (GIN-8379) and 9760±150 BP (GIN-8379a), indicating the beginning of a lake transgression which flooded the site. Bone and antler artefacts (Fig. 8.5) include long narrow arrowheads (Fig. 8.5, 1–4), primarily with slots for insets along one or both sides, fragments of massive lance heads (*e.g.* Fig. 8.5, 5), an antler dagger with grooves for insets along both sides, massive antler tine points, broad knives made of elk shoulder blade, awls, a bone polisher, beaver mandible tools, antler axe and adze blades (Fig. 8.5, 7 and 11), two adze sleeves (one with a fragment of a wooden shaft in a shaft-hole (Fig. 8.5, 6), an antler pressure flaker and wedge, preforms and cut bones.

The next occupation of the site took place during a short lake regression, with an early Mesolithic cultural layer (IIIa) formed during this period. Pollen analysis places this layer in the first half of the Pre-Boreal, just before its climatic optimum at *c.* 9600 BP. A wooden stake, sharpened with a stone axe or adze, in cut 3, is radiocarbon dated to 9620±60 BP (GIN-8377). Two more similar stakes, found in a horizontal position in trench 2, below cultural layer III (equivalent to level IIIa of the cultural layer) has been radiocarbon dated to 9620±50 BP (GIN-8374) and 9590±40 BP (GIN-8376). A sample of peat with gyttja, overlying cultural layer (III), in the central part of cut 3 is radiocarbon dated to 9560±40 BP (GIN-10125 II) and 9480±120 BP (GIN-10125 I) respectively, thereby indicating that the submergence of the site occurred during the middle Pre-Boreal.

Bone artefacts include a massive coarse arrowhead with irregular bi-conical head, a gouge, a fragment of a knife and a tubular bone with a deep groove. A fragment of basket-like fish trap was also found in this layer, with a large stone sinker beside it. Only the lower part of the fish trap, including the mouth, was preserved. The rim is composed of an intact branch 15–20 mm in diameter, with the ends of longitudinally split, long willow branches about 7–10 mm in diameter twisted over it, making the split branches perpendicular to the rim. These branches were carefully woven transversely by similar split branches at intervals of about 70–80 mm. Such fish traps, usually about 2 m in length, were widely used in prehistory and are still used for fishing in the villages of the Volga-Oka region. On the basis of the radiocarbon determinations, it appears that this find represents the oldest fish trap in Europe known today.

The middle Mesolithic layer in cut 3 is dated by pollen analysis to the second half of the Pre-Boreal period (*c.* 9300–9000 BP), with peat samples radiocarbon dated to 9280±240 BP (GIN-10122 I), 9090±400 BP (GIN-10124) and 8610±40 BP (GIN-10122 II). The youngest of these dates probably reflects some later mixing of the sample, while the other two are in good agreement with the pollen data. A wooden stake, sharpened with a stone adze, is dated to 9220±60 BP (GIN-8375), which delineates the occupation phase of the site more accurately. An elk bone from this layer in the western part of the cut was dated to 8850±90 BP (GIN-11093a), marking the latest date for activity in this area. Numerous wooden stakes, driven from this layer into the basal layers, and especially into the river bed, are most probably the remains of fish weirs. Fragments of wooden fishing equipment were also recovered at this location, including disks 0.10–0.12 m diameter, 10–20 mm thickness which were perforated in the centre. These were hafted on to sticks and probably used to drive fish into nets.

Various bone and antler artefacts were found (Fig. 8.6), among them numerous arrowheads. These included needle-shaped forms, with bi-conical head or base, those with a leaf-shaped head, which are narrow and flat, and mostly with slots for inserts along one or both sides, along with asymmetric two-winged forms with slots for inserts and with barbs near the tip or along one side. Slotted spearheads and massive lance heads were also found. Barbed points were scarce, and where evidenced these had either sparse or dense fine teeth. Fragments of harpoons were also scarce. Straight or slightly curved daggers, with micro-blade inserts mounted in one or two slots using a grey glue (as described above), were recovered. Fragments and waste from fishing hooks indicate their wide use, while numerous bone and antler tools were used for various domestic activities. These included knives, scrapers, perforated plates for dragging sinew (*e.g.* Fig. 8.6, 11), awls, a needle case, chisels, knives and scrapers made of beaver mandibles, antler axe and adze blades and sleeves for their mounting, narrow bone chisels, punches and a pressure flaker. One adze sleeve was found with a chert core adze still preserved in the groove at its working edge (Fig. 8.6, 17). The butt end of a broken antler adze was preserved in the groove of another, and a sleeve for an axe had a fragment of a wooden handle in a shafting hole. Many items are decorated with engraved geometric designs, and personal ornaments include various teeth pendants and flat rectangular perforated pendants.

The equivalent layer (III) in cut 2, situated closer to the mineral lake margins, is younger in date, being placed in the second quarter of the Boreal period (about 8800–8600 BP) by pollen analysis. Radiocarbon

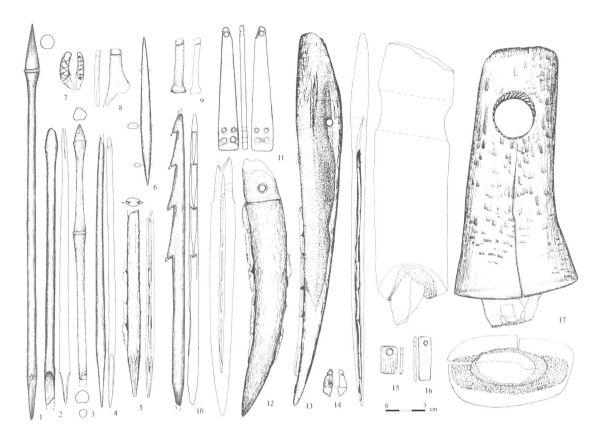

Fig. 8.6 Stanovoje 4, upper Mesolithic layer (III), cut 3, bone and antler artefacts. 1–6, 10: arrowheads; 7–9, 14, 15–16: pendants; 11: plate for dragging sinew; 12–13: daggers; 17: antler sleeve with chert adze blade in a socket.

determinations on peat containing cultural remains give dates of 8930±40 BP (GIN-10109 II), 8640±60 BP (GIN-10110 II), 8540±80 BP (GIN-10109 I) and 8500±150 BP (GIN-10110 I). Three wooden stakes from this layer, sharpened with a stone adze, produced dates of 8700±70 BP (GIN-8854), 8670±50 BP (GIN-8856) and 8540±60 BP (GIN-8853), which help determine the period of occupation. Bone and antler tools (Fig. 8.7) are numerous and include various arrowheads (Fig. 8.7, 1–7, 11 and 12) of needle-shape, with bi-conical head and a long stem, needle-shaped forms with a bi-conical base and narrow and flat profile or with slots along one or both sides, and asymmetric two-winged types with slots for inserts. Fragments of harpoons or barbed points and spearheads were found, accompanied by massive lance heads made of elk long bones and daggers with slots for inserts along one or both sides.

It is worth noting that some of the slotted artefacts had regular un-retouched micro-blades preserved in their original position in the slots, fixed by the same dark grey glue as seen on objects in cut 3. Many bone artefacts were ornamented by engraved geometric designs. Fishing tools include intact hooks and sinkers, while bone and antler tools from domestic activities include various knives (*e.g.* Fig. 8.7, 16), scrapers, awls, a needle and needle case, and chisels, knives and scrapers made of beaver mandibles (Zhilin 1998b).

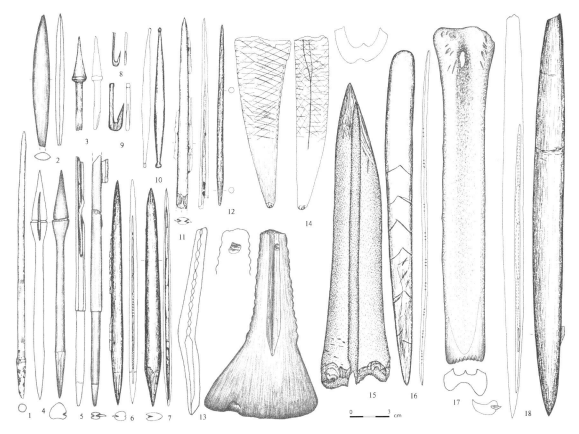

Fig. 8.7 Stanovoje 4, upper Mesolithic layer (III), bone and antler artefacts. 14: cut 1; the rest: cut 2; 1–7, 11, 12: arrowheads; 8–9: fishing hooks; 10: sinker; 13, 17: endscrapers; 14: antler punch; 15: lance head; 16: fish scaling knife; 18: dagger.

Also recovered were antler axes, adzes and sleeves for their mounting, bone chisels and gouges, digging tools, punches and pressure flakers. Personal ornaments are represented by various animal teeth pendants with grooves, and flat rectangular perforated pendants made of split ribs.

THE SAHTYSH PEAT BOG

Sahtysh is situated *c.* 40 km southwest of Ivanovo, being connected by the small River Koika to the Klyazma, a tributary of the Oka. The site is located in a peat bog at the foot of a sloping promontory of a terrace of the Late-glacial lake. During the Mesolithic it occupied a lower lake shoreline position, near a river outlet. The site was discovered in 1999, and 65 square metres were excavated in 1999–2000. Five Mesolithic cultural layers were investigated, and sampling and analysis for dating are on-going. The first layer has produced mammalian and bird bones, a core and several flakes, and it appears that the excavations are in a peripheral zone of a site. It is preliminarily dated by palynology to the Boreal period. Significant quantities of mammalian, bird and fish bones, coprolites, and lithic and bone artefacts were found in the second layer, which is also preliminarily dated to the Boreal period by palynology. The bone and antler tools include a needle-shaped slotted arrowhead with grey glue preserved in the slot, a massive needle-shaped spearhead, a side scraper made of elk rib, broad elk scapula knives and a knife made of split long bone, chisels, knives and scrapers made of beaver mandibles, pendants made of brown bear and small predator canines, and a flat oval pendant with ten perforations along its perimeter.

Cultural layer IIIa has produced numerous bones of forest mammals, birds and fish, coprolites, and chert and flint flakes and blades. Bone and antler artefacts are represented by a short massive needle-shaped arrowhead with glue preserved at the base, a fragment of a flat intact dagger, fragments of a slotted dagger, broad knives made of elk shoulder blades, a fragment of a knife made of elk long bone, a denticulate end scraper, various beaver mandible tools, an antler adze, a hollow edge gouge, a bone chisel, fragments of shallow spoons, one of which was ornamented, an antler punch, elk teeth pendants and a flat rectangular perforated pendant.

The next cultural layer (IIIb) yielded, besides faunal remains and coprolites, wooden stakes sharpened with a flint adze, and a wooden hook made from a curved branch without bark, cut at both ends. Bone artefacts include needle-shaped arrowheads, a knife made of long bone, a chisel-knife made of beaver mandible, fragments of shallow spoons, longitudinally cut elk bones, a bear canine pendant, and fragments of tools and pre-forms.

The basal cultural layer (IV) has produced bone tools, including a needle-shaped arrowhead, an antler dagger, an axe blade, a fragment of a massive lance head, an intact fishing hook, fragments of broad elk scapula knives, a fragment of a side scraper made of tubular bone, awls, hollow edge gouges, a beaver mandible chisel-knife, elk tooth pendants, pre-forms and fragments of tools.

ENVIRONMENTAL CHANGE AND THE MESOLITHIC ECONOMY

Wetland sites with good preservation of organic materials provide ideal conditions for the study of environmental change with the help of scientific methods such as palynology, radiometric dating and osteological analyses. The beginning of the Mesolithic in Central Russia is traditionally connected with the transition from the Pleistocene to the Holocene, which is dated on the Upper Volga to approximately 10,000 BP. According to Spiridonova and Aleshinskaya (1999), the climate and vegetation changed several times during the Mesolithic, with the early Pre-Boreal (*c.* 10,000–9600 BP) being marked by a gradual spread of birch-pine forests. Initially these were sparse, occupying mainly river valleys and lake depressions, while certain elements of the periglacial landscape were still preserved on the interfluves. At around 9600 BP, forest became the main element in the landscape, but the species composition shifted.

The period *c.* 9500–9300 BP marks the climatic optimum of the Pre-Boreal, when pine forests of the taiga type dominated in the Upper Volga region. During the Pre-Boreal the landscape was characterised by a mosaic of vegetation types, with meadows being apparently rather significant in the uplands, and bogs being widespread in depressions. The end of the Pre-Boreal and beginning of the Boreal period (*c.* 9300–8800 BP, with the Pre-Boreal/Boreal boundary at *c.* 9000 BP) is marked by some climatic deterioration, characterised by increasingly cold and damp conditions when compared with the previous optimal period. Birch dominated the forests and the extent of bogs increased. At *c.* 8800–8600 BP the climate gradually improved, and eventually this led to the Boreal optimum, dated to 8600–8200 BP. Forests of the southern taiga type, with pine as the dominant species, were widespread in the region, and the extent of bogs decreased. Some broadleaf species, such as elm, lime and oak, occurring firstly in the early Boreal, occupy a somewhat subordinate place in these forests. The transition from the Boreal to the Atlantic period at *c.* 8000–7800 BP is marked by the dominance of pine and a steady increase in broadleaf species. At *c.* 7800 BP taiga gives way to the mixed forests of the temperate zone, which dominated in the landscape until the

Atlantic optimum. The emergence of pottery around 7100 BP marks the end of the Mesolithic in Central Russia.

Faunal remains from habitation sites show that throughout the Mesolithic various forest mammals were hunted. Elk was the most important of these, being exploited from the earliest period up until the middle Neolithic. Beaver was the second most important, being hunted primarily for its meat, judging from the numerous butchery marks preserved on its bones. The earliest Mesolithic layer of Stanovoje 4 (IV) also produced bones of brown bear, musk-rat, hare and domestic dog. The early Mesolithic layer (IIIa) of this site yielded only elk and beaver bones, suggesting specialised hunting activities. Mammalian remains from the early Mesolithic layer (IV) of Ivanovskoje 7 again show a dominance of elk and beaver, with other species including pine marten, domestic dog, hare, musk-rat, water mouse, hamster, brown bear, fox, badger, reindeer, otter, wolf, mink, squirrel and roe deer. It is worth noting that reindeer bones comprise only 0.46% of faunal remains, while pine marten occurs at 11%, hamster at 1.16%, musk-rat at 1.16%, elk at 38% and beaver at 36%.

The distribution of mammalian bones indicates that taiga forest surrounded settlements from the beginning of the Mesolithic, and that these developed into the mosaic landscapes that characterised the Pre-Boreal period. Forest hunting was supplemented by fowling, with 26 species of waterfowl being identified in the lower layer of Ivanovskoje 7. These were supplemented by wood-grouse, black grouse and also some predatory species such as hawk and white-tailed eagle. Fishing also played an important role during the Mesolithic, with pike being the most significant species exploited, supplemented by perch, perch-pike, whitefish, eel-pout, crucian and roach. In the first half of the Boreal period, wild pig appears synchronously with the emergence of broadleaf trees, though its role in the economy remains small even in the early Atlantic period. At this time sheath-fish also begin to play some role, represented mainly by large examples of 2 m length and more, and both of these species reflect the amelioration of climate in the Atlantic period.

Food gathering was also important, with both dryland and water plants being exploited. Hazelnut shells and seeds of yellow water lilies have been recovered from the early Mesolithic layer (IV) of Ivanovskoje 7. The latter are often found in coprolites, together with small fish bones. Snowball-tree fruits, waternut shells and seeds of berries were also found in the terminal Mesolithic layer of Ozerki 5, and at some other sites. The shells of various aquatic molluscs, found in cultural layers, indicate their use as another food source.

The exceptionally well preserved archaeological and environmental records from the sites considered above suggest that the complex economy of forest hunters, fishers and gatherers of the temperate zone was well developed in Central Russia at the beginning of the Mesolithic. It was perfectly adapted to the forest environment, being both conservative and flexible, and on the basis of the available evidence, the economy of the Mesolithic communities appears to have survived without any substantial change until the late Neolithic. Also, there is nothing in these records to suggest that any crisis, either of a sustained or periodic nature, occurred in the hunting, fishing and gathering economies of these populations during the periods considered.

ACKNOWLEDGEMENTS

I am grateful to E.A. Spiridonova and A.S. Aleshinskaya for pollen analyses, to L.D. Sulerzhitsky and N.E. Zaretskaya for radiocarbon analyses, and to I.V. Kirillova, A.A. Karhu and E.K. Sytchevskaya for the identification of mammalian, bird and fish remains.

REFERENCES

Koltsov L. V. & M. G. Zhilin 1999. Tanged point cultures in the upper Volga Basin, in. S. Kozlowski, J. Gurba & L. L. Zaliznyak (eds) *Tanged Points Cultures in Europe*, 346–60. Lublin: Wydawnictwo UMCS.

Spiridonova E. A. & A. S. Aleshinskaya 1999. Результаты палинологического изучения мезолита Волго-Окского междуречья (Results of pollen studies of the Volga-Oka Mesolithic), in Л. В. Кольцов & М. Г. Жилин (eds) *Мезолит Волго-Окского междуречья*, 137–153. Москва: Наука.

Zhilin M. G. 1996. The Western Part of Russia in the Late Palaeolithic – Early Mesolithic, in L. Larsson (ed.) Earliest Settlement of Scandinavia. *Acta Archaeologica Lundensia* 80, 273–84.

Zhilin M. G. 1998a. Technology of the manufacture of Mesolithic bone arrowheads on the Upper Volga. *European Journal of Archaeology* 1 (2), 149–75.

Zhilin M. G. 1998b. Artifacts, made of animals' teeth and jaws in the Mesolithic of Eastern Europe, in M. Pearce & M. Tosi (eds) *Papers from the European Association of Archaeologists Third Annual Meeting at Ravenna 1997*, 26–31. Oxford: British Archaeological Reports S716.

Zhilin M. G. 1999. New Mesolithic peat sites on the Upper Volga, in S. Kozlowski, J. Gurba & L. L. Zaliznyak (eds) *Tanged Points Cultures in Europe*, 295–310. Lublin: Wydawnictwo UMCS.

Chapter 9

The Ritual Use of Wetlands During the Neolithic: A Local Study in Southernmost Sweden

Lars Larsson

One sees great things from the valley; only small things from the peak. (G. K. Chesterton 1911)

INTRODUCTION

Prehistoric societies can be studied from many different perspectives. To analyse them using an historical-geographical approach, with the landscape as a functioning, active agent, has been shown to provide an important contribution to their understanding (Bradley 1993, 2000, Barrett 1994, Topping 1997, Edmonds 1999). This approach emanates from the potential of humans to understand the landscape they are living in. However, the landscape might be theorised in many different ways, and such an approach therefore challenges our capacity to understand past societies, with their rules and ideas far distant from those of modern 'civilisation'.

If society and landscape are studied from a long chronological perspective, it is of primary importance to view these entities as both dynamic and interactive. For example, farming practices during recent millennia have had a major effect on both social structures and natural landscapes, and one therefore needs to consider that the pre-conditions to study past societies might change considerably, particularly due to processes acting since the Neolithic. This is especially the case in terms of those archaeological remains deposited in extinct wetlands that are related to the ritual activities of Neolithic societies. A taphonomic reconstruction (Benes and Zvelebil 1999, 75) is of importance in understanding how the remains of different behaviours by a society, and recovered in the landscape context by archaeologists, have been deposited but also distorted.

Recent discourse on the relationship between society and landscape has shown that interpretation should not be attempted solely through consideration of artificial monuments; it also needs to involve the natural landscapes (Bradley 2000). Societies with monuments such as barrows, megalithic tombs, causewayed enclosures and henges are obviously stimulating to investigate because of the amount of available information. We might even talk about a 'Wessex syndrome' (*i.e.* the over-emphasis by prehistorians on this region's monuments), but what happens outside such cultural regions? How do past peoples express their world-view? In marginal regions, the symbols may be less impressive, but their determination is still of importance if we are to attempt to understand a Neolithic society. In fact, less obvious symbolism may be of even greater importance than monuments, as this might represent the fundamental expression of social interaction. If one believes that societies with monuments are the reaction to specific stress conditions caused by a concentration of people and resources, societies at a stage without developed monumentality might well give valuable insights into the rationale or pre-conditions under which monuments are subsequently introduced.

THE DEVELOPMENT AND REDUCTION OF WETLANDS

Wetland finds and sites play a very important role in prehistoric research in southern Scandinavia in general, and from the Stone Age in particular (Fig. 9.1). The importance of wetlands as a resource for information on prehistory in Scandinavia is due to the large number of water basins, from large lakes to small kettle holes, formed during the deglaciation of southern Scandinavia.

Most of the lakes in the southernmost part of Sweden have been prone to eutrophic conditions and rapid infilling with organic material during the earlier Holocene. Due to the nature of the bedrock and overlying glacial deposits, the preservation of organic material in wetlands varies. A considerable number of lakes were substantially filled with organic material and became bogs during the Atlantic period. In the Sub-Boreal period, which corresponds to the start of the Neolithic at 6000 cal. BP, many bogs dried out and were covered with forest. At the start of the Sub-Atlantic period, which marks the transition between the Bronze Age and the Iron Age at *c.* 2600 cal. BP, there was increased precipitation, which resulted in the development of raised bogs. In many cases these eco-systems developed on top of the former wetlands (von Post and Granlund 1926); ponds and small lakes were once again partly filled with water, even those which had been relatively dry for millennia. However, this description of wetland transformation is obviously very generalised, and wetland development was dependent upon a range of local factors such as topography, hydrology and catchment size. Even wetlands in the same landblock will therefore have the potential to exhibit diverging developmental histories.

In Scandinavia, population expansion and early forest clearance caused fuel shortages, and from the late eighteenth century peat cutting was introduced on a large scale (Kristiansen 1974, 1985). In order to gain more arable land, many wetlands were also drained, a process that started in the second half of the nineteenth century and is still occurring today. This process has consequently caused a radical change in the extent of the wetland landscape (Bengtsson-Lindsjö *et al.* 1991) (Fig. 9.2).

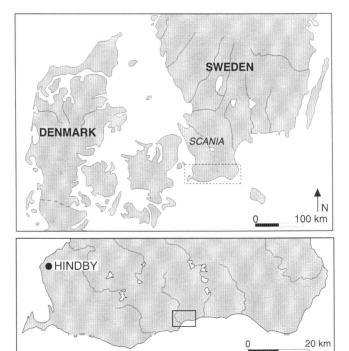

Fig. 9.1 The southern part of Scandinavia and the southern part of Scania with sites mentioned and the research area specially marked.

With the availability of cartographic sources, that are extremely rich by international standards, we can gain a good idea of the extent of wetlands during the period immediately preceding the large-scale drainage endeavours of the last 150 years. Obviously this picture does not agree entirely with conditions prevailing in the Stone Age, but the difference was probably not very great. If we compare the situation in the late eighteenth century, when our map sources are particularly numerous and detailed, we see that very few wetlands have survived in today's landscape.

A study of a small river catchment in the western part of Scania (the southernmost part of Sweden) shows that in the early nineteenth century, wetland covered 29% of the drainage system (Wolf 1956). In the 1950s the wetland area was reduced to about 3%, and today the figure is even lower. The wetlands that have survived are the lakes, as their drainage would have been more difficult and economically unviable.

Depositions of artefacts in bogs were recognised during peat digging in the eighteenth and nineteenth centuries, and some of these constitute the oldest finds in

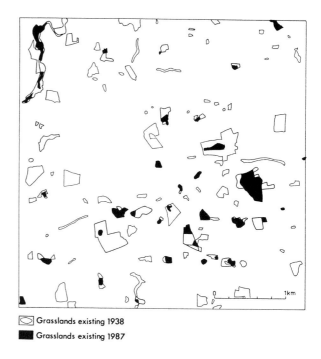

☐ Grasslands existing 1938
■ Grasslands existing 1987

Fig. 9.2 An area within Västra Nöbbelöv parish (part of the research area) showing changes of semi-natural grassland mostly equals to former wetlands (after Bentsson-Linsjö et al. *1991).*

museum collections (Nielsen 1985). In the late nineteenth and early twentieth century, a large number of artefacts were retrieved in southern Sweden. During the two world wars and shortly afterwards, peat cutting was intensive. In the late 1940s most of the cutting was carried out manually, which meant that artefacts and sites were easily recognised. Today peat cutting is of a minor extent and totally mechanised. Finds are rarely made except by archaeologists who are responsible for surveying those areas where sites have been found previously, or where the environment indicates a reasonable chance of finding prehistoric sites (Larsson 1998, 1999). Because of intensive drainage, most of the Neolithic finds are made during ploughing in former wetlands.

The widespread drainage has had far-reaching consequences for prehistoric remains in wetlands. In the 1980s and 90s, Scania was the subject of a total area survey, during which it became evident that a large majority of all the Stone Age objects discovered in the last two decades were found in wetlands, or exhibit a yellow to brown moss patina that indicates that they were deposited in former wetlands. Earlier finds in museum collections were to a large extent found on elevated areas beyond the wetlands, and consequently these lack a moss patina. They are thus most probably the remnants of burials either disturbed or destroyed by arable activities.

WETLAND DEPOSITS

When considering the Stone Age finds in southern Scandinavia in relation to wetlands, the first impression might be that they are linked to bog sites. This is a well known phenomenon because of the good preservation, mostly of bone and antler but also of wooden artefacts, that is encountered in bogs (as attested earlier in Chapter 8).

From the Pre-Boreal onwards, many sites were indeed located in bogs, often as small camp sites that included a hut structure, formed on layers of gyttja and peat at the edge of a lake. These sites have occasionally been identified as becoming overgrown by bog expansion and transgressive episodes of the lake, and are therefore regarded as temporary camps of a seasonal nature (Larsson 1983, 1990). From the Late Mesolithic and Early Neolithic onwards, very few comparable bogs sites are found, primarily because they are often located in what is now the deepest, most central section of the former lake area. The post-activity stages of the lake infill process at these sites produces a situation whereby these activity foci are especially difficult to trace.

However, the finds from most bogs have a state of preservation which clearly indicates that they are not the remains from settlement areas or ordinary domestic refuse contexts. Among the numerous finds from the Mesolithic, some occur in a combination and location which makes it very plausible that they derive from activity of a ritual nature (Larsson 1978a, 1999). Some of the Mesolithic finds exhibit considerable similarities to Neolithic votive contexts. For instance, in the refuse layer of a site from central Scania dating to *c*. 9000 cal. BP, at the edge of a large bog, a collection of more than 30 microliths, some in a fragmentary state, were found (Larsson 1978b). They had been stored in a container, but were broken before they were deposited, so this context rules out the interpretation of the deposit as a cache for future use.

The finds at an inland site dated to the Late Mesolithic at Bökeberg in southwestern Scania are of importance in this discussion (Karsten 2001). The site was located at the shore of a former lake. Artefacts, among which antler tools with decoration were found in the refuse deposits, were positioned in a way that strongly indicates intentional depositions of special importance.

The role of wetlands as a place of contact with the spiritual world was clearly important during the Neolithic. A study of the find contexts of more than 600 Neolithic hoards in Denmark has shown that 80% were found in former wetlands (Nielsen 1985). The Danish study is based on depositions which included two or more flint or stone objects, but a large number of single artefacts have also been found in wetland contexts (Karsten 1994). Additionally, in a study of depositional contexts in Scania, *c*. 370 hoards have been identified, and the proportion found in wetlands is similar to the figure for Denmark (Karsten 1994). Again, however, in addition to the hoards, there are more than 900 single finds recorded.

Peat cutting and drainage have uncovered a large number of objects, mainly of flint but also of bone, antler and amber. The most common category of find is flint axes, and while in most cases the deposition seems to be an isolated expression of a votive ritual, in some instances a number of artefacts were deposited within a limited area over a period of time, often amounting to hundreds of years. Intact tools are the most common find, but in some cases these objects might simply represent the most visible element of depositions that may have included a range of additional artefact types of less durable or obvious nature. Recent Quaternary geological investigations have shown that most of the bogs in which votive deposits were found were only seasonally waterlogged when used for votive offerings.

A case study

The information obtained from museum collections as well as surveys provides a general overview of depositional practices during the Stone Age. In order to obtain a more detailed and hopefully more holistic understanding of votive deposition in Neolithic society, an area in the southernmost part of Sweden was chosen for further study, including the neighbourhood of the author's residence in southern Scania. The area contains a complex range of landscape features, the understanding of which is fundamental to the analysis of wetland depositions. The landscape is mainly undulating and consists of clay, with a high lime

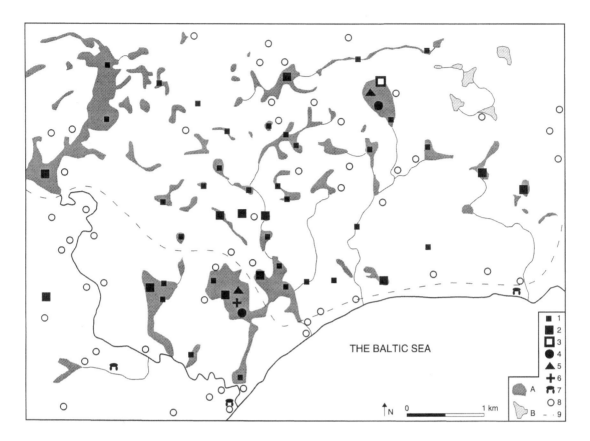

*Fig. 9.3 Votive depositions from the Neolithic within the research area in southernmost Scania, southern
Sweden. 1: deposition of a single artefact; 2: deposition of at least two artefacts; 3: deposition of several
artefacts; 4: deposition of artefacts made of antler or bone; 5: depositions from the Bronze Age; 6:
depositions from the Iron Age; 7: megalithic tombs; 8: settlement; 9: the extent of the hummocky area; A:
wetlands; B: lakes.*

content, that almost reaches the coastline, and is delimited on both sides by sandy plains (Fig. 9.3). The
sandy coastal area, at some points just a couple of hundred meters wide, is a glaciofluvial formation from
the end of deglaciation (Fig. 9.3). Due to the nature of the topography, wetlands occur extensively in this
region, and consequently land suitable for settlement is limited. The number of Neolithic artefacts in
museum collections are low from this area in comparison with those from the bordering sandy plains
(Karsten 1994), as also are the monuments from this period. Just one megalithic tomb is recognised within
the survey area, while several tombs are known to have existed on the adjacent plains (Fig. 9.3).

As can be seen from Fig. 9.3, the wetlands have been considerably reduced in both area and number
during recent decades. Also, an important data-set is preserved in a military reconnaissance map from
around 1815, which pre-dates the implementation of large-scale drainage activity and therefore gives a
unique insight into the size and distribution of wetlands, several of which have been totally drained today.
In most of these cases the former wetland area can still be monitored as these areas are highlighted by

darker soil, which remains visible as a result
of the high organic content of the plough-
zone.

Because of the undulating topography, the
view of the landscape varies considerably,
with excellent views from hilltops and a
restricted field of vision in the lower areas.
This change can take place within less than
100 m, and can significantly affect a person's
perception of the landscape. These characteristics of the landscape imbue it with a kind
of monumentality in itself, an element which
contrasts markedly with the surrounding
plains. In this topographic setting the
wetlands are usually rather small in size, or
they are long and narrow, and occupy the
discrete areas between the hills (Fig. 9.4). In
several cases it is possible to see from one
wetland to another, but lines of sight are
restricted and controlled by the higher
ground. Even in areas close to the coast, the
Baltic Sea is not visible from any of the
wetlands.

The settlement remains of the Neolithic
period are situated on small hills, especially
on those formed of well-drained material
(Fig. 9.3). Unfortunately, ploughing has
heavily damaged all of these settlement sites.
However, they are all of limited extent,
usually encompassing an area of less than
1000 m².

Although the wetlands form an important
part of the landscape, their functional use
appears to have been limited. There is no
evidence of wetland having been used as
pasture, a practice which is first evident
during the middle Bronze Age (Berglund *et
al.* 1991). Hunting and fishing could have
been of some importance, even though the
latter would have been a seasonal activity, as
large areas of the wetlands were almost dried
out during the summer.

The hummocky landscape has yielded a

*Fig. 9.4 View of the hummocky landscape within the
research area, with wetlands in different stages of
preservation.*

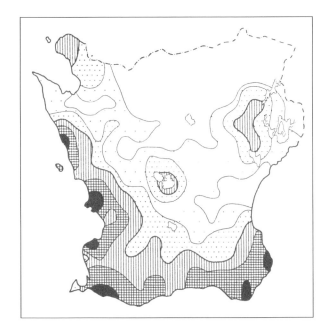

*Fig. 9.5 The distribution of thin-butted flint axes from
the Late Early Neolithic and Early Middle Neolithic
(after Malmer 1957).*

small number of finds compared to most of the southern coastal zone of Scania (Fig. 9.5). However, field
survey by the present author, along with information gained from several farm collections, means that the
find locations of a considerable number of archaeological deposits are now known (Fig. 9.3), with

information on over 80 flint artefacts having been recorded. Most of these artefacts have been found during ploughing, with a small number recovered during digging for drainage. Judging by the yellow to red patination of artefacts in the farm collections, a large number were found in wetlands, though it should be noted that no information on the circumstances of discovery is available.

Axes, mainly of flint but also of stone, are the dominant artefact type in most of the farm collections, and chronologically, because during the latest part of the Neolithic axes are replaced by daggers, we have a *terminus ante quem* for these objects. Most of the finds from museums, as well as from farm collections, include intact tools. However, during surveys, several fragments of axes and also daggers have been found. While some of these might have been intentionally split, some axe fragments have a pattern of breakage consistent with their having been broken during use (Olausson 1983, fig. 26).

DURATION OF DEPOSITION

Another feature of deposition is a mixture of tool types which is often associated with a marked chronological gap occurring within the same small bog, or within the same part of a somewhat larger wetland. One example of this phenomenon was recovered from a small wetland, some 20 m across, where a battle-axe from the Early Neolithic was found along with a flint axe from the Late Neolithic, representing a minimum of two depositions occurring over a millennium apart (Fig. 9.6). In some cases, finds have been made in small wetlands that are not recorded on the map from around 1815. One of these bogs has an area of less than fifteen metres across. On different occasions, a flint axe, a fragmentary stone axe, a flint core with a shape resembling an axe, and a small polishing stone have been recovered (Fig. 9.7). However, in this instance, as the axes are of the same type, the artefacts might have been deposited simultaneously and dispersed later by ploughing.

The simplicity of the artefacts makes it difficult for the layman to identify some of the objects as part of intentional votive depositions, and it is possible that, except where they occur in more easily recognised forms such as flint axes or daggers, such deposits have not been recognised in some bogs. The excavation of a small bog at Hindby, in southwestern Scania (Fig. 9.1), illustrates this situation (Svensson 1993). Here the remains of votive practices were found running through the Late Mesolithic, most of the Neolithic and into the Bronze Age (5000–1500 cal. BC), and a number of tree trunks had been placed as a short trackway in order to facilitate access to the central part of the bog. There are examples of axes deposited in pairs, but it is more common to find combinations of tools, some of which were broken up before deposition. In addition, two human bones and three pig 'eye-teeth' have been recovered. These remains show that in some cases sorting of the votive material was carried out, as well as deliberate fragmentation by cracking and burning before the deposition was made. While these depositions are the most difficult to recognise, the possibility exists that they may well have been amongst the most common, and that the failure to identify them has severely distorted the record. During the survey of the research area several bones

Fig. 9.6 A flint axe from the Late Neolithic and a battle-axe from the Early Neolithic, found in a small wetland.

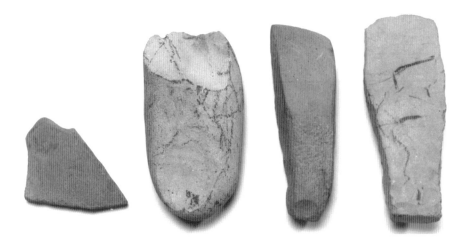

Fig. 9.7 From the left: a small polishing stone, a flint core with a shape resembling an axe, a fragmentary stone axe and a flint axe, found in a small wetland.

were found, but as the species identified does not give any indication of the chronological age, these objects may very well be part of later depositional practices.

From the perspective of the Neolithic period, almost every wetland was used for some form of votive deposition. Compared to Continental Europe, where depositions in rivers and lakes are rather well known, the number in southern Scandinavian contexts is small (Forssander 1933). Only two instances have been reported where artefacts were found in small streams within the research area.

WETLANDS AND ARTEFACT DEPOSITION

Most of the wetlands considered here contain only a few artefacts, whereas in the larger wetlands on the plain a considerable number of artefacts have been found according to the evidence from some of the farm collections studied. In most cases these were recovered during peat cutting which took place over a number of years, but which occurred within a limited area of the wetland. There are observations of tree trunks and concentrations of bark and branches in connection with the discovery of artefacts, but it is not possible to state whether these are natural finds of fallen trees or parts of trackways or platforms associated with the act of deposition.

The largest votive deposit found within the undulating area comprises a total of seventeen axes, with a shape and size that strongly indicates that they were made by the same flint-knapper (Rydbeck 1918, fig. 42–3) (Fig. 9.3). Other artefacts have been found in the same wetland, which is the largest within the hummocky area. However, the most numerous deposits seem to be related to the plains – areas that are also marked by a larger number of settlement sites and megalithic tombs.

In some cases the remains of settlements are situated only a short distance from a wetland in which artefacts have been found. In an area with intensive excavations about 10 km west of the survey region, deposits of axes appear in close connection with settlement sites (Larsson 1992). They were deposited a short distance from the shoreline area, which was marked by decaying stumps of bushes and small trees. When relating the find spots with the surrounding landscape, it is remarkable how most deposition sites have a very delineated setting, with most occurring in close proximity to steep hillsides.

In view of the fact that we are dealing with wetland deposits, it is conceivable that people stood on the bank and threw the objects out into the water. However, in those cases where a more detailed account is given about the precise context of axes that were found close to each other, the finder has observed that they were often carefully placed in a special arrangement within the wetland. Axes have been found close to, and sometimes on top of, each other, with the edges directed downwards or in a circle (Rech 1979). This substantiates the observation that such artefacts were carefully placed in the water rather than being thrown out into the basin.

LANDSCAPE AND SOCIETY

In certain respects, the cosmology which is related to wetland offerings was active throughout most of the Neolithic, and also in later periods. Some wetlands appear to have been imbued with ideas of a sacred character over many centuries, and in some cases millennia (Stjernquist 1997). Artefacts were often deposited within a delimited area of the bog, even though they may represent material from a considerable timespan. The dating of the finds from the surveyed area highlights an interesting variety in the intensity of deposition that is in good agreement with the record from the Scania region (Karsten 1994) (Fig. 9.8). A small number of artefacts belong to the Early Neolithic (EN I) but a much larger number of tools are dated to the latest part of the Early Neolithic or the Early Middle Neolithic (EN II–MNA II). A much smaller number can be dated to the later part of the Middle Neolithic (MNA III–V), but the latest part of the Middle Neolithic (MNB) and the Late Neolithic (LN) are also well represented. The earliest concentration of deposition is contemporaneous with the building of megalithic tombs and causewayed enclosures. Thereafter, there appears to be a hiatus in Scania, during which the wetland locations that acted as foci for votive deposition were seldom exploited. In spite of the time interval, however, there is a close spatial relationship between depositions dating from the late Early Neolithic/early Middle Neolithic and those belonging to a late part of the Middle Neolithic, and also continuity into the Late Neolithic. This observation appears to be of major importance for understanding the world-view of Stone Age societies and how it can be observed in material culture.

Depositing artefacts within a delimited area of a wetland, over intervals of several centuries, means that knowledge of the ritual importance of the site survived for generations. Knowledge of the physical as well as the metaphysical components of the landscape was passed on over long timescales, and during certain periods the relation of these components to the wider world-view is marked by ritual practice involving the material culture. However, it is apparent that for long intervals the knowledge was passed on without any visible reaction by the Neolithic societies in question, in terms of active votive deposition, and this observation perhaps best reinforces the conclusion that these locations held significance beyond the spiritual.

In this context, there may have been changes occurring in the society, when people had a need to disrupt as well as establish links with much earlier societies. Bringing the old offering sites back into use during the later part of the Middle Neolithic might have been a way of re-establishing contact with earlier

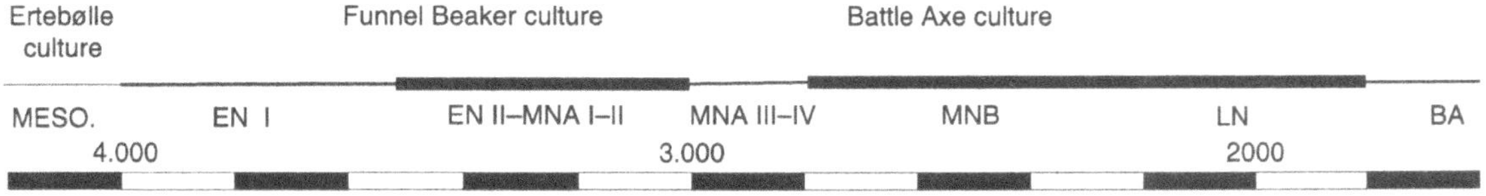

Ertebølle culture		Funnel Beaker culture		Battle Axe culture			
MESO.	EN I	EN II–MNA I–II	MNA III–IV	MNB		LN	BA
4.000			3.000			2000	

Fig. 9.8 The intensity of deposition pattern in wetland within the research area during the Neolithic.

generations. The sites represent a connection with societies of the past based on legends, and could function as a means of legitimisation, as well as emphasising a different value system from the society which had been replaced.

As repeated depositions of artefacts occurred within a limited area of the wetland, which was shallow and in some cases seasonally desiccated, there would have been some knowledge of the excellent status of the tools used in offerings, which could have been several generations old. The deposition or transformation of artefacts in water would presumably confer special value to them, and this must surely have been the primary intention behind such offerings. Where wetland depositions became visible during dry summers, when the harvest might have been adversely affected, the appearance of such deposits might well have been an important stimulus to reinforce connections with the metaphysical world.

In the cosmology of certain societies the cosmos involves three worlds – the underworld, the earth and the sky (Jacobson-Widding 1979, Helskog 1999). The underworld is usually connected to water, so wetland might have been regarded as the liminal zone between the underworld and the physical world. Water is life-giving for all organisms, and wetland depositions might therefore be related to underground spirits, connected with fertility, where the wetland was regarded as a point of bodily access to the hidden soul, *i.e.* to the underworld. Votive offering in wetlands might have been a regular practice, where the desired effects were long-lasting but not immediately noticeable. We can envisage a situation where change is not visible, unless the offerings cease. Some of the wetland offerings may be viewed as abandoned projects, especially where the deposition period stretches over several generations (Barrett 1994, 13).

According to palaeoecological studies, most Neolithic wetlands were partially covered or encircled by small trees and bushes. The depositional context of a small wetland, with a dense vegetation of bushes and trees surrounded by steep hills, focusing on a small part of the sky, produces a location tailor-made for secret depositions made by individuals or a small group of people. The situation is quite different in larger wetlands on a plain, where the deposition could be witnessed by a large group of people. The first example is a ritual act of which the most important part is to create contact with members of the underworld or upper world. In the second case, it should be anticipated that at least some depositions are made in order to allow several people to take part in an act of votive offering which was initiated and performed by certain members of the society. This latter form of ceremony might be related to the activities of burial in megalithic tombs, or involve offerings made to the monuments themselves. Such ceremonies are well attested by the large quantity of pottery and tools found outside the entrance to such tombs (Strömberg 1968, 1971, Tilley 1996).

Votive depositions within the hummocky landscape are usually rather small-scale, with the more substantial depositions being found in the larger wetlands on the plain, but in a context that exhibits a close connection with the adjacent hummocky landscape. This might indicate that wetlands in different kinds of landscapes could have special importance in the metaphysical world, and the fact that they are still used during later parts of the prehistoric period also indicates their continued ritual and symbolic importance.

The numerous small wetlands within the hummocky landscape will have had a limiting effect on population size and the potential for communities to expand their resource exploitation areas. This might mean that conflicts between any new settlers and the existing habitants were minimised, thereby reducing the potential for stress in the society. The need for marking the relationship of people with monuments was therefore unnecessary in this more marginal landscape, but this does not mean that ceremonies of different kinds were not taking place. As stated earlier, ceremonies will have taken the form of both more individual votive offerings in small wetlands, and larger societal expressions of ritual behaviour in the wetlands of the plains. Wetlands clearly functioned as natural monuments within the landscape in regions where monumental architecture was either an impractical or unnecessary form of societal ritual expression and legitimisation.

REFERENCES

Barrett, J. 1994. *Fragments from Antiquity: an archaeology of social life in Britain, 2900–1200 BC*. Oxford: Blackwell.

Benes, J. & M. Zvelebil 1999. A historical interactive landscape in the heart of Europe: the case of Bohemia, in P. J. Ucko & R. Layton (eds) *The Archaeology and Anthropology of Landscape*, 73–93. New York: Routledge.

Bentsson-Lindsjö S., M. Ihse & G. Olsson 1991. Landscape patterns and grassland plants species diversity in the 20th century, in B.E. Berglund (ed.) *The Cultural Landscape during 6000 years in Southern Sweden*, 388–96. Lund: Ecological Bulletins 41.

Berglund, B. E., N. Malmer & T. Persson 1991. Landscape-ecological aspects of long-term changes in the Ystad area, in B. E. Berglund, (ed.) *The Cultural Landscape during 6000 years in Southern Sweden*, 405–24. Lund: Ecological Bulletins 41.

Bradley, R. 1993. *Altering the Earth: the origins of monuments in Britain and continental Europe*. Edinburgh: Society of Antiquaries of Scotland 8.

Bradley, R. 2000. *An Archaeology of Natural Places*. London: Routledge.

Chesterton, G. K. 1911. *The Innocence of Father Brown, the Hammer of God*. Leipzig: B. Tauchnitz.

Edmonds, M. 1999. *Ancestral Geographies of the Neolithic: landscapes, monuments and memory*. London: Routledge.

Forssander, J-E. 1933. En fyndplats från stenåldern i Sege å . *Meddelande från Lunds universitets historiska museum* 1933, 24–41.

Helskog, K. 1999. The shore connection: cognitive landscape and communication with rock carvings in northernmost Europe. *Norwegian Archaeological Review* 32, 73–94.

Jacobson-Widding, A. 1979. *Red–White–Black as a Mode of Thought: a study of tradic classification by colours in the ritual symbolism and cognitive thought of the people of the lower Congo*. Uppsala: Almqvist & Wiksell International, Uppsala Studies in Cultural Anthropology 1.

Karsten, P. 1994. *Att Kasta Yxan i Sjön: en studie över rituell tradition och förändring utifrån skånska neolitiska offerfynd* (To Throw the Axe in the Lake: a study of ritual tradition and change from Scanian Neolithic votive offerings). Stockholm: Almqvist & Wiksell International, Acta Archaeologica Lundensia, Series in 8°, No. 23.

Karsten, P. 2001. *Dansarna från Bökeberg: om jakt, ritualer och inlandsbosättning vid jägarstenålderns slut* (The dancers at Bökeberg: hunting, rituals and inland settlement during the end of the Mesolithic).. Malmö: Riksantikvarieämbetets Arkeologiska Undersökningar Skrifter 37.

Kristiansen, K. 1974. En kildekritisk analyse af depotfund fra Danmarks yngre bronzealder (periode IV–V): et bidrag til den arkæologiske kildekritik (A source-critical analysis of hoards from the Danish Late Bronze Age (periods IV–V): a contribution to archaeological source-criticism). *Aarbøger for Nordisk Oldkyndighed og Historie* 1974, 119–60.

Kristiansen, K. 1985. Economic development in Denmark since agrarian reform: a historical and statistical summary, in K. Kristiansen (ed.) *Archaeological Formation Processes: the representativity of archaeological remains from Danish prehistory*, 41–62. København: Nationalmuseet.

Larsson, L. 1978a. Mesolithic antler and bone artefacts from central Scania. *Papers of the Archaeological Institute University of Lund* 2, 28–67.

Larsson, L. 1978b. *Ageröd I:B – Ageröd I:D: a study of early Atlantic settlement in Scania*. Lund: Acta Archaeologica Lundensia, Series in 4°, No. 12.

Larsson, L. 1983. *Ageröd V: an Atlantic bog site in central Scania*. Lund: Acta Archaeologica Lundensia, Series in 8°, No. 12.

Larsson, L. 1990. The Mesolithic of southern Scandinavia. *Journal of World Prehistory* 4, 257–309.

Larsson, L. 1992. Neolithic settlement in the Skateholm area, southern Scania. *Papers of the Archaeological Institute University of Lund* 9, 5–43.

Larsson, L. 1998. Prehistoric wetland sites in Sweden, in K. Bernick (ed.) *Hidden Dimensions: the cultural significance of wetland archaeology*, 64–82, Vancouver: UBS Press.

Larsson, L. 1999. Settlement and palaeoecology in the Scandinavian Mesolithic, in J. Coles, R. Bewley & P. Mellars (eds) *World Prehistory: studies in memory of Grahame Clark*, 87–106. London: Oxford University Press, Proceedings of the British Academy 99.

Malmer, M. P. 1957. Pleiombegreppets betydelse för studiet av förhistoriska innovationsförlopp. *Finska Fornminnesföreningens tidskrift* 58, 160–184.

Nielsen, P. O. 1985. Neolithic hoards from Denmark, in K. Kristiansen (ed.) *Archaeological Formation Processes: the representativity of archaeological remains from Danish prehistory*, 102–9. København: Nationalmuseet.

Olausson, D. 1983. Lithic technological analysis of the Thin-butted Flint Axe. *Acta Archaeologica* 53, 1–87.

Rech, M. 1979. *Studien zu Depotfunden der Trichterbecher und Einzelgrabkultur des Nordens* (Studies of Hoards from the Funnel Beaker Culture and the Single Grave Culture of Northern Europe). Neumünster: Karl Wachholtz Verlag, Offa-Bücher 39.

Rydbeck, O. 1918. *Slutna mark- och mossfynd från stenåldern i Lunds Universitets Historiska Museum, deras tidsställning och samband med religiösa föreställningar* (Field- and bog-finds from the stone age in Lund University's Historical Museum, the chronology and relation to ritual beliefs). Lund: Gleerup, Från Lunds Universitets Historiska Museum.

Stjernquist, B. 1997. *The Röekillorna Spring: spring-cults in Scandinavian prehistory*. Stockholm: Almqvist & Wiksell International, Regia Societatis Humaniorum Litterarum Lundensis LXXXII

Strömberg, M. 1968. *Der Dolmen Trollasten in St. Köpinge, Schonen* (The dolmen Trollasten in St. Köpinge, Scania). Lund: Acta Archaeologica Lundensia. Series in 8°. No. 7.

Strömberg, M. 1971. *Die Megalithgräber von Hagestad: zur Problematik von Grabbauten und Grabriten* (The Megalithic Tombs of Hagestad: problems concerning construction and rituals). Lund: Acta Archaeologica Lundensia, Series in 8°, No. 9.

Svensson, M. 1993. Hindby offerkärr – en ovanlig och komplicerad fyndplats (The bog for offering at Hindby – an unusual and complex site). *Fynd* 1993, 5–11.

Tilley, C. 1996. *An Ethnography of the Neolithic: early prehistoric societies in southern Scandinavia*. Cambridge: Cambridge University Press.

Topping, P. (ed.) 1997. *Neolithic Landscapes*. Oxford: Neolithic Studies Group Seminar Paper 2, Oxbow Monograph 86.

von Post, L. & E. Granlund 1926. *Södra Sveriges Torvtillgångar I* (The Peat Resources of Southern Sweden I). Stockholm: Nordstedt, Sveriges Geologiska Undersökning Ser. C, no 2.

Wolf, P. 1956. *Utdikad Civilisation* (Drained Civilisation). Malmö: Gleerups.

Chapter 10

Prehistoric Wetland Environments and Human Exploitation of the Foulness Valley, East Yorkshire, UK

Peter Halkon

INTRODUCTION

The River Foulness (Fig. 10.1) is fed by a series of smaller streams, which rise on the western escarpment of the Yorkshire Wolds in eastern England. It runs southwestwards across the lowlands of the southeastern part of the Vale of York, turning in a large arc towards the River Humber, one of England's major estuaries. At present, the Foulness joins the Humber via the Market Weighton canal (constructed in AD 1772); before then it entered the Walling Fen, which was at various times a tidal estuarine inlet of the Humber (Halkon and Millett 1999). At points along the river, extensive areas of wetland, all now drained, are evident in aerial photographs and on geological survey maps (*e.g.* Geological Survey of England and Wales sheet 71 [Selby] and sheet 72 [Beverley]). Of particular importance in this context are those wetlands located at Everingham Carrs, Hasholme and Bursea. However, these wetlands can only be properly understood in the context of the better drained ridges of aeolian sand and other Quaternary sediments which border them.

What follows is an outline of a landscape archaeology project initiated in 1980 by the present author, with members of the East Riding Archaeological Society (ERAS). Initial field walking was concentrated to the south of the parish of Holme-on-Spalding Moor (HOSM), where ERAS had excavated a series of Roman pottery kilns and an underlying Iron Age settlement (Hicks and Wilson 1975). This work followed on from that of Corder (1930), whose excavation at Throlam had provided a 'type-site' for this regionally important Roman pottery production centre.

This primary phase of field survey (Halkon 1983) revealed a much greater density of past activity than had previously been supposed. There was an obvious correlation between sites and soil types, as scatters and concentrations of Romano-British greyware pottery, iron slag and other material, presumed to indicate ploughed-up settlement and industrial sites, were largely located on the ridges of sand close to the River Foulness. Flint tools, debitage, and cracked and burnt stones recovered during field walking also highlighted an earlier prehistoric presence within this landscape. The potential for the survival of organic material within the wetlands of the Foulness system was evident, as bog oaks were often encountered during drainage and cultivation in the extensive carrs flanking the river.

In 1983 the author joined forces with Professor Martin Millett, and a multidisciplinary approach to the Foulness landscape was adopted. This incorporated excavation, field walking, aerial photography and

Peter Halkon

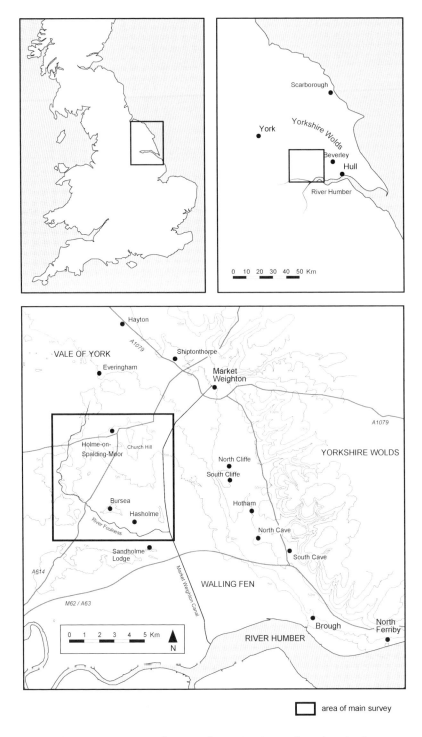

Fig. 10.1 Location map showing the main sites referred to in the text.

geophysical survey. The results of this work have now been fully published (Halkon 1987, Halkon and Millett 1999). With the enthusiastic co-operation of local farmers, a further technique employed was the examination of drainage schemes on the carrs near the river. This led directly to the discovery of the Iron Age Hasholme logboat in July 1984. As well as recovering Britain's largest surviving logboat (Millett and McGrail 1987), this rescue excavation also provided an opportunity for detailed palaeoenvironmental research, particularly on the pollen and minerogenic aspects of these deposits (Jordan 1987, Turner 1987), thereby laying the foundation for subsequent work, which is reviewed below.

This integrated approach to landscape archaeology continues to be central to the project, which has subsequently examined a Roman roadside settlement at Shiptonthorpe (Millett and Halkon 1986, Millett 1992, Taylor 1995) and is currently investigating a 3 × 3 km area around Hayton (Halkon and Millett 1996, Halkon *et al*. 1997, 1998, 2000), further up the Foulness valley. Although not wetland archaeology as it is usually perceived, both excavations were found to contain well preserved waterlogged deposits.

Largely in response to chance finds by farmers and workers on the land, the research rationale has been expanded temporally to include earlier prehistory and the Medieval period, with a full report on prehistoric aspects currently in preparation (Halkon *et al*. forthcoming). The area of research has also been enlarged spatially to cover the whole catchment of the River Foulness.

It must be emphasised that this account represents the results of the contribution made by several hundred people – students, specialists and volunteers of all ages. Though now firmly embedded within a university research regime, public participation, education and access remain core components of the research agenda. The following review aims to provide a chronological overview of this research, with particular regard to the wetlands and their drier margins during prehistory.

GEOLOGY AND EARLY PREHISTORY

Within the Foulness valley it is possible to trace wetland development from before Isotope Stage 7 (245,000–186,000 yr BP). Chance discoveries of the bones of straight tusked elephant (*Palaeoloxodon antiquus*), (Halkon 1999, Schreve 1999, Halkon *et al*. forthcoming), combined with a re-assessment of the important mammaliferous deposits at Bielsbeck Farm (Lamplugh *et al*. 1910, Harcourt 1829, Schreve 1999), have demonstrated the presence of an infilled valley of this age, *i.e.* a proto-Foulness. This feature was cored in 1999, and alluvial clay deposits were found to be *c*. 9 m deep, with palaeoenvironmental material well preserved within them. A molluscan fauna (Lamplugh *et al*. 1910, Schreve 1999), which included terrestrial, marsh and freshwater species, complements the coleopteran and larger faunal evidence in indicating the presence of a substantial river, bordered by marsh, woods and some open grassland. Compared to deeply buried early sites elsewhere, such as Boxgrove in West Sussex (Roberts and Parfitt 1998), those at Bielsbeck, North Cliffe are close to the modern ground surface. It is possible to trace this Pleistocene alluvial deposit from the air as a crop mark, as well as determining the later, Holocene streams that followed it (Fig. 10.2).

There may also be evidence for Palaeolithic human activity along this river system, as a pointed Acheulian handaxe has been found at the top of Austin's Dale, overlooking the lowlands (Roe forthcoming). The unworn condition of this item showed that it had not been subjected to glacial erosion and may have been a casual loss *en route* between the Foulness lowlands and the adjacent uplands of the Wolds.

THE MESOLITHIC PERIOD

The presence of former lakes and relict streams can be traced from aerial photographic evidence, especially at Bursea and Hasholme (Fig. 10.3). Coring and subsequent palaeoenvironmental analysis along the southern stretch of the Foulness valley between Welham Bridge and Sandholme (Long *et al*. 1998, Innes

Fig. 10.2 Aerial photograph showing the Foulness relict streams near Bielsbeck Farm, North Cliffe. The pond marks the location of the 1829 fossil finds.

Fig. 10.3 Aerial photograph showing relict streams near Bursea. Early Mesolithic finds cluster on the sand ridges near these streams.

et al. forthcoming) has revealed the existence of reedswamp environments of high biomass potential, thereby complementing chance finds of elk and red deer at various locations in the valley system. It is on the sand ridges adjacent to the alluvium at Bursea and Hasholme and the now largely destroyed Howe Hill on the edges of Everingham Carrs, however, that the majority of evidence for human activity is to be found (Fig. 10.4). Several core tools of axe/adze type have been recovered by chance and during systematic survey, along with evidence for flint working (Halkon 1999, 2003), and there is a distinct possibility that other early Mesolithic material lies buried under alluvium in some places.

A range of later Mesolithic material has also been found within the study area and the flint distribution shows expansion into the wider landscape, away from the wetlands, though some locations near streams are still favoured. Innes (Long *et al.* 1998, Innes *et al.* forthcoming) has shown that water levels rose during the Mesolithic, with reedswamp environments being replaced by a series of freshwater lakes. Further upstream a naturally fallen oak dated to 4834–4576 cal BC (2 sigma) (GU5001) was found during the construction of the Market Weighton by-pass, where the road crossed Market Weighton Beck (Halkon *et al.* forthcoming). Bones of dog and red deer were also found during the digging of a borrow pit nearby, in the same deposits as the tree.

Coring at Sandholme shows that wetland conditions were altered further by a marine transgression which created a tidal inlet at around 4789–4352 cal BC (2 sigma) (SRR-4894) (Innes *et al.* forthcoming), with estuarine conditions extending as far north as Hasholme.

THE NEOLITHIC PERIOD

A combination of palaeoenvironmental evidence, especially from Sandholme (Innes *et al.* forthcoming) and Hasholme Grange (Turner 1987, Heath and Wagner forthcoming), shows that the Foulness valley in this period was a diverse mosaic of vegetation environments. Further away from the estuarine inlet were freshwater reedswamp and fen areas, with extensive woodland differing in composition along the valley. As elsewhere in Britain (Whittle 1999), there is little evidence for a dramatic change in human activity in the early Neolithic. The discovery of a number of characteristic leaf-shaped arrowheads, especially close to what is now Market Weighton Beck (Halkon *et al.* forthcoming), implies that hunting continued to be an integral element of food procurement strategies in this region during the earlier part of the Neolithic. This is reinforced by the evidence from Howe Hill, Everingham, a site intimately associated with the wetland zone, where early and later Mesolithic material has been recovered. Neolithic activity at this location is attested by finds of a Neolithic arrowhead in association with Grimston ware pottery (Manby 1988, Halkon *et al.* forthcoming), suggesting a continuity of wetland exploitation strategies into the Neolithic.

Elsewhere along the valley, pollen and beetle remains show some evidence of human activity within the woodland bordering the wetlands during the Neolithic. Some form of woodland management may have been in operation at this time, as 17 stone and flint axes and adzes have been found within a 6 × 9 km area, with a total of 30 such finds within the larger area of the Foulness valley. Most of these tools were found on the light, better-drained soils close to the alluvium (Fig. 10.5) (Halkon *et al.* forthcoming).

There is little evidence to indicate the products of such management. Perhaps the timber used in the construction of the long barrow at Market Weighton Wold, excavated by Rolleston (Greenwell 1877, Kinnes 1992), may have originated in these lowlands; such a structure would surely have required considerable quantities of timber. Manby (1999) postulates the presence of a timber façade, comparable with Willerby Wold and Street House, which, like the Market Weighton long barrow, were deliberately burnt. The location of the monument itself is also likely to be of considerable significance, as it lies near the head of a prominent dry valley – a route-way linking the chalk uplands of the Yorkshire Wolds with the lowlands and wetlands of the Foulness valley. A continued spatial, and possibly ritual, relationship can also

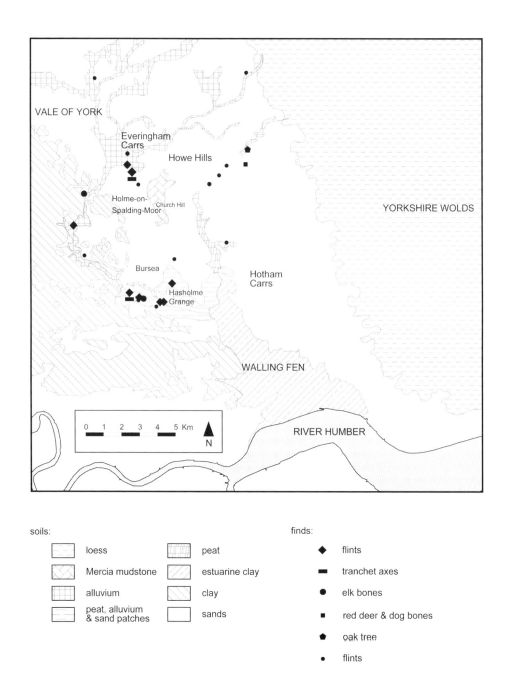

Fig. 10.4 Mesolithic sites in the Foulness Valley.

be postulated in the Bronze Age, with a large round barrow cemetery at Market Weighton Wold and the famous square barrow appearing to take account of this line of movement through the landscape.

The findspots of some of the finest polished tools (Fig. 10.6) from the Foulness valley may also warrant a ritual/votive explanation, especially the magnificent flint adze of Duggleby-Seamer type from Hasholme Carr Farm (Manby forthcoming). The mint condition and high quality of this object make it unlikely that it was ever used in wood-working, and it may be a special deposit similar in type to those known from the wetlands of other areas (Bradley 1990, 66–7, Bradley and Edmonds 1993, 49, Larsson *this volume*).

Some caution is needed, however, in applying this interpretation to all such discoveries. Haughey (2000) has recently pointed out that some of the Neolithic objects within the floodplain of the middle Thames valley, previously interpreted as votive offerings in watery places, were placed, or simply lost, in areas that subsequent palaeoenvironmental analysis has shown were dry at the time of deposition. At Hayton, however, to the north of the present study area, three pits excavated in 1999, close to a relict stream associated with Hayton Beck, did contain material suggestive of ritual deposition. This included Grooved ware of Woodlands style (Cleal and Macsween 1999, T. Manby pers. comm., Halkon *et al.* 2000), flint assemblages, animal bone and hazelnuts. Deposits of similar composition have been found in southern Sweden, which Larsson (this volume) suggests are indicative of votive deposition at the margins of wetland areas.

Within the Foulness valley and its hinterland, later Neolithic flint finds demonstrate a greater intensity of activity, especially at South Cliffe Common, Bursea and Hasholme (Halkon *et al.* forthcoming). Once again all these sites are located on the drier, sandy ridges at the wetland margins. At South Cliffe Common 19 scrapers were found, as well as a sickle fragment and a saddle quern, which are likely to be of Neolithic date and which may provide evidence for settlement of the valley margins (Healey forthcoming).

THE BRONZE AGE

Deposits cored at Sandholme demonstrate that by *c.* 2880–2667 cal BC (SRR-4743) the Foulness valley landscape was characterised by the withdrawal of estuarine conditions and the continuation of peat formation (Innes *et al.* forthcoming). Woodland coverage remained extensive, though there is inter-site variation, with thicker afforestation nearer the Humber (Buckland *et al.* 1990,143, Lillie and Gearey 1999, 93, Heath and Wagner forthcoming). Red deer bones from Hasholme suggest that hunting may have continued to be important within this environment. As well as pollen, a large quantity of timber survives, particularly in wetland deposits within the lower Foulness valley. Bog oaks, such as those observed in the excavation of drains and an irrigation pond at Hasholme Grange (Hillam 1987, Halkon *et al.* forthcoming), demonstrate

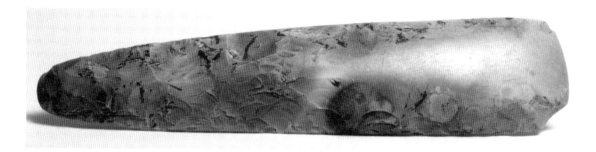

Fig. 10.5 Polished flint adze and Group XVIII axe from Hasholme Carr Farm.

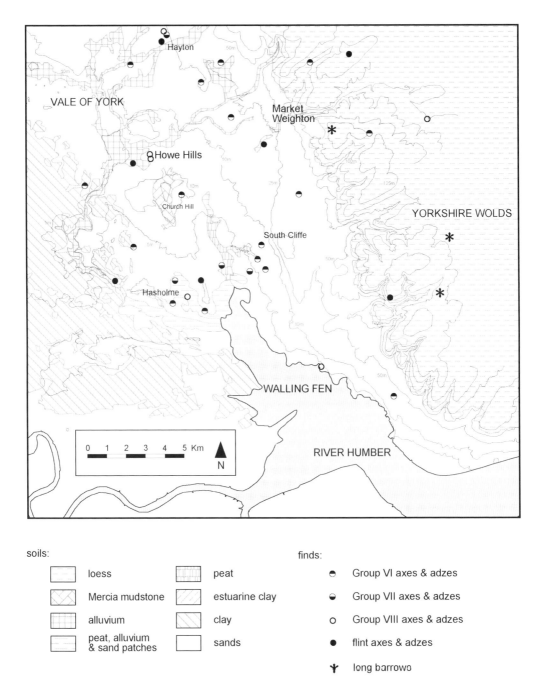

Fig. 10.6 Neolithic axes and adzes in the Foulness Valley.

the presence of heavy woodland. These woodlands would have provided a valuable raw material for the construction of the various wooden structures that have been recorded on the Humber foreshore; the most spectacular of all of these finds are the Bronze Age boats from North Ferriby (Wright 1990). The most recent radiocarbon dating from AMS re-calibration places three of the vessel remains within the range 2030–1780 cal BC at 95% to 1680 cal BC) (http://www.english-heritage.org.uk/news-events/news), though previous assays (Switsur and Wright 1989) have produced dates as late as 400 BC.

Also observed by Wright in the 1970s, recorded by ERAS members in 1984 (Crowther 1986) and further investigated by the Humber Wetlands Survey (Fletcher *et al.* 1999, 230) were a series of possible trackways on the Humber foreshore near Melton, one of which (Melton 35) was dated to *c.* 1440–1310 cal BC. Many of the foreshore timbers bore tool marks, which may provide a context for several notable hoards and single finds of bronze axes such as those from Hotham Carrs (Schmidt and Burgess 1981, Halkon *et al.* forthcoming). The distribution of perforated stone axe-hammers of the earlier Bronze Age may also relate to woodland management or exploitation, although their find spots close to streams within the Foulness valley means that the possibility of ritual deposition of both artefact types cannot be ignored (Halkon *et al.* forthcoming).

The Melton trackways themselves may have been associated with 'the exploitation of the saltmarsh environment for pasture' (Van de Noort *et al.* 1999, 276). Further evidence for the grazing of large ungulates within the first millennium BC has been identified through coleopteran evidence at Skelfrey Beck, Shiptonthorpe, one of the northern feeder streams of the Foulness valley (Wagner 1999, 168, Heath and Wagner forthcoming). Here the remains of insects such as *phylloperthus horticola* also indicated the presence of permanent grassland.

THE IRON AGE

The beginning of the Iron Age in the Foulness valley was marked by a significant period of environmental change, with a major marine transgression *c.* 815–405 cal BC (2 sigma) (Long *et al.* 1998, Innes *et al.* forthcoming). This event transformed a Bronze Age forested, freshwater environment into an estuarine inlet, and its tributary stream valleys into tidal creeks (Halkon 1987, 1990, Millett and McGrail 1987). The Foulness therefore became integrated more closely within the fluvial regime of the Humber estuary, which consequently had a considerable impact on the development of the wider region (Halkon and Millett 1999, 2000a). Woodland management remained crucial on the wetland margins, as large quantities of charcoal were needed to fuel an iron industry. The wetland edges also provided considerable deposits of bog iron ore. During field survey, 19 smelting sites of presumed Iron Age date were located, with almost all of them being situated on the sand ridges close to the river Foulness (Fig. 10.7). A slag heap at Moore's Farm, Welham Bridge was fully excavated and found to contain *c.* 5 tonnes of slag (Fig. 10.8), enough for the manufacture of *c.* 800 bars of trade iron. It was dated to 450–250 cal BC (2 sigma) (HAR-9234) and 600–380 cal BC (2 sigma) (HAR 9235) (Millett and Halkon 1986, Halkon 1997a, Halkon and Millett 1999), and is therefore contemporary with the East Yorkshire Arras culture, with its chariot burials and square barrows (Stead 1979, 1991). It is tempting (though to date unproven) to see a link between the iron industry and these burials, some of which contained iron swords, mirrors and chariot fittings; the Arras cemetery itself lies at the head of the dry valley of Sancton Dale, which may have formed a natural route-way between the Foulness wetlands and the Yorkshire Wolds.

The low-lying carr land adjacent to the River Foulness continued to provide summer grazing for cattle, and enclosures and droveways which show up as crop marks on the sand ridges close by may have been associated with stock control (Halkon and Millett 2000b). Cattle bones formed part of the contents of the 12 m long logboat from Hasholme (Fig. 10.9) (Millett and McGrail 1987), which was hewn from an 800 year

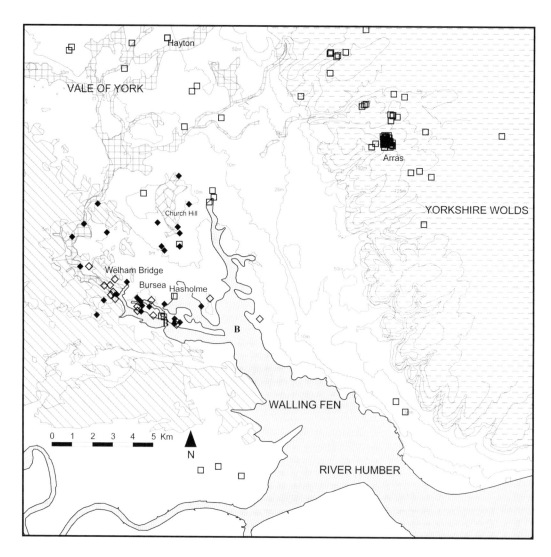

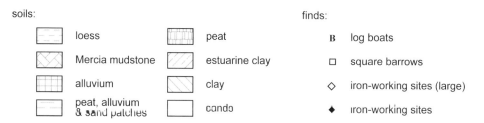

Fig. 10.7 Iron Age sites in the Foulness Valley.

Fig. 10.8 The Iron Age slag heap at Welham Bridge.

old oak cut down *c*. 322–277 BC and is therefore contemporary with the iron industry and the Arras culture burials. The parent log of the boat itself is of great significance, as it would thus indicate that closed canopy woodland characterised the region during the latter part of the Bronze Age at around 1100–1000 BC.

A further, now destroyed logboat is recorded from South Carr Farm (Halkon 1997b), a few kilometres to the east of Hasholme, which would appear from the description provided by the drainage contractors who found it, to have sunk in the same strata of estuarine clay as the Hasholme vessel. This boat, though longer and containing decorated internal ribs, was otherwise typologically similar to the Hasholme vessel. It also lies within a kilometre of an Iron Age settlement and smelting site at North Cave, excavated by Dent (1989). In September 2000, a local farmer reported to the present author that another boat had been found and destroyed by drainage activities near Faxfleet, an observation which has yet to be investigated. There is, therefore, as in the Bronze Age, ample evidence for waterborne transport within the Humber wetlands during the Iron Age.

In the later Iron Age, it is clear that waterborne trade continued, with evidence for cross-Humber contacts (Halkon and Millett 2000b). At Bursea House, Holme-on-Spalding Moor (Halkon and Millett 1999, 123) fine ware pottery with cordons, closely paralleled in Ceramic stages 10 and 11 at the Iron Age site of Dragonby to the south of the Humber (Elsdon 1989, 37, 1996 *cf*. fig. 19.38 no. 340, 473), was found. This discovery makes the Bursea House finds one of the most northerly find spots of this kind of pottery. A similar vessel was also excavated at Brantingham (Dent 1989), a site overlooking the eastern side of the Foulness tidal inlet, where Iron Age Corieltauvian coins were also discovered. A gold stater of inscribed VEP CORF type (Van Ardsell 1989, type 930) was also found at Holme-on-Spalding Moor, and similar finds of coins and imported pottery have also been made in the valley of the River Hull, to the east of the Yorkshire Wolds (Didsbury 1988, 26, Halkon and Millett 2000b, 90).

THE ROMAN PERIOD

Landscape factors that determined settlement and industry in the prehistoric period remained influential during the Roman period and beyond. The Walling Fen tidal inlet remained important, with the Roman fort and probable *Civitas* capital at Brough positioned at the interface of the inlet and the Humber estuary. The route of the main Roman road, which took the higher, drier ground close to the Wolds, was also determined by the inlet, and roadside settlements developed at Hayton and Shipton-thorpe, where the road crossed tributary streams of the Foulness system. These sites were to become the main markets for the pottery industry around HOSM, which largely replaced iron production. During investigation of the 8 × 8 km HOSM landscape block, 16 certain and 18 possible pottery kilns were found, along with around 70 other sites indicated by sherds of Roman pottery and other material (Halkon and Millett 1999). Again most activity was located on the sand ridges close to the River Foulness (Fig. 10.10).

Coppiced woodland continued to provide a source of fuel, as it had done for the iron industry before. Palmer (1999, 131–141) has shown, through examination of the charred botanical remains and the plant impressions on kiln fabric from Bursea House, that the wetlands provided other resources such as clay and sedges, with species associated with wet habitats, such as *Litorella uniflora*, having been iden-tified.

A further, 'post-processual' link between watercourses and industry

Fig. 10.9 The Hasholme Boat.

may be derived from the face pots and smith's tool pots discovered at various locations in the Foulness valley (Halkon 1992). At Shiptonthorpe, for example, sherds and a near complete vessel bearing the bearded face of the Vulcan/Celtic smith god have been found, one overlying a pond or waterhole at the junction of the main Roman road and a possible side road leading towards HOSM (Millett 1992). This feature, presumably after it fell out of use as a source of water, became a focus for infant and animal burial (Fig. 10.11). The fill of the waterhole also contained other elements which can be interpreted as being of ritual significance, such as shoes, further animal bones (including dog) and bundles of mistletoe. In these anaerobic conditions, a wax recessed type writing tablet was also preserved, which may have significant implications for assessing the level of Romanisation of these sites.

CONCLUSIONS

From this chronological survey it is possible to draw together strands of continuity and change over a long timespan in a limited area. Whilst woodland management was particularly important in the Iron Age and Roman periods, for fuelling furnace-based industries, a combination of artefact distribution and palaeoenvironmental examination shows that it was also important during earlier periods, and indeed, coppiced woodland must have been important in fuelling the later medieval Humber Ware pottery industry which focused on HOSM.

There does seem to have been special deposition of artefacts and animals close to the Foulness streams and within wells and ponds in both the prehistoric and Roman periods. If the face and smith tool pots are, as has been suggested, linked to Vulcan or his Celtic equivalent, what better deity could the people of this area of furnace-based industries have chosen to venerate?

In the Neolithic, Bronze Age, Iron Age and even into the Anglo-Saxon period, the dry valleys linking the lowlands and Wolds appear to have been marked by burial monuments. Good evidence for waterborne communication either on or beside watercourses can also be demonstrated from boat finds in the Bronze Age and Iron Age, and the riverine focus of site distribution during most periods.

The wet deposits along the whole length of the Foulness valley provide a vital resource for examining changes in climate and environment over a long timespan in a compact, well defined area, and there is still a great deal to be explored. Of particular importance is the evidence for marine transgression and regression and its effects on

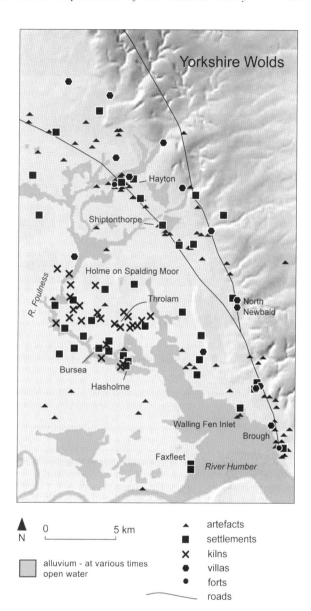

Fig. 10.10 Roman sites in the Foulness Valley.

human activity within the landscape. Anaerobic wet deposits have also led to the preservation of remarkable objects such as prehistoric boats and more mundane, but in many ways no less important, palaeo-environmental remains, through which a more holistic picture of human-landscape interaction is developing.

Finally, the Foulness valley project demonstrates that the wetlands cannot be studied in isolation. It is,

Fig. 10.11 The waterhole at Shiptonthorpe during excavation.

after all, on the drier ridges of the wetland and watercourse margins that cropmark and artefact distributions are concentrated, and these margins are presumably where many past populations lived and worked. Although the poorer preservation qualities of these acidic sandy deposits are apparent, the consideration of wetland landscapes alone within the Foulness valley would only provide a partial picture of past human activity within this dynamic landscape block.

REFERENCES

Bradley, R. 1990. *The Passage of Arms: an archaeological analysis of prehistoric hoards and votive deposits.* Cambridge: Cambridge University Press.

Bradley, R. & M. Edmonds 1993. *Interpreting the Axe Trade: production and exchange in Britain.* Cambridge: Cambridge University Press.

Buckland, P. C., C. J. Beal & S. V. E. Heal 1990. Recent work on the archaeological and palaeoenvironmental context of the Ferriby boats, in S. Ellis & D. Crowther (eds) *Humber perspectives: a region through the ages*, 131–46. Hull: Hull University Press.

Cleal R. & A. Macsween 1999. *Grooved Ware in Great Britain and Ireland.* Oxford: Oxbow Books.

Corder, P. 1930. *The Roman Pottery at Throlam, Holme-on-Spalding Moor, East Yorkshire.* Hull: Roman Malton and District Report 3.

Crowther, D. 1986. Some recent fieldwork by the Archaeology Department of Hull City Museums and Art Galleries, in V. Fairhurst (ed.) *Bringing the Past up to Date: some recent archaeological work in Yorkshire and Humberside*, 14–17. Hull: East Riding Archaeological Society.

Dent, J. S. 1989. Settlements at North Cave and Brantingham, in P. Halkon (ed.) *New Light on the Parisi: recent discoveries in Roman and Iron Age East Yorkshire*, 15–22. Hull: East Riding Archaeological Society and Department of Adult and Continuing Education, University of Hull.

Didsbury, P. 1988. Evidence for Romano-British settlement in Hull and the Lower Hull Valley, in J. Price, P. Wilson & S. Briggs (eds) *Recent Research in Roman Yorkshire*, 21–37. Oxford: British Archaeological Reports British Series 193.

Elsdon, S. M. 1989. *Later Prehistoric Pottery in England and Wales.* Aylesbury: Shire Archaeology.

Elsdon, S. M. 1996. *Iron Age Pottery in the East Midlands: a handbook.* Nottingham: Department of Archaeology, University of Nottingham.

Fletcher, W., H. Chapman, R. Head, H. Fenwick, R. Van de Noort & M. Lillie 1999. The archaeological survey of the Humber estuary, in R. Van de Noort & S. Ellis (eds) *Wetland Heritage of the Vale of York: An archaeological survey*, 204–41. Hull: Humber Wetlands Project, University of Hull.

Greenwell, W. 1877. *British Barrows: a record of the examination of sepulchral mounds in various parts of England.* Oxford: Clarendon Press.

Halkon, P. 1983. Investigations into the Romano-British landscape around Holme-on-Spalding Moor, East Yorkshire. *East Riding Archaeologist* 7, 15–24.

Halkon, P. 1987. *Aspects of the Romano-British Landscape around Holme-on-Spalding Moor, East Yorkshire.* Unpublished MA thesis, University of Durham.

Halkon, P. 1990. The archaeology of the Holme-on-Spalding Moor Landsdcape, in S. Ellis and D. R. Crowther (eds) *Humber perspectives a region through the ages*, 147–157. Hull: University Press.

Halkon, P. 1992. Romano-British face pots from Holme-on-Spalding Moor and Shiptonthorpe, East Yorkshire. *Britannia* 23, 222–8.

Halkon, P. 1997a. Fieldwork on early iron working sites in East Yorkshire. *Historical Metallurgy* 3, 12–16.

Halkon, P. 1997b. A log boat from South Carr Farm, Newport, East Yorkshire. *East Riding Archaeologist* 9, 7–9.

Halkon, P. 1999. The early landscape of the Foulness valley, East Yorkshire, in D.R. Bridgland, B.P. Horton & J.B. Innes (eds), *The Quaternary of North East England Field Guide*, 173–6. London: Quaternary Research Association.

Halkon P. 2003 Researching an Ancient Landscape: The Foulness Valley, East Yorkshire, in T. Manby, S. Moorhouse and P. Ottaway (eds) *The Archaeology of Yorkshire An Assessment at the Beginning of the 21st Century*, 261–74. Leeds: Yorkshire Archaeological Society.

Halkon, P. & M. Millett 1996. Fieldwork and excavation at Hayton East Yorkshire 1995. *Universities of Durham and Newcastle upon Tyne Archaeological Reports* 19, 62–8.

Halkon, P. & M. Millett 1999. *Rural Settlement and Industry: studies in the Iron Age and Roman archaeology of lowland East Yorkshire.* Leeds: Yorkshire Archaeological Report 4.

Halkon, P. & M. Millett 2000a. Foulness, valley of the first Iron Masters. *Current Archaeology* 169, 20–21.

Halkon, P. & M. Millett 2000b. The Foulness valley: investigation of an Iron Age landscape in lowland East Yorkshire, in J. Harding & R. Johnston (eds) *Northern Pasts: interpretations of the later prehistory of Northern England and Southern Scotland*, 81–93. Oxford: British Archaeological Reports British Series 302.

Halkon, P., E. Healey, J. Innes & T. Manby (eds) forthcoming. Change and continuity within the prehistoric landscape of the Foulness valley, East Yorkshire. *East Riding Archaeologist.*

Halkon, P., M. Millett & J. Taylor 1997. Fieldwork and excavation at Hayton, East Yorkshire, 1996. *Universities of Durham and Newcastle upon Tyne Archaeological Reports* 20, 39–41.

Halkon, P., M. Millett & J. Taylor 1998. Fieldwork and excavation at Hayton East Yorkshire, 1997. *Universities of Durham and Newcastle upon Tyne Archaeological Reports* 21, 71–73.

Halkon, P., M. Millett, E. Easthaugh, J. Taylor & P. Freeman 2000. *The Landscape Archaeology of Hayton.* Hull: University of Hull.

Harcourt, W. V. 1829. On the discovery of fossil bones in a marl pit near North Cliffe. *Philosophical Magazine* 6, 225–32.

Haughey, F. 2000. From prediction to prospection: finding the prehistory of London's floodplain. Abstracts of conference papers, The alluvial archaeology of North-west Europe and the Mediterranean, University of Leeds, School of Geography.

Healey, E. forthcoming. Other worked flint artefacts from the survey area (excluding axes), in P. Halkon, E. Healey,

J. Innes & T. Manby (eds) Change and continuity within the prehistoric landscape of the Foulness valley, East Yorkshire. *East Riding Archaeologist.*

Heath, A. & P. Wagner forthcoming. Coleopteran evidence from the Foulness valley, in P. Halkon, E. Healey, J. Innes & T. Manby (eds) Change and continuity within the prehistoric landscape of the Foulness valley, East Yorkshire. *East Riding Archaeologist.*

Hicks, J. D & J. A. Wilson 1975. Romano-British kilns at Hasholme. *East Riding Archaeologist* 2, 49–70.

Hillam, J. 1987. The tree-ring dating, in M. Millett & S. McGrail. The archaeology of the Hasholme logboat. *Archaeological Journal* 144: 79–84.

Innes J., A. Long & I. Shennan forthcoming. Stratigraphic and pollen analyses in the lower Foulness Valley, in P. Halkon, E. Healey, J. Innes & T. Manby (eds) Change and continuity within the prehistoric landscape of the Foulness valley, East Yorkshire. *East Riding Archaeologist.*

Jordan, D. 1987. The investigation of the minerogenic deposit, in M. Millett & S. McGrail, The archaeology of the Hasholme Logboat. *Archaeological Journal* 144, 90–7.

Kinnes, I. A. 1992. *Non-megalithic long barrows and allied structures in the British Neolithic.* London: British Museum Occasional Paper.

Lamplugh, G. W., J. W. Stather, T. Anderson, J. W. Carr, W. Lower Carter, A. R. Dwerryhouse, F. W. Harmer, J. H. Howarth, W. Johnson, P. F. Kendall, G. W. B. Macturk, E. T. Newton, H. .M. Platnauer, C. Reid & T. Sheppard 1910. Investigation of the fossiliferous Drift deposits at Kirmington, Lincolnshire, and at various locations in the East Riding of Yorkshire. *Report of the British Association, Winnipeg for 1909*, 177–80.

Lillie, M. & B. Gearey 1999. The palaeoenvironmental survey of the Humber estuary, incorporating an investigation of the nature of warp deposition in the southern part of the Vale of York, in R. Van de Noort & S. Ellis (eds) *Wetland Heritage of the Vale of York: an archaeological survey*, 79–108. Hull: Humber Wetlands Project, University of Hull.

Long, A. J., J. B. Innes, J. R. Kirby, J. M. Lloyd, M. M. Rutherford, I. Shennan & M. J. Tooley 1998. Holocene sea-level change and coastal evolution in the Humber estuary, eastern England: an assessment of rapid coastal change. *The Holocene* 8, 229–47.

Manby, T. G. 1988. The Neolithic period in Eastern Yorkshire, in T. G. Manby (ed.) *Archaeology in Eastern Yorkshire: essays in honour of T.C.M. Brewster*, 35–88. Sheffield: Department of Archaeology and Prehistory, University of Sheffield.

Manby, T. G. 1999. Market Weighton Wold long barrow. Unpublished note.

Manby, T. G. forthcoming. Flint and stone axes from the survey area, in P. Halkon, E. Healey, J. Innes & T. Manby (eds) Change and continuity within the prehistoric landscape of the Foulness valley, East Yorkshire. *East Riding Archaeologist.*

Millett, M. 1992. Excavations and Survey at Shiptonthorpe, East Yorkshire 1991. *Universities of Durham and Newcastle Archaeological Reports* 15, 29–33.

Millett, M. & P. Halkon 1986. Excavations at Shiptonthorpe, Welhambridge Farm and East Bursea Farm, East Yorkshire 1985. *Universities of Durham and Newcastle-upon-Tyne Archaeological Reports* 10, 40–3.

Millett, M. & S. McGrail 1987. The archaeology of the Hasholme logboat. *Archaeological Journal* 144, 69–155.

Palmer, C. 1999. The charred botanical remains, in P. Halkon & M. Millett (eds) *Rural Settlement and Industry: studies in the Iron Age and Roman archaeology of lowland East Yorkshire*, 131–140 Leeds: Yorkshire Archaeological Report 4.

Roberts, M. & S. Parfitt 1998. *Boxgrove: a Middle Pleistocene hominid site at Eartham Quarry, Boxgrove, West Sussex.* London: English Heritage.

Roe, D. forthcoming. A Lower Palaeolithic handaxe from Hotham, East Yorkshire in P. Halkon, E. Healey, J. Innes & T. Manby (eds) Change and continuity within the prehistoric landscape of the Foulness valley, East Yorkshire. *East Riding Archaeologist.*

Schmidt, P. K. & C. B. Burgess 1981. *The axes of Scotland and Northern England.* Munich: Prahistoriche Bronzefunde 9.

Schreve, D. 1999. Bielsbeck Farm, East Yorkshire, in D. R. Bridgland, B. P. Horton & J. B. Innes (eds) *The Quaternary of North East England Field Guide*, 176–9. London: Quaternary Research Association.

Stead, I. M. 1979. *The Arras Culture*. York: Yorkshire Philosophical Society.

Stead I. M. 1991. *The Iron Age Cemeteries from East Yorkshire*. London: English Heritage.

Switsur, V. R. & E. V. Wright 1989. Radiocarbon ages and calibrated dates for the boats from North Ferriby, Humberside – a reappraisal. *Archaeological Journal* 146, 58–67.

Taylor, J. 1995. Surveying small towns: the Romano-British roadside settlement at Shiptonthorpe, East Yorkshire, in A. E. Brown (ed.) *Roman Small Towns in Eastern England and Beyond*, 39–52. Oxford: Oxbow books.

Turner, J. 1987. The pollen analysis, in M. Millett & S. McGrail, The archaeology of the Hasholme Logboat. *Archaeological Journal* 144, 85–8.

Van Arsdell. R. D. 1989. *Celtic coinage of Britain*. London: Spink.

Van de Noort, R., M. Lillie, B. Gearey, H. Fenwick, H. Chapman, W. Fletcher & R. Head 1999. Conclusions, in R. Van de Noort & S. Ellis (eds) *Wetland Heritage of the Vale of York: an archaeological survey*, 269–79. Hull: Humber Wetlands Project, University of Hull.

Wagner, P. 1999. The Skelfrey Beck section, in P. Halkon & M. Millett (eds) *Rural Settlement and Industry: studies in the Iron Age and Roman archaeology of lowland East Yorkshire*, 170. Leeds: Yorkshire Archaeological Report 4.

Whittle, A. 1999. The Neolithic period, *c.* 4000–2500/2200 BC: changing the world, in J. Hunter & I. Ralston (eds) *The Archaeology of Britain*, 58–76. London: Routledge.

Wright, E.V. 1990. *The Ferriby Boats: seacraft of the Bronze Age*. London: Routledge.

Chapter 11

Medieval Coastal Landscape Evolution – the Example of the Lincolnshire Marsh, England, UK

Helen Fenwick

INTRODUCTION

The coastal margins of Britain are an intrinsic part of the character of this island country, but the study of its coastal marshes is a relatively young discipline within wetland archaeology. These are areas which have seen major changes over the centuries, and millennia, with sea-level fluctuations often making them an unpredictable environment in which to settle. Despite this, these areas have seen varying degrees of settlement through time, and their study can provide valuable insights into human-environment interactions in terms of the adaptation of local communities to changing factors such as economic, social and environmental conditions (Rippon 2000, 10).

The Lincolnshire coastal zone is one such area that has witnessed extensive changes over the last 10,000 years, both physically and culturally, and these changes are reflected in the way in which the landscape of the area was exploited and settled. One important resource of the area that has influenced the nature and extent of settlement has been salt. The earliest evidence for salt-production in Lincolnshire has been dated to the Late Bronze Age (Palmer Brown 1993), with extensive evidence from the Iron Age and Roman periods, and an even larger body of evidence for the Medieval salt industry.

This paper summarises preliminary results on the development of the landscape of the Lincolnshire Marsh, in an area from the mouth of the Humber estuary near Cleethorpes, southwards to Skegness (Fig. 11.1). This is an extension of initial research which was undertaken during the English Heritage funded Humber Wetlands Project (Ellis *et al.* 2001, Fenwick 2001).

MARSHLANDS AS MARGINAL ZONES

Coastal zones, prone to flooding, can be classed as marginal areas in which individuals need to persevere to survive in the ever-changing environmental conditions, but in some cases the rewards outweigh the initial hardships. The occupation of marginal zones can relate to a range of push and pull factors. Those that push the population into such a zone include increased population pressure, land shortage and social pressures, while those that exert a pull on the population include improved physical conditions and the abundance of natural resources such as food, fuel, building materials and salt. In settling coastal areas, the

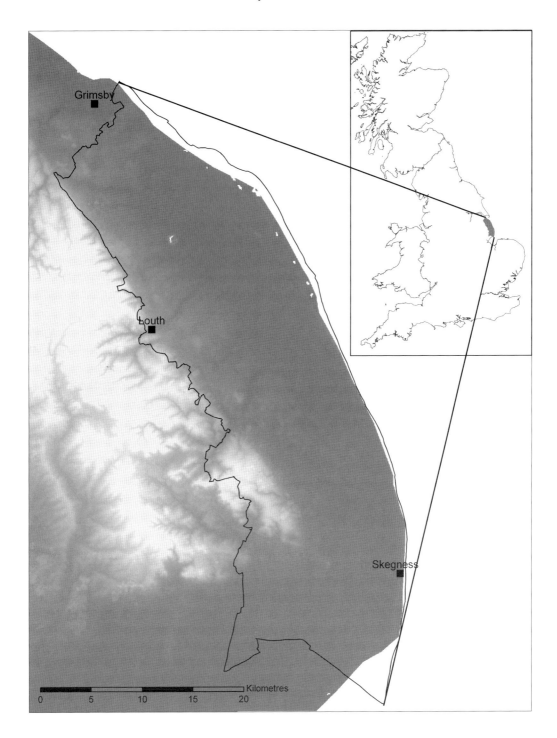

Fig. 11.1 The Lincolnshire Marsh.

population invariably changes and develops these regions so that they become less inhospitable. In some cases this can be controlled change, while in others it is a by-product of subsistence or economic activity in the region.

Many of the ideas of marginality that have been developed are over-simplistic, and past communities are often labelled as marginal due to the position of their habitat in the modern world rather than in the past (Coles and Mills 1998). As Rippon (2001, 153) notes: 'coastal wetlands are 'marginal' only if human communities perceived them to be so'. When classing the coastal margins of Lincolnshire as a marginal zone, this is an important point to note, and only by studying the present and past landscape can we hope to consider whether the past communities perceived it as such.

MEDIEVAL EXPLOITATION OF MARSHLANDS

The main research that has been carried out on the use of marshland areas in the Roman and post-Roman periods is that of Rippon (2000), who has suggested three broad ways in which coastal marshes have been used by communities in the past – exploitation, modification and transformation – which show different intensities and reflect different strategies of landscape change. Exploitation is simply the use of the natural resources of the region such as willow, reeds, foodstuffs and peat. There is also the possible extraction of salt and the use of saltmarshes for seasonal grazing. Modification of the landscape is designed to increase its natural productivity. This is most often seen through the control of water by the digging of ditches and the construction of banks, although the areas may still be threatened by flooding. The third strategy sees the eventual transformation of an area by the construction of permanent banks and defences, thus changing the nature of the landscape from one of seasonal flooding to one that is no longer inundated.

Factors such as the type of natural resources available and the socio-economic position of the coastal hinterland result in different strategies being adopted in different coastal zones (Rippon 2001). In certain areas, such as the Severn Estuary, it has been suggested that a sea-bank was first built which was followed by settlement, whereas areas on the Norfolk coast have shown a more gradual intensification of activity, with settlement appearing before the construction of sea walls (Rippon 2001).

THE LINCOLNSHIRE MARSH

The Lincolnshire Marsh is a strip of coast, bounded to the west by the Lincolnshire Wolds, and consisting of two zones – Middle Marsh and Outmarsh. Middle Marsh is an undulating landscape extending from the edge of the Wolds to the Outmarsh, whereas Outmarsh is a flat, low-lying area bordering the coast and constituting alluvial deposits. Fluctuating sea-levels have resulted in an ever-changing coastline. During the Roman period the coastline was much further inland than at present, and during prehistory it was further to the east. This very mobile interface of land and sea has been the cause of changing settlement patterns, but has also hidden the majority of the pre-Medieval remains under a blanket of alluvium. Only modern development, agricultural activities, drainage works and the natural erosive powers of the sea allow insights into these earlier periods. This chapter will examine the development of the Lincolnshire Marsh and the ultimate reclamation of these areas along the east coast of England, which occurred as a result of the exploitation of the region's natural resources.

Early investigations into Medieval settlement along the Lincolnshire coast suggested that permanent settlement was only possible in this wetland environment with the creation of the sea-bank defences (Owen 1975), but further investigations indicated that the crucial factor in the development of the area was in fact the salt industry (Owen 1984). It was shown that this industry was not only the cause of settlement in the area, but that it actually made further settlement possible with the establishment of spoil-heaps which, due to their elevated situation, facilitated the building of subsequent settlements away from areas of potential flooding (Owen 1984).

Owen's work on the development of these settlements has emphasised certain key features that assist in the understanding of both the salt industry and settlement evolution. In essence the salt workers had to work close to the high-water mark of spring tides, and access was needed to the salt water or mud at all times. The process of salt production resulted in large spoil-heaps known as saltern mounds, and there could be no sea-bank on the seaward side of the saltern while it was in use. The elevated topography of the salterns facilitated expansion of settlement further into what had previously been an intertidal zone. This feature of settlement development was also highlighted in the work of Morris (1989), who linked the development of the churches and hence the settlements in the area, to the salt industry. The churches in the marsh area are often sufficiently large to suggest a degree of wealth in the local community, similar to the 'wool churches' of East Anglia. It was only after such studies that a more integrated approach to landscape evolution was developed, with The Royal Commission on Historical Monuments of England linking the salt-working features to other landscape elements and considering the development of the landscape as a whole, including aspects such as coastal change and the sea-banks and storm beaches of the area (Grady 1998).

MEDIEVAL SALT PRODUCTION IN THE LINCOLNSHIRE MARSH

'The round groundes at the Easte end of Marshchappell are called mavres and are firste framed by layinge together of great quantities of moulde for the making of Salte. When the mavres grow greate the Salt makers remove more easte and come nearer to the Sea and then the former mavres become in some fewe years good pasture groundes.' This account comes from a map of the parishes of Fulstow and Marshchapel dated to 1595, drawn by W. Haiwarde (Walshaw 1935). The text and the detail on the map highlight the way in which much of this stretch of coast was reclaimed from the saltmarsh using the waste products of the salt industry. The area to the east of the main road (which may occupy the first sea-bank in the area) is shown with many of the saltern mounds still under construction (Fig. 11.2). Salt-cotes occur on the most recently abandoned mounds, and it is very easy to see how the land had been reclaimed in the past with the surfaces of the mounds being exploited for arable farming.

The Medieval method of salt production has been described elsewhere (Bridbury 1955, Grady 1998) but a summary is provided here to aid in the understanding of the development of the saltern mounds. Salt can be produced from natural brine springs, as are found in the west of the country at places such as Nantwich and Droitwich, but elsewhere it is produced from seawater. In some areas, water could be channelled into large lagoon type structures and the sun used for evaporation. These sites, known as sunworks, have been postulated at Sutton-on-Sea in Lincolnshire, where the great storm of 1953 removed sand from the beach, revealing rectangular structures, although these may have been clay extraction pits (Grady 1998). However, as this form of salt production was heavily reliant on the sun, it was not widely used in this country to any great success. The salt produced in the region currently under study was undertaken by the method of sand-washing. This involved the filtering of salt from salt-laden sand and boiling of the resultant liquid. Three salt-working sites have been excavated in south Lincolnshire, at Bicker Haven (Healey 1999), Wainfleet St Mary (McAvoy 1994) and Wrangle Toft (Bannister 1983), and these show different elements of the sand-washing processes.

The method of salt production inferred from these sites and from documentary sources is as follows: sand and silt, laden with salt crystals, was collected from the coast after spring tides and taken to an area, often associated with a small building called a salt-cote, where a process of filtration was carried out to separate the salt from the sand. The result was a salty water and waste material comprising sand and silt which was then disposed of close to the processing area, resulting in the mounds seen in the landscape today. The salty

Fig. 11.2 Extract from Haiwarde's map of 1595.

water then went through various boiling processes to produce the salt. A major resource required for this method of production was fuel, although the method did require less fuel than that of the open-pan processes seen in earlier periods. It is assumed that peat was the main source of fuel; although the local supplies were limited, many of the monastic houses that held rights to salt production on the coast also held turbury (peat-cutting) rights elsewhere.

MEDIEVAL LANDSCAPE DEVELOPMENT OF THE LINCOLNSHIRE MARSH

The initial results of the research into landscape development of the region have postulated three separate developmental zones along the Lincolnshire coast (Fig. 11.3). The area has been considered in terms of the nature of the historic environment, the positions of settlements, their relation to the geology of the area, the location of drainage features such as moated sites, and the associated field systems.

The first area consists of two near-parallel sets of settlements running roughly north-south. The first line of settlement is situated on the edge of the Outmarsh region, on the 10 m contour, and corresponds to the edge of the glacial deposits, where they meet the alluvial area of the Marsh. The settlements are recorded in the Domesday Book, and some have earlier origins, with Saxon evidence recovered from Tetney and North Thoresby. Many of these settlements have moated sites, often located next to tongues of alluvium stretching into the area of glacial deposits. In contrast, the settlements in the second line are not mentioned in the Domesday Book and would appear to have developed in a later period. They also lack features such as moated sites. The majority of these settlements have developed away from those further inland as 'daughter villages' of the earlier settlements.

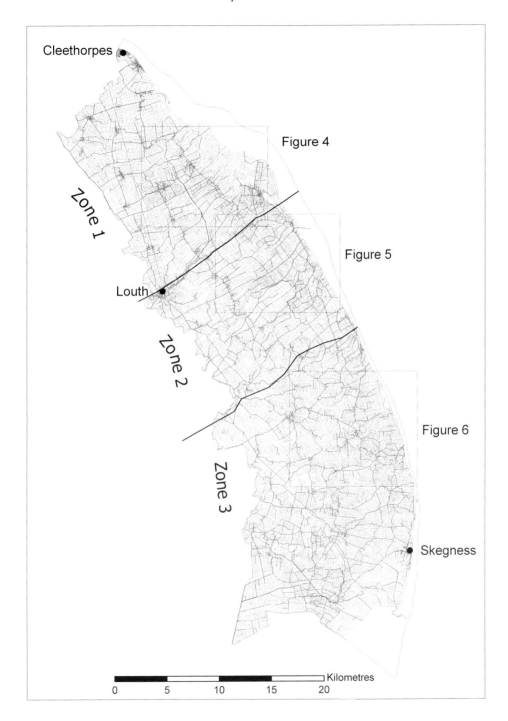

Fig. 11.3 The three landscape blocks.

In the past this area has received the greatest amount of study and has the best evidence of the salt industry, with aerial photographic data and the British Geological Survey having plotted many such sites across the area. The evidence from excavations to the west of Marshchapel indicates that Domesday period salt-making activity in the region was located away from the possible Medieval sea-bank, which it is suggested runs along the line of the main Grimsby road (Fenwick *et al.* 2001). It is clear that settlements gradually moved eastwards with the abandonment of salterns, and that the construction of sea-banks occurred only after this abandonment. The resultant landscape thus contains an 'image' of the former salt workings (Fig. 11.4).

Preliminary investigation of this landscape block indicates initial exploitation for its natural resources, in particular salt, with the debris from this exploitation increasing the amount of land raised above the area of flood influence.

The second landscape area stretches from Saltfleet to Mablethorpe. Although this exhibits a similar development to the area to the north (considered above), the evidence of landscape change takes a different form. The settlements in many cases are again daughter settlements of those on the drier areas, but the waterways which cross the area have played an important additional role in its development. The Great and Long Eau rivers traverse the region in a west-east direction, and their importance in the Medieval period is highlighted by the presence of motte and bailey castles where both of these waterways enter the area of glacial till and become more constricted *e.g.* at Toothill, Withern and Castle Carlton.

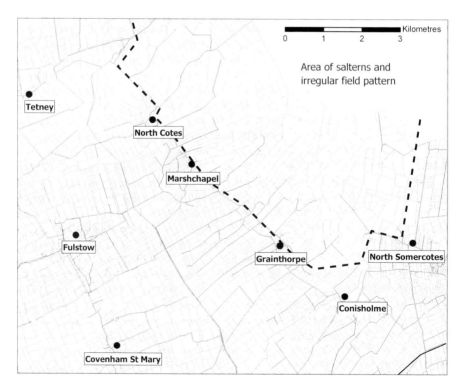

Fig. 11.4 Landscape features of the first landscape block.

The evidence from aerial photographs for salterns is limited, but the churches at Skidbrooke, Saltfleetby St Peters and Saltfleetby All Saints have been postulated as being situated on salterns. This area does not have the characteristic 'moonscape' of the area further north around Marshchapel, where the mounds are still very visible in the flat alluvial landscape. A more striking feature of this landscape is its regimented field systems and road network; the first region has a very 'organic' field system, with the watercourses and marshy areas surrounding the saltern mounds producing an irregular field pattern (Fig. 11.5). Although the majority of this more controlled landscape in this second area will date to the later Medieval period, it would appear that it was initiated earlier; many of the settlements include moated sites which have been shown to be an integral element of the colonisation process of other wetland areas in the Humber region (Fenwick 1997, 1999).

The most common landscape evidence from aerial photographs in this area is for Medieval field systems and the associated ridge and furrow. In some locations this can be quite substantial and can possibly be viewed as being more akin to the 'ridge and vurrow' seen in other regions such as the Gwent Levels, and which was used to help drain land to be used for cultivation (Rippon 1996). If this is the case in this part of the Lincolnshire Marsh, it may suggest that the intensity of salt production here was not as great as in the northern area, or that later agricultural and drainage activity has obscured the evidence.

This initial investigation would suggest that this landscape block was the result of landscape modification, or even transformation, with drainage and regular field systems being established, but further investigation is required to clarify this issue.

The final landscape block lies between Mablethorpe and Skegness. This area has not been studied in any

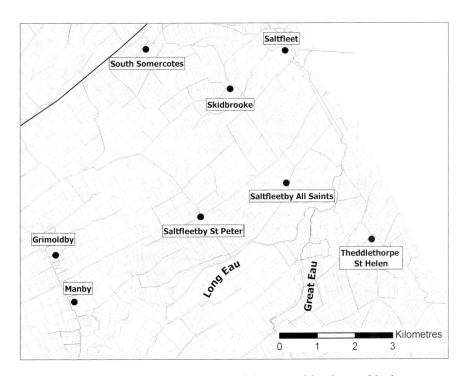

Fig. 11.5 Landscape features of the second landscape block.

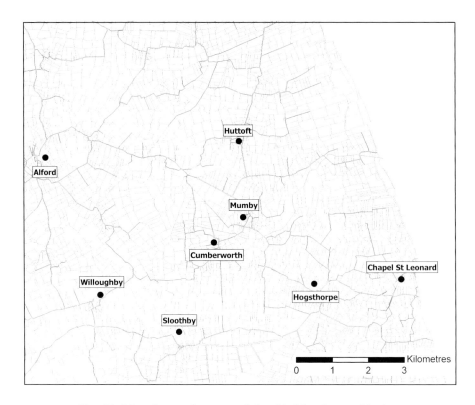

Fig. 11.6 Landscape features of the third landscape block.

great detail in relation to the Medieval salt industry, but it does include the area of Ingoldmells and Hogsthorpe where evidence of intensive Iron Age and Roman salt production has been recovered. The natural geology of this area differs from that of the other areas, being a mixture of outcrops of glacial till and gravels surrounded by alluvium. The areas of till and gravels provide a more stable environment for settlement than other areas of the Lincolnshire Marsh, and many of the settlements in the region are located on them (Fig. 11.6). Some of the settlements have early origins dating to the Saxon period. The proximity of the till outcrops to the coastline has meant that the area of tidal influence is smaller than in the regions to the north. This area was protected at some time in the past by a large sea-bank, a feature that remains to be dated.

This final area of the Lincolnshire Marsh, in essence, is not a coastal marshland. The geology of the region and the area which actually came under the influence of coastal processes is different from that of the regions to the north. Areas of high, drier ground occur naturally within the wetland zone. These outcrops of gravels and sands would have provided areas from which the surrounding alluvium could have been exploited, without the need for habitation sites located within the coastal zone.

CONCLUSIONS

Coastal zones are excellent landscapes in which to study the different mechanisms that have influenced the way in which humans have interacted with their environment. Initial study in the region of the Lincolnshire coast has highlighted the importance played by the natural resources of the marshlands. Three landscape blocks have been postulated, with the northernmost area exhibiting exploitation, and as a by-product of the industrial activity, modification of the landscape. The middle landscape block also shows landscape modification, if not wholesale transformation, with landscape reclamation highlighted in the field system and drainage patterns. The final region is different from the former two zones, with the natural landscape enabling exploitation from dryland areas, and with transformation of the coastal zone occurring in the form of a major sea-bank, which as yet, remains undated.

Finally, can the coastal zone of the Lincolnshire Marsh be described as a 'marginal' zone, or does the evidence point towards a 'struggle' against the elements in order to settle in the area? The available evidence would suggest that the region was far from being considered a marginal zone and was seen as an area with a profitable resource to be exploited. The abundant evidence for salt production from the Late Bronze Age onwards signifies the importance of this resource in the coastal zone, and the exploitation of salt not only enabled permanent settlement in the region, but would also appear to have been a causal element of this settlement, especially in the first two zones considered here.

REFERENCES

Bannister, R. T. 1983. Wrangle Toft. *Lincolnshire History and Archaeology* 18, 104–5.

Bridbury, A. R. 1955. *England and Salt Trade in the Later Middle Ages*. Oxford: Clarendon Press.

Coles, G. & C. M. Mills 1998. Clinging on for grim life: an introduction to marginality as an archaeological issue, in C. M. Mills & G. Coles (eds) *Life on the Edge: human settlement and marginality*, vii–xii. Oxford: Oxbow Monograph 100.

Ellis, S, H. Fenwick, M. Lillie & R. Van de Noort (eds) 2001. *Wetland Heritage of the Lincolnshire Marsh: an archaeological survey*. Hull: Humber Wetlands Project, University of Hull.

Fenwick, H. 1997. The wetland potential of Medieval moated sites in the Humberhead Levels, in R. Van de Noort & S. Ellis (eds) *Wetland Heritage of the Humberhead Levels: an archaeological survey*, 429–38. Hull: Humber Wetlands Project, University of Hull.

Fenwick, H. 1999. Medieval moated sites in the Vale of York: distribution, modelling and wetland potential, in R. Van de Noort & S. Ellis (eds) *Wetland Heritage of the Vale of York: an archaeological survey*, 255–68. Hull: Humber Wetlands Project, University of Hull.

Fenwick, H. 2001. Medieval salt-production and landscape development in the Lincolnshire Marsh, in S. Ellis, H. Fenwick, M. Lillie & R. Van de Noort (eds) *Wetland Heritage of the Lincolnshire Marsh: an archaeological survey*, 231–41. Hull: Humber Wetlands Project, University of Hull.

Fenwick, H., H. Chapman, W. Fletcher, G. Thomas & M. Lillie 2001. The archaeological survey of the Lincolnshire Marsh, in S. Ellis, H. Fenwick, M. Lillie & R. Van de Noort (eds) *Wetland Heritage of the Lincolnshire Marsh: an archaeological survey*, 99–202. Hull: Humber Wetlands Project, University of Hull.

Grady, D. M. 1998. Medieval and Post-Medieval salt extraction in north-east Lincolnshire, in R.H. Bewley (ed.) *Lincolnshire's Archaeology from the Air*, 81–95. Lincoln: Occasional Papers in Lincolnshire History and Archaeology 11.

Healey, H. 1999. A Medieval salt-making site at Bicker Haven, in A. Bell, D. Gurney & H. Healey (eds) *Lincolnshire Salterns: excavations at Helpringham, Holbeach St Johns and Bicker Haven*, 82–101. Sleaford: East Anglian Archaeology 89.

McAvoy, F. 1994. Marine salt extraction: the excavation of salterns at Wainfleet St Mary, Lincolnshire. *Medieval Archaeology* 38, 134–43.

Morris, R. 1989. *Churches in the Landscape*. London: Dent and Sons.

Owen, A. E. B. 1975. Hafdic: a Lindsey name and its implications. *Journal of the English Place-Name Society* 7, 45–56.

Owen, A. E. B. 1984. Salt, sea banks and Medieval settlement on the Lindsey coast, in N. Field & A. White (eds) *A Prospect of Lincolnshire*, 46–9. Lincoln: Field & White.

Palmer Brown, C. 1993. Bronze Age salt production in Tetney. *Current Archaeology* 136: 143–45.

Rippon, S. 1996. *Gwent Levels: the evolution of a wetland landscape*. York: Council for British Archaeology Research Report 105.

Rippon, S. 2000. *Transformation of Coastal Wetlands*. London: British Academy.

Rippon, S. 2001. Reclamation and regional economies of Medieval marshland in Britain, in B. Raftery & J. Hickey (eds) *Recent Developments in Wetland Research*, 139–58. Dublin: Seandalaoiocht Monograph 2 & WARP Occasional Paper 14.

Walshaw, G. R. 1935. An ancient Lincolnshire map. *Lincolnshire Magazine* 2, 196–206.

Chapter 12

The Palace Centre of Sago City: Utti Batue Site, Luwu, Sulawesi, Indonesia

F. David Bulbeck, Doreen Bowdery, John Field and Bagyo Prasetyo

INTRODUCTION

Archaeological remains rarely preserve well in humid equatorial regions, apart from those in a few favoured locations such as anaerobic swamps. One of the best known of these sites is Kuala Selinsing, in Peninsular Malaysia, an important offshore trading and industrial centre occupied from *c*. 2000–1000 yr BP (Nik Hassan Shuhaimi 1991). In this paper we briefly describe a second site of comparable significance – Utti Batue, which evidently served as the palace centre of Luwu, a prominent Bugis kingdom, during its pre-Islamic heyday in the fifteenth and sixteenth centuries. The site has major implications for historians of Sulawesi in view of Luwu's widely held reputation as the oldest and most prestigious Bugis kingdom (*e.g.* Pelras 1996). It also has great archaeological significance as it is, to our knowledge, the best preserved Southeast Asian palace centre which dates to the 'early modern period'. Following its abandonment *c*. AD 1600 and the shift of Luwu's capital to Palopo between 1610 and 1630 (Bulbeck and Caldwell 2000, 70), Utti Batue became sealed beneath approximately 1 m of sediment, and its material remains were virtually free of degradation from subsequent habitation or agricultural disturbance.

Utti Batue is located at 2° 47' 50" S, 120° 24' 00" E in Kecamatan Malangke Barat, Kabupaten Luwu Utara, in the province of South Sulawesi (Fig. 12.1). The site lies at 1–2 m above sea-level, in an environment of oppressively high temperatures and humidity, and an average annual rainfall of 2500–3000 mm. It was first recorded in 1997 during a preliminary survey by the 'Origin of Complex Society in South Sulawesi' (OXIS) project (Bulbeck and Prasetyo 1998), and in 1999 a borehole survey estimated its extent as 4 ha. During this survey, test pits were excavated to the north of the site, adjacent to the drainage ditch which had previously exposed its existence; this paper describes the artefacts and ecofacts recovered through excavation.

One of the hypotheses formulated by the OXIS project is that the dominant plant food of the Utti Batue inhabitants was sago (Bulbeck and Caldwell 2000, 14–5); the coastal swamps surrounding the site are one of the main sources of sago in South Sulawesi, exporting over 1000 tons of wet sago in 1984 (Osazawa 1986). It is also estimated that the population stood at about 10,000 inhabitants in the fifteenth century, and 10–15,000 in the sixteenth century, on the basis of looted pre-Islamic cemeteries in the Malangke site complex, and the size of the two main settlements of Utti Batue and Pattimang Tua (Fig. 12.2) (Bulbeck and

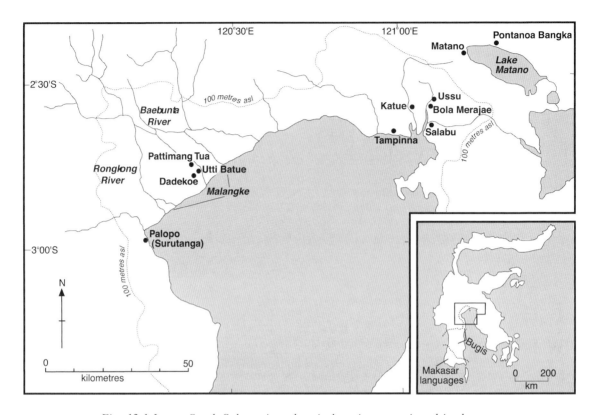

Fig. 12.1 Luwu, South Sulawesi, and main locations mentioned in the text.

Caldwell 2000, 73–6). These urban population levels, in conjunction with a monumental structure with brick foundations at Tampung Jawa, a centralised economy and the well-established literary tradition in pre-Islamic Luwu, would satisfy all four of Fagan's (1995, 343) criteria for a 'civilisation'. Utti Batue would accordingly rank with the famous fifteenth century entrepôt of Melaka, in Peninsular Malaysia, whose inhabitants also established the capital of a civilisation in an environment too brackish for wet-rice cultivation, and instead turned to sago for their staple diet (Wheatley 1961, 311–2).

SURVEY AND EXCAVATION

Immediately north of Utti Batue lies the site referred to as the Dato Sulaiman Islamic Cemetery. It features the tombs of the first sultan of South Sulawesi, Matinroe ri Ware Sultan Muhammad Wali Muzahir, and his teacher Dato Sulaiman who converted his royal pupil to Islam in 1605. This major Islamic site, noted by Caldwell (1993), was subsequently shown to abut a rich, looted, late pre-Islamic cemetery (Bulbeck 1996–7). The tombs face a large fishpond, excavated by the local inhabitants in 1997 and containing a lock-controlled drainage ditch in which buried posts of massive dimensions were unearthed. Pottery and other debris were also reported.

The OXIS team was able to survey the bottom of the ditch, and in addition to observing an abundance of earthenware pottery, an iron knife with a curved blade and other iron fragments were collected. A water-

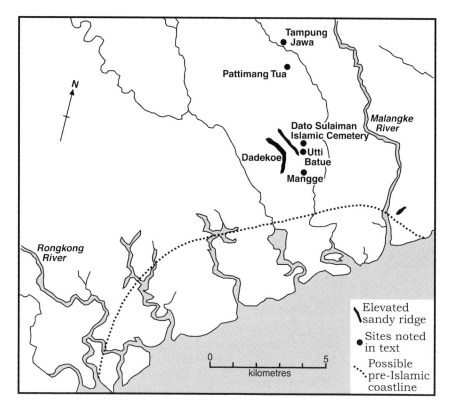

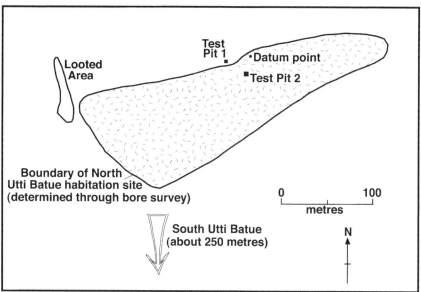

Fig. 12.2 Malangke (above) and Utti Batue (below).

buffalo bone, a pottery fragment with dammar gum adhering to it,[1] and imported ceramic shards dating to the fifteenth and sixteenth centuries were also retrieved (Bulbeck and Prasetyo 1998, 1999, 25). The site's proximity to the tomb of Muhammad Wali Muzahir strongly suggested that it represented Luwu's last pre-Islamic palace, and the range of organic and barely corroded iron remains indicated excellent preservation conditions, brought about by the deposition of flood sediments since the site's abandonment.

To determine the extent of the main habitation area at Utti Batue 35 boreholes were sunk along transects running through the mandarin orchard at the site, and the surrounding *rawa* forest. The boreholes were drilled to a depth of 1.5–2.0 m, or until evidence of habitation, in the form of pot shards or food remains, was encountered (between 0.6 and 1.9 m depth). A boundary drawn around the productive borehole locations produced a shoehorn-shaped site, called North Utti Batue, covering approximately 3 ha (Fig. 12.2). A second area a short distance to the south, reportedly yielded a stone board with engraved squares, evidently designed for playing *macang* (a Bugis game similar to chess), and an extraordinary concentration of imported ceramic shards from a 2 × 2 m illegal excavation, totalling 29 g of glass and 31.4 kg of imported ceramics. It was tentatively inferred that this area, designated South Utti Batue, would have covered around 1 ha and probably served as an elite compound.

An initial test pit (Selatan 4–5/Barat 25–26) (Fig. 12.2) failed to reveal any archaeological traces, but a second, 9 m² test pit (Selatan 23–25/Barat 2–4) yielded an abundance of materials (Fig. 12.3). The top 0.7 m of this pit was removed by shovel until the first archaeological trace, an earthenware shard, was exposed (Fig. 12.3). Subsequent deposits were removed in spits of approximately 5 cm depth. The first three such

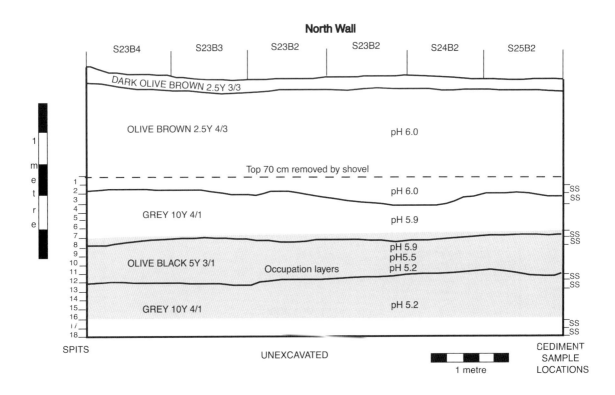

Fig. 12.3 Stratigraphic section of Utti Batue Test Pit 2 (northern wall).

spits, which yielded very sparse remains, covered the entire exposure. Spits 4 to 18, which were excavated through the main cultural deposit to sterile deposits at about 1.6 m depth, were restricted in area to the 4 m² in the centre of the test pit. Most of the pit was excavated at levels below the water table, and excavation was only possible by regularly pumping out the groundwater.

In describing the materials retrieved, along with the more interesting finds from the surface survey, we must stress that the work at Utti Batue provides only a very partial view of a site whose total material contents may be in the order of 10,000 times larger than the sample that OXIS was able to extract; based on the second test pit's excavated finds of approximately 45 kg, the total artefactual and ecofactual content of Utti Batue could conceivably lie in the region of 400–500 tonnes.

Site environment and sediments

In the 1980s, Utti Batue lay in a mixed forest of *Rhizophora apiculata*, *Avicennia marina* and *Xylocarpus granatum*, within a 12,000 ha expanse of mangrove forest which was, at the time, the largest in Sulawesi (*cf.* Whitten *et al.* 1987, 125, 190). Intuitively it is assumed that at the time of occupation the Utti Batue site also lay in a mangrove setting, but closer to the shoreline which may then have reached several kilometres to the west of its present position. The colour contrast in the excavated deposits, from predominantly grey in the lowest metre to olive brown in the upper layers (Fig. 12.3), would be consistent with the transition from a hydromorphic alluvium to a gley humus, which Whitten *et al.* (1987, 110) associate respectively with the seaward and terrestrial zones of the Malangke mangrove forest.

The majority of sediment samples (labelled SS in Fig. 12.3) comprised clay-rich, silty fine sands, and all are acidic, increasing in acidity with depth (*cf.* Whitten *et al.* 1987, 110). Clay samples from spits 2 and 3 (post-occupation), 7 and 8 (late occupation), 12 and 13 (middle occupation), and 17 and 18 (pre-occupation) were analysed by X-ray diffractometry, and the results from a representative sample (spit 2) are illustrated in Fig. 12.4. Mineralogically the samples are virtually identical, containing orthoclase and plagioclase feldspar, quartz, mica (biotite and muscovite), amphibole, chlorite, kaolinite, smectite, vermiculite and inter-layered clay (in a biotite/chlorite phase). These minerals are consistent with derivation of the sediments from the granodiorite hills which abut the coastal plain where the Rongkong and Baebunta rivers leave the ranges before flowing towards Malangke. The lack of variation in the samples suggests that the same sedimentation regime has prevailed from the time of initial settlement until the present, and chronological indications are consistent with this scenario. The *c.* 0.5 m of sediment deposited during the period of occupation of the site, *i.e.* the 200 years between AD 1400 and 1600, and the overlying 1 m of sediment deposited during the 400 following years, both correspond to an average deposition rate of 0.25 cm yr⁻¹.

Consequently, the abandonment of Utti Batue cannot be attributed to a disastrous flood or other major natural event, and is more likely due to socio-political changes, as the local oral history would suggest (Bulbeck and Caldwell 2000, 70). Moreover, the excellent preservation of organic materials at the site would appear to reflect anaerobic, waterlogged conditions at the time of occupation, rather than the creation of such conditions through the deposition of a sealing cap of sediment at the time of the site's abandonment.

Site contents

The primary means of dating the period of site occupation come from consideration of the 466 shards of imported stoneware and porcelain ceramics identified from the site. While the vast majority of these pieces (434) were recovered from the illegal excavation (noted above) in South Utti Batue, they are entirely consistent with the smaller body of finds recovered from North Utti Batue. Sixteenth century wares dominate the assemblage (Table 12.1). The single fourteenth century shard could have arrived at the site later than its identified age of manufacture, and is thus insufficient evidence of occupation before the fifteenth

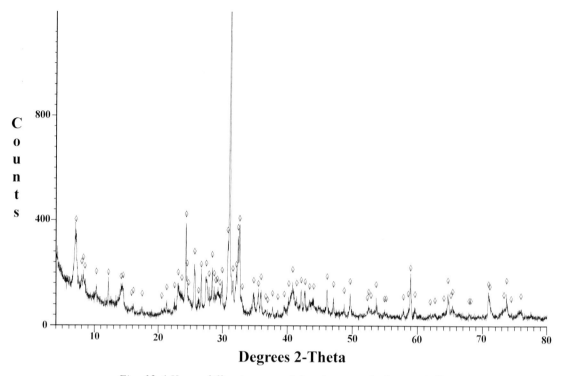

Fig. 12.4 X-ray diffractogram of the clay sample from spit 2.

century. As no wares definitely post-date the sixteenth century, the total period of occupation can be bracketed between *c*. AD 1400 and 1600, suggestive of sudden abandonment at the end of the site's sixteenth century heyday. The variety of imported wares is great (Table 12.1), and includes all manner of large and small jars, jarlets, vases, plates, bowls and covered bowls, with many of the pieces being of excellent quality. The assemblage gives the impression that imported wares served as a regular component of the domestic chattels, although the excavated finds show that earthen vessels greatly outnumbered imported ceramics in daily use (Table 12.2).

A few of the surface finds collected near the test pit in 1998 merit attention. One is a fragment of a brick 16.4 cm wide and 3.85 cm thick, with a length in excess of 19 cm. Its breadth and thickness fall in the range recorded on brick fragments from Tampung Jawa, which are interpreted to have been used in erecting a thirteenth to fifteenth century monumental structure at Malangke (Bulbeck and Prasetyo 1999, 27). Another find is a beautifully polished 'axe' with a square cross-section and dimensions of 149 × 37 × 35 mm, made of a hard, mottled pink and reddish brown stone. Even though it was surely a ceremonial piece (with no traces of use), its appearance at Utti Batue appears anachronistic, and it must have been brought on to the site as a curio or heirloom. The two other surface finds of polished stone are a heat-cracked river pebble (from a hearth or earth oven?), subsequently utilised as a pestle, and a soft river pebble with use wear striations on one face and transverse fractures along its sharper edge (probably used as a pestle-cum-scraper). Stone tools are rare at this site and the only other identified example is a small polished stone from spit 17 of the test pit. This peg-like chaplet or pin measures 48 mm long × 28 mm maximum diameter, and has been shaped to only a slight degree from a suitably proportioned pebble blank (Fig. 12.5, UTB.2.17.4).

	14th	14th–15th	15th	15th–16th	15th–18th	16th	16th–17th	Undated
Chinese whiteware	1	1	0	13	0	17	16	0
Jizhou black-and-white	0	3	0	0	0	0	0	0
Chinese blue-and-white	0	2	1	5	0	239	8	0
Chinese lead-glazed	0	0	2	0	0	0	0	0
Chinese celadon	0	0	1	7	0	1	0	0
Chinese monochrome	0	0	0	4	0	10	0	0
Sawankhalok	0	0	0	18	0	0	0	0
Sukothai	0	0	0	2	0	0	0	0
Vietnamese	0	0	0	46	0	41	0	0
Pegu (Burma) plainware	0	0	0	0	2	0	0	0
Coarse stoneware jar	0	0	0	1	0	1	0	4
Chinese blueware	0	0	0	0	0	1	0	0
Chin. overglaze enamels	0	0	0	0	0	16	3	0
Total	1	6	4	96	2	326	27	4

Table 12.1 Imported stoneware and porcelain identified from Utti Batue. Summarised from Bulbeck and Prasetyo (1999), and Bulbeck and Caldwell (2000: 119–20), apart from a few coarse stoneware shards subsequently identified among the excavated earthen pottery.

Spit	Gravels and misc. pebbles	Ironstone pebbles	Baked/indurated earth	Earthen shards	Ceramic shards	Glass bead
5	–	–	–	1.1 g	–	–
6	4.0 g	–	–	34.1 g	–	–
7	1.0 g	–	–	541.5 g	–	–
8	25.0 g	–	–	596.1 g	–	–
9	20.0 g	–	5.0 g	1082.0 g	–	–
10	210.0 g	–	117.0 g	3015.0 g	17.0 g	–
11	160.0 g	86.0 g	–	2937.8 g	210.9 g	–
12	564.0 g	–	174.0 g	3120.9 g	2.7 g	–
13	120.0 g	–	196.7 g	3305.3 g	19.0 g	–
14	247.2 g	–	105.0 g	2512.0 g	86.5 g	–
15	26.0 g	–	–	1263.1 g	–	0.1 g
16	6.0 g	–	6.6.g	681.3 g	37.4 g	–
17	–	–	–	38.0 g	–	–
18	–	–	–	9.6 g	–	–
Total	1.4 kg	86.0 g	604.3 g	19.8 kg	379.8 g	0.1 g

Table 12.2 Weight of inorganic finds from the Utti Batue test pit. Earthenware total includes 0.7 kg of shards collected from the wall, and the total of imported ceramic shards includes one from the wall.

Various pebbles and gravels, weighing 1.4 kg in total, were collected during the excavation. Their occurrence rises and falls with that of the artefacts and ecofacts (Tables 12.2 and 12.3), indicating that they had been transported manually on to the site. Some of the rocks show traces of exposure to heat and others may be spalls from used pebbles, confirming their role in the material culture of the occupants. A single jagged pebble of ironstone (Table 12.2) is of interest for its possible suggestion that iron ore had been imported to Malangke. The 600 g of baked and/or indurated earth point to some lighting of fires in the

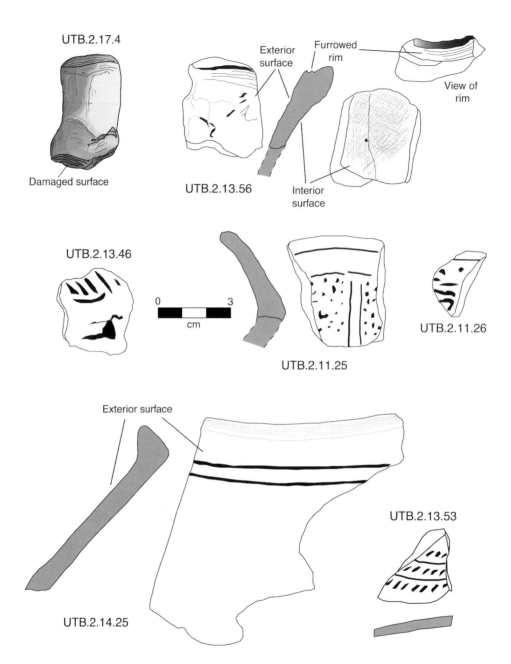

Artifacts excavated from Utti Batue. UTB.2.17.4: polished stone chaplet. UTB.2.13.56: "soft pink" rim. UTB.2.13.46: pottery disc. UTB.2.11.25-26: high-fired pottery with Islamic decorations. UTB.2.14.25: imitation martavan (high-fired pottery). UTB.2.13.53: high-fired cover sherd with geometric motifs.

Fig. 12.5 Representative artefacts excavated from Utti Batue Test Pit 2.

vicinity of the settlement, notwithstanding the waterlogged terrain that surrounded the site. The most prominent class of finds is earthenware shardage which, at nearly 20 kg, accounts for almost half of all recorded finds. The excellent preservation conditions probably account for the observation (detailed below) that a dammar coating (see note 1) has been better preserved on the local pottery at Utti Batue than at other excavated sites in Luwu.

The organic finds include over 1 kg of charcoal, about 13 kg of timber, over 330 g of other plant macrofossils, about 6.6 kg of faunal remains, and a lump of dammar gum (Table 12.3). Closer study of the plant macrofossils (which were identified only in the field) would quite likely reveal a greater variety than those shown here, but the direct evidence for the consumption of canarium nuts and coconuts is of interest in dietary reconstruction. Many of these fragments were recorded as burnt, as were all the faunal fragments. The wood had been deposited in large blocks of up to 2.7 kg in weight, and two pieces were observed to show clearly planed surfaces, as would be expected if the timber had been used for constructing houses on the site. According to the workers at a sawmill near the Makassar Archaeology Office in Sudiang, the type of wood represented is locally named *taluttu*, a mangrove timber of fine quality for carpentry and construction.

In view of the incomplete nature of the reference collection and the lack of a specialist analyst to undertake faunal identification, we attempted to be conservative in our identifications (Table 12.4). However, it is clear that bovids account for almost two-thirds of the faunal remains by weight. Large mammals, including pigs and deer, dominate the assemblage, while the contribution from medium-sized mammals, apparently including monkeys and dogs, but possibly also cuscus, was modest.

Few small fauna were recorded, and of these the rats were probably unwelcome guests. Fish and turtle make up a restricted but recognisable marine component of the assemblage, but no shellfish were recorded. This contrasts with the archaeological excavation of shellfish middens in Luwu, which are both

Spit	Charcoal	Canarium nut husks	Carbonised seed	Coconut shell	Wood	Dammar lumps	Faunal fragments	Organic lumps
2	8.5 g	8.0 g	–	–	–	–	–	–
3	1.4 g	–	–	–	–	–	–	–
5	3.5 g	–	0.3 g	–	–	–	–	–
6	0.6 g	15.0 g	–	–	–	–	–	–
7	77.0 g	2.0 g	–	–	–	–	124.7 g	–
8	53.0 g	1.2 g	–	–	2.7 kg	–	104.5 g	–
9	76.0 g	9.0 g	–	–	4.5 kg	–	700.6 g	15.0 g
10	193.0 g	7.0 g	–	30.0 g	–	–	1629.9 g	11.0 g
11	112.0 g	33.0 g	–	33.0 g	3.1 kg	–	641.4 g	26.0 g
12	240.0 g	27.0 g	–	–	–	–	677.6 g	3.0 g
13	293.7 g	33.3 g	–	1.0 g	2.1 kg	1.3 g	387.5 g	9.0 g
14	162.0 g	68.0 g	–	–	118 g	–	853.4 g	18.0 g
15	106.0 g	35.0 g	–	4.0 g	148 g	–	177.9 g	5.0 g
16	69.0 g	19.0 g	–	2.0 g	100 g	–	59.6 g	8.0 g
17	10.0 g	1.0 g	–	–	62 g	–	26.7 g	2.0 g
18	–	–	–	–	131 g	–	–	–
Total	1.4 kg	260.5 g	0.3 g	70.0 g	13.0 kg	1.3 g	6.95 kg	97.0 g

Table 12.3 Weight of organic finds by spit from the Utti Batue test pit (spits 1 and 4 were sterile). Totals include 12.4 g of charcoal, 2 g of canarium fragments, 6.6 g of wood, and 1.55 kg of bone fragments collected from the test pit wall.

Identification	Weight (g)	NISP	Identification	Weight (g)	NISP
Bovid	4234.2	272	*Bovid?*	112.4	15
Pig	39.8	8	Pig?	21.8	5
Deer	299.0	1	Large	912.9	302
Large/Medium	680.5	422	Medium	87.1	49
Macaque?	12.4	2	Dog?	2.0	1
Medium/Small	14.3	8	Small	3.1	7
Rat	1.0	7	Rat?	2.0	5
Bat?	1.8	1	Turtle/tortoise	5.7	4
Fish	110.8	52	Unidentified	413.1	352
			TOTAL	**6953.9**	**1513**

Table 12.4 Recorded weights and numbers of individual specimens (NISP) of fauna from the Utti Batue excavation.

contemporary with Utti Batue (Salabu, Tampinna) and younger than it (Surutanga, in Palopo) (Bulbeck and Caldwell 2000). The identifications at Utti Batue suggest an animal protein diet dominated by domestic animals, especially bovids, along with some exploitation of resources from the adjacent sea. It is assumed that the bovids were mainly water-buffalo, which would have thrived in the wet conditions at Utti Batue, rather than cattle.

The term 'organic lumps' is used for the 97 g of rounded cylindrical objects (Table 12.3) which the excavators tentatively identified as coprolites, but this was later demonstrated not to be the case at the Palynology Laboratory of Texas A&M University, although their identification remains uncertain. X-ray mineralogical analysis showed that one specimen consisted entirely of amorphous material, a second was predominantly of amorphous matter but included some quartz, possibly accompanied by some chloritic material and a little kaolin, and a third specimen consisted of an outer skin, which is mineralogically identical to the surrounding sediments, and inner matter which produced the same essentially amorphous X-ray spectrum as observed for the other two samples (Fig. 12.6). The most plausible identification currently comes from Malcolm Lillie at the University of Hull, who observed pockets of oil in the matrix and inferred that the specimens sent to him could be some sort of seed or fruit.

Phytolith analysis

Analysis of phytoliths (plant biogenic silica microfossils) was carried out on nine sediment samples from the test pit, in particular to determine whether sago palm (*Metroxylon sagu*) was part of the vegetative landscape when occupation at the site was flourishing. Illustrated references of phytoliths extracted from tropical material are limited. Nonetheless, leaf material from a small phytolith reference collection of tropical economic plants indicates that sago palm is a high volume producer of small, spherical echinate phytoliths in three sizes, *c.* 5, 8 and 10 mm diameter. However, as small spherical echinate and similar phytoliths are also identified in other palms (*e.g.* coconut), use of a higher magnification scanning electron microscope would be required to distinguish phytoliths of less than 10 mm. In this instance an assumption has been made that the presence of high numbers of phytoliths within the size range noted in sago reference material, and the current presence of sago palms throughout this area, indicate a probable continuous presence through time, as reflected by the high numbers of phytoliths with small palm type morphology. The sago palm also produces starch grains which can and do migrate and cluster on the slide, but not necessarily in the count transect. While phytoliths classified to a morphological group are not necessarily identical, their size, shape and ornamentation place them within the various groups.

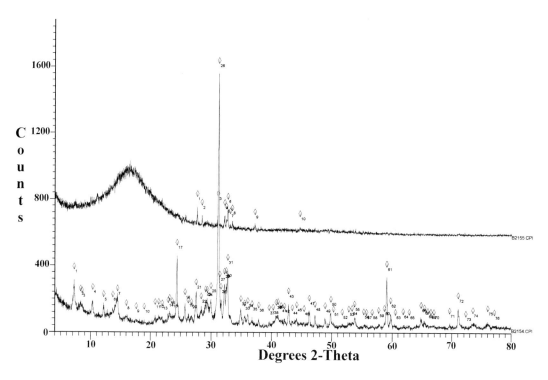

Fig. 12.6 X-ray diffractogram of an 'organic lump' sample contrasting the outer skin (with peaks) and the inner matter (amorphous spectrum).

Given the high rainfall, high water table, low relief and likely regular inundation of the area through time, it is suggested that the excavation area has always experienced various degrees of wetness, similar to the existing swampy conditions. In the Utti Batue phytolith assemblage (Table 12.5) two morphologies are indicative of wet conditions. These are the spherical echinate palm types of less than 10 mm (Group 1) and grass arc/triangle cells (Groups 69–71).

Where possible, plant names are tentatively identified to the groups. Groups 1–8 are palms with Group 1, which dominates the phytolith assemblage, probably sago palm and Group 7 probably coconut palm. Also tentatively identified are Groups 29 (*Barringtonia*), 33 (wild banana) and 34 (ginger). These plants, along with a variety of palms and other flora, are all to be found in disturbed areas of Indonesia (Whitten *et al.* 1987). Group 25, Cuticle 2 is identified (Fahn 1989) as from the Iridiaceae family, whose members are swamp plants in their native habitat (Dykes 1974). Rectangles (Groups 36–48) are ubiquitous and unidentified to plant in this assemblage. 'Grasses' (Groups 50–71) follow a classification scheme devised by Bowdery (1998) to identify short cell morphology. Group 55 and Angular groups 58–67 in this instance are tentatively associated with bamboo. The arc/triangle morphology of Groups 69–71 identifies hydrophylic grasses such as *Imperata*, *Phragmites* and bamboo. It is suggested that, given the short chronology of the deposit, morphologies occurring in all nine samples (final column, Table 12. 5) may be considered as part of the phytolith 'noise' for the site. Omitting Group 72 ('others'), this gives 27 morphologies occurring as noise.

Table 12.6 gives relative frequencies of phytoliths obtained from a count of one 10 mm transect for each

PHYTOLITH MORPHOLOGY	Gp #	5	8	9	10	11	12	13	14	18	Occurrence in seds (max. 9)
Spit #		5	8	9	10	11	12	13	14	18	
dbs cm		105	120	125	130	135	140	145	150	170	
Slide ID		713	723	714	715	716	717	718	719	720	
Spherical echinate (palm)	-										
- <10μ (sago)	1	x	x	x	x	x	x	x	x	x	9
- <20μ	2	x	x	x	x	x	x	x	x	x	9
- <30μ	3	x	x	x	x	x	x	x	x	x	9
- <40μ	4	x	x	x		x	x	x	x		7
- <50μ	5			x							1
- <75μ	6			x	x						2
- Se1 (coconut)	7	x	x	x	x	x	x	x	x	x	9
- Se 2	8	x	x	x	x	x	x	x	x	x	9
psilate (arboreal)											
- <10μ	9	x	x	x	x	x	x	x	x	x	9
- <20μ	10	x	x	x	x			x			5
- <30μ	11	x	x	x				x	x		5
- <40μ	12		x	x		x	x	x			5
- <50μ	13		x	x		x	x	x	x		6
- <60μ	14							x	x	x	3
- <75μ	15				x			x		x	3
- <100μ	16			x							1
- <100μ	17					x					1
3D chunks	18	x	x	x	x	x	x	x	x	x	9
Sheet	19	x	x	x	x	x	x			x	7
Facetted	20			x	x						2
Crescent	21				x	x					2
Perforated - clear	22		x	x	x	x	x	x	x	x	8
- brown opaque	23	x	x	x	x	x	x	x	x		8
Cuticle - C1	24		x	x	x	x	x	x	x		7
Cuticle - C2 Iridaceae	25		x	x	x	x	x	x			6

x = presence

Table 12.5 Utti Batue phytolith presence.

Type	No.	Total
Cuticle - 3	26	4
Cuticle - C4	27	4
Cuticle C5 - other	28	7
Barringtonia?	29	8
Ornamented. rect. - arboreal	30	9
Ornamented rect. - grasses	31	9
Adiantaceae type	32	5
Wild banana?	33	9
Ginger	34	9
Trichome/hair	35	9
Rectangles		
- short - narrow	36	9
- medium	37	9
- wide	38	9
- medium - n	39	9
- medium	40	9
- wide	41	9
- long - narrow	42	8
- medium	43	9
- wide	44	2
- tapered	45	9
- sinuous	46	8
- rounded	47	8
- square	48	9
- oval	49	6
GRASSES		
Bilobe group - 1	50	8
- B2	51	8
- B3	52	5
- B4	53	8
- B5	54	5
- B6	55	9
- B7	56	8
- B8	57	4

x = presence

Table 12.5 continued.

	ID									n
Angular group -1	58	x	x	x	x	x	x	x	x	9
- A2	59	x	x	x	x	x	x	x		6
- A3	60	x	x	x	x	x	x	x	x	8
- A4	61	x	x	x	x	x	x	x	x	9
- A5	62	x	x	x	x	x	x	x	x	9
- A6	63	x	x		x	x	x			5
- A7	64	x	x	x	x	x	x	x		8
- A8	65		x							2
- A9	66	x	x	x	x	x	x	x	x	8
- A10	67	x	x	x	x	x	x	x	x	8
- A other	68	x	x	x	x	x	x	x	x	9
Arc/triangle - AT1	69	x	x	x	x	x	x	x	x	9
- AT2 (bamboo)	70		x	x	x	x	x	x	x	8
- AT3 (*Phragmites/Typha*)	71	x	x	x	x		x	x	x	7
OTHERS - not classified	72	x	x	x	x	x	x	x	x	9
TOTAL GROUPS (71): Presence		44	61	59	58	56	56	52	49	48
TOTAL GROUPS (71): Table 6		29	44	42	44	47	46	39	35	36
OTHER BIOGENIC SILICA MICROFOSSILS										
Diatoms - D1		x	x	x	x	x	x	x	x	
- D2		x	x	x		x	x	x	x	
- D3		x	x	x			x			
- D4		x	x			x	x			
Sponge spicules		x		x	x	x	x	x	x	
OTHER MICROFOSSILS										
Nematoda		x	x	x	x	x	x	x	x	x
Starch grains		x	x	x	x	x	x	x	x	x

x = presence

Table 12.5 continued.

Spit #		5		8		9		10		11		12		13		14		18	
dbs cm		105		120		125		130		135		140		145		150		170	
Slide ID		713		723		714		715		716		717		718		719		720	
PHYTOLITH MORPHOLOGY			%		%		%		%		%		%		%		%		%
Spherical																			
- echinate (palm)																			
sago		392	76.3	215	43.3	245	39.9	181	40.8	351	37.6	185	33.2	250	63.3	40	19.7	97	38.3
coconut		1	0.2	4	0.8	18	2.9	6	1.4	15	1.6	10	1.8	1	0.3	2	1.0	1	0.4
palm type other		20	3.9	10	2.0	10	1.6	16	3.6	25	2.7	12	2.2	14	3.5	9	4.4	5	2.0
- psilate (arboreal)		9	1.8	10	2.0	15	2.4	6	1.4	12	1.3	10	1.8	7	1.8	3	1.5		
Cuticles				3	0.6	11	1.8	2	0.5	12	1.3	4	0.7	1	0.3	5	2.5		
Rectangles		14	2.7	63	12.7	80	13.0	67	15.1	123	13.2	80	14.4	24	6.1	23	11.3	27	10.7
Grasses																			
Bilobe		2	0.4	29	5.8	26	4.2	24	5.4	49	5.2	34	6.1	5	1.3	4	2.0	6	2.4
Angular		23	4.5	48	9.6	78	12.7	56	12.6	112	12.0	73	13.1	18	4.6	17	8.4	18	7.1
Arc/triangle				6	1.2	16	2.6	1	0.2	24	2.6	7	1.3	8	2.0	23	11.3	10	4.0
Grass others		2	0.4	6	1.2	13	2.1	8	1.8	14	1.5	9	1.6	4	1.0	23	11.3	7	2.8
Others - identified		13	2.5	42	8.5	43	7.0	20	4.5	74	7.9	56	10.1	7	1.8	16	7.9	12	4.7
Others - not identified		38	7.4	61	12.3	59	9.6	57	12.8	123	13.2	77	13.8	56	14.2	38	18.7	70	27.7
TOTAL PHYTOLITHS		514	100.0	497	99.9	614	100.0	444	100.0	934	100.0	557	100.0	395	100.0	203	100.0	253	100.0
OTHER BIOGENIC SILICA MICROFOSSILS																			
Diatoms - D1		x																	
- D2		x																	
- D3		x																	
- D4		x																	
Sponge spicules		x		x												x			
OTHER MICROFOSSILS																			
Nematoda				x				x				x		x					
Starch grains		c91		c188		c52		c143		c378		c93		c268		c302		c56	

x = presence

Table 12.6 Utti Batue phytolith relative frequency.

sampled spit. Counts were made using the morphologies listed in Table 12.5; these were then consolidated into broader groups. The number of groups observed during the count for each spit is shown in Table 12.5 in the row headed 'TOTAL GROUPS observed in count'. These numbers are lower than those observed for each spit, recorded in the adjacent row headed 'TOTAL GROUPS Presence'. Of importance is the number of groups not shown by the transect count. In a possible presence of 639 groups (over all sampled spits), inspection of the entire slide returned 483 groups (75.6%) compared to the 362 groups (56.7%) observed in the transects – a difference of 18.9 percent of the morphologies not reflected by the counts. For this reason it is suggested that phytolith presence as reported in Table 12.5 is a more reliable indication of phytolith diversity than the 10 mm transect counts. The difference may be attributed to how phytolith morphology affects migration across the cover slip when mounted. The small palm type has an even distribution amongst all other shapes and sizes and this may cause bias in a count. It will be seen from Table 12.6 that palms (groups 1–8) dominate the assemblage with sago palm (group 1) in turn dominating the palms.

Of the other microfossils, diatoms, sponge spicules and *Nematoda* are associated with water. At least four species of diatoms were noted. Their presence in spits 13, 8 and 5 suggests that these were relatively wetter periods. There were few sponge spicules or *Nematoda*. Starch grain size fell within the range noted in reference material, *i.e.* up to 32 mm in diameter.

The cultural sediments in spits 8–17 are positively correlated with phytolith diversity. Group numbers rise from spit 14 to spit 8, with the highest diversity of phytoliths, most indicative of a human presence, occurring in spit 8. The phytolith presence data in spits 12–10 suggest that these sediments were the driest in the column. This inference is confirmed by the excavated materials, at least in spits 10 and 12, where the finds include over 100 g of baked or indurated earth (Table 12.2), but no wood (Table 12.3). The data from Tables 12.5 and 12.6 are combined below to provide a summary of phytolith presence through the assemblage.

> *170 cm, Spit 18*: Non-cultural spit. Lowest phytolith diversity, low palm and coconut. High bamboo presence. Few other grasses represented. Possibly no standing water but high precipitation.

> *150 cm, Spit 14*: Similar to 170 cm but with increased phytolith diversity related to cultural presence. First appearance of eight phytolith groups including various palms. The appearance of diatoms, sponge spicules and *Nematoda* indicates an increase in water availability. Wetter than spit 18.

> *145 cm, Spit 13*: Increase in phytolith diversity but with less bamboo (bilobes and angular types). First appearance of Iridaceae cuticle. Complete specimens of diatoms (all four species), and sponge spicules with two terminations intact, indicate little movement likely to cause damage, *i.e.* probably standing water.

> *140 cm, Spit 12*: Further increase in phytolith variety. The increase in tree/shrub morphologies and grasses other than bamboo possibly indicates some clearance and drier conditions. Fewer small palm and starch grain numbers.

> *135 cm, Spit 11*: Highest number of small palm phytoliths in the cultural levels. All grass groups present, and many cuticle fragments. *Canarium* trichome tentatively identified.

> *130 cm, Spit 10*: Grass presence maintained, increase in *Barringtonia* (tree) morphology, fewer cuticles.

> *125 cm, Spit 9*: Highest coconut count. Grass presence maintained, increase in cuticle types. Increase in water indicated by an increase in hydrophylic AT cells and three species of diatoms noted.

> *120 cm, Spit 8*: Highest phytolith diversity. Many arboreal ornamented rectangles. Fewer grass groups represented with many hydrophylic AT, all groups.

> *105 cm, Spit 5*: Non-cultural. Drop in phytolith diversity to less than that of spit 18 at the base of the excavation. Highest number of sago phytoliths present in the column. Possible extension of sago palm cultivation with increasingly swampy conditions.

From the summary above it can be seen that phytolith presence changed continuously throughout the column. At the same time the available water changed from still, to drier to wetter (for example, spits 13,

12–10, and 9 respectively). However, the phytolith assemblage indicates that this variation is modest, and the site has always been wet to varying degrees. The sago palm dominates at all levels and increases to its highest numbers in spit 5 above the habitation layers. It may be the case that some clearing of bamboo occurred during a drier interval corresponding to spit 12, and this clearance enabled expansion of sago palm, by natural or anthropogenic means, as water availability subsequently increased (represented by the higher spits). This argument would support the hypothesised importance of sago to the inhabitants of Utti Batue. The continuing high numbers of sago phytoliths may have possibly been maintained by natural regeneration of *Metroxylon* sp. in the increasingly swampy conditions after the abandonment of the site.

Pottery analysis

Owing to time constraints, the local pottery was only recorded in detail in spit 14 and especially spit 13. These two assemblages appear to be dominated by shards from storage vessels, some of which may have been used in storing water and victuals at the site, but many of which were perhaps related to cartage of goods. In support of this claim, only 33 shards (22.9 g) of the >5.8 kg of earthen pottery bore traces of charring attributable to cooking fires, and an apparent ceramic disc from spit 13 may have functioned as a marker of merchandise (Fig. 12.5, UTB.2.13.46). When this incised disc is excluded, along with shards bearing simple ridges and horizontal incisions, and rims with marginally decorated notches, the number of clearly decorated shards is reduced to two – a tiny number possibly more compatible with outdoor utilitarian wares than with domestic utensils.

A wide variety of fabrics was encountered, and these were grouped into five 'wares': soft white, soft pink, fine, coarse, and high-fired. Soft white grades into soft pink, soft pink into fine, fine into coarse and high-fired, and coarse into high-fired. The soft pottery (whose shards typically have water-rounded edges) was recorded at about the same frequency as the high-fired pottery, and either the fine or the coarse pottery was the most commonly observed class (Table 12.7). Dammar is the suspected resinous agent on dark-surfaced ware (see note 1), based on reports of its continued use in coating locally made vessels at Lake Matano (field observation), and its historically attested role as a major forest product from Luwu (Caldwell 1993, 1995, but see Bulbeck 1993).

Fabric class (spit 13)	No.	Weight (g)	Fabric class (spit 14)	No.	Weight (g)
Soft white	27	37.3	Soft white	37	54.1
Soft pink	262	542.8	Soft pink	143	167.0
Utti Batue fine	433	1597.8	Utti Batue fine	204	544.6
Utti Batue coarse	67	511.7	Utti Batue coarse	169	1035.0
High-fired*	16	615.7	High-fired*	3	113.5
Surface effects (13)			**Surface effects (14)**		
None	429	1084.1	None	386	887.0
Textile-impressed	66	218.1	Textile-impressed	16	20.7
Red-slipped	27	161.6	Red-slipped	20	83.9
Dark-surfaced	265	1194.8	Dark-surfaced	126	758.4
Dark-surfaced and decorated*	11	598.0	Dark-surfaced and decorated*	1	110.8
Other decorated	7	48.7	Other decorated	6	52.9
Total	**805**	**3305.3**	**Total[#]**	**608**	**2512.0**

*Table 12.7 Pottery counts and weights from spits 13 and 14. *Includes many shards which may belong to a single large jar represented in both spits, and so are not enumerated separately. #Total includes 54 rim and base shards (~598 g) not classified into sub-classes.*

Textile impressions were first observed on the spit 13 pottery curated in Makassar, and studied in more detail on the spit 14 pottery brought to Canberra. Some of the cloth impressions were of a plain weave with a single warp and weft, *i.e.* 1/1 tabby. Some shards were impressed with coarse fibres and others with fine fibres. A number of shards were impressed with fine netting, made from a single set of interlinked elements. The more angled and curvilinear patterns (Fig. 12.5, UTB.2.13.56) could reflect irregularities in the weaving, or creasing and distension of the textiles. The more regular patterns resemble the textile impressions on the interior of shards from a late prehistoric context at Gunung Kidul, Java, related by van der Hoop (1941, fig. 96) to palm cloth. Although cotton has evidently replaced the traditional bast fibres used by local groups in South Sulawesi, matting and textiles woven from *Metroxylon sagu* (sago palm), *Corypha* palm and *Borassus flabellifer* are still being produced on nearby islands. Analysis is continuing, with impressions from woven palm textiles being used for comparative purposes.

Brian Vincent of the University of Otago, New Zealand, has suggested that the cloth may have been employed to cover the surface of a vessel to keep it separate from clay that was packed on to the cloth. The inner vessel would thus have acted as the mould for the new vessel formed around it. After shaping, the new vessel could have been bisected to dislodge it, and fired after the two halves were luted back together. Vincent's suggestion would explain the occurrence of the impressions only on the interior face of the vessel, which rules out cord-marking or any other decoration as the cause, as well as the curved and angled orientation of the thread impressions. The manufacture technique he proposed would also tally with the irregular appearance of the shards' exterior face, the dominance of geometrically simple forms (lids, inverted jars and, possibly, boxes) and the use of easily moulded clay with remarkably little of the temper generally required for structural support. The vessels in question could have been rapidly produced containers designed to traffic their contents, with little consideration given to use-life beyond the intended destination. In support of this interpretation, 'soft' pottery has been observed in *c.* fifteenth century contexts at Bola Merajae and Matano, two sites which are suspected to have been linked to Malangke in the exportation of Luwu's iron (Bulbeck and Caldwell 2000).

In an attempt to quantify the impressions, Bulbeck calculated the average warp and weft on 65 shards by measuring the extent of the affected area in both directions, and dividing the linear distances by the number of thread impressions. The resulting 130 observations showed that the average distance between threads was 0.74 mm, with a standard deviation of 0.20 mm, and a range from 0.4 to 1.3 mm (apart from an outlier at 2.1 mm). The average distance between threads of the example of particularly fine palm cloth illustrated by van der Hoop (1941, fig. 96) is about 0.3 mm. The comparable measurement for a fragment of fine, historical cotton (Bulbeck and Caldwell 2000, 25) from the Pontanoa Bangka cemetery in Luwu, studied by Cameron (2000), is 0.4 mm. Thus the Utti Batue impressions have a distance between threads which is equal to or greater than the between-thread distance of fine palm cloth and cotton. This result is readily understood because old, less valuable cloth that had lost its structural resilience, but which was easier to stretch and distend, would probably have been used in the vessel manufacturing technique suggested by Vincent. Further, coarser cloth of lesser value might have been used for this utilitarian task, and Bulbeck's counts of the number of threads may be too low when the impressions are faint. Cameron's opinion (pers. comm.) that cotton is probably responsible for similar textile impressions on the 'soft orange' pot shards from Bola Merajae, suggests that cotton was surely one of the textiles which left its impressions at Utti Batue. However, local palm cloth was probably involved too.

Some of the shards from the higher spits of the excavation, identified as 'special', deserve comment. Spits 6 and 10 both produced single examples of shards from 'cogwheel pots'. This is the name given by Clune and Bulbeck (1999) for serving dishes with nipples spaced evenly around the circumference, at an interval that is ideal for clasping the vessel with the fingers. A complete example has been recorded by Bulbeck from the looted Mangge pre-Islamic cemetery in Malangke. Previously, cogwheel pots had been documented solely in fifteenth to sixteenth century contexts from the Makasar-speaking area along the south coast of South Sulawesi, and not from any Bugis haunts (Clune and Bulbeck 1999, 46). The Malangke specimens are distinct in being high-fired, dark-surfaced wares, in having the cogwheel band luted on rather than moulded, and in the position of the band along the shoulder. Along with a shard from a vertically gouged pot excavated in spit 10 at Utti Batue, which is a decoration characteristic of the southeast lowlands of the peninsula (Clune and Bulbeck 1999, Table 12.3), the cogwheel pots hint at a link between pre-Islamic Luwu and its Makasar counterparts.

Spits 8 and 11 yielded three shards from what is probably the same vessel, an everted jar with arabesque decorations in vertical and horizontal panels along the inside of its mouth (Fig. 12.5, UTB.2.11.25/26). This apparently local vessel suggests influences from one of the Islamic trading polities already established widely across the archipelago by the sixteenth century. One possible source of inspiration would be Banten (see Ambary *et al.* 1993, 136–42) or another of Java's north-coast ports. Spit 11 produced another high-fired, dark-surfaced shard, the only decorated cover knob yet recorded from Utti Batue. More prosaically, spit 10 yielded a shard from what was probably the cover of a heavy pottery stove, made of Utti Batue 'coarse' fabric. Stoves such as these have been commonly carried around on sailing vessels in the archipelago (Bellwood 1997, 227), so the shard is consonant with the inferred maritime focus of Utti Batue.

In summary, the local Utti Batue pottery is highly variable, and rich in its potential to yield socio-economic information on the site. It provides evidence on textiles unknown through other avenues, and may hint at the importance of dammar resins in the local economy. Links are suggested with the Makasar belt along the south coast of South Sulawesi, as insinuated in the historical texts (Caldwell and Druce 1998), and possibly with early Islamic polities in the region. The assemblage as a whole testifies to systematic maritime trade, and the soft ware reiterates the case for a connection via Ussu to the iron-producing centre of Lake Matano.

DISCUSSION AND CONCLUSIONS

Sago is identified as a major plant species at Utti Batue before, during and, particularly, after the site's period of occupation. While there is little indication that arboricultural practices had ever promoted the growth of sago at Utti Batue, its natural abundance would have provided an immediately available, major source of food. Coconuts and canarium nuts were also evidently available, but presumably of much lesser significance than sago. Rice and other imported plant foods may also have been consumed, but they were not identified among the macro-botanical remains, and the phytolith record would be unlikely to register their presence as processed imports. Water-buffalo seems to have been the major source of animal protein, based on the high rate of bovid identifications and the wet conditions that prevailed at the site. Marine resources such as fish and turtle are also reflected in the faunal assemblage. Overall, the ecofactual evidence is consistent with the claim that the subsistence regime at Utti Batue, and presumably more widely in Malangke, had relied heavily on sago. The subsistence economy at Malangke would appear broadly similar to that at Melaka in Peninsular Malaysia where, according to eye-witness accounts, the dietary staple of sago was supplemented with vegetables, sugar cane and various tree crops such as bananas and jackfruit, as well as marine foods provided by the fishermen attached to the entrepôt (Wheatley 1961, 311–2). Malangke and Melaka pair together as trading settlements with the social complexity of a 'civilisation', and a diverse subsistence economy centred on sago, as witnessed by the archaeological record in one case and the textual record in the other.

Large quantities of timber were recorded only between spits 13 and 8, though not in spits 12 and 10 which may correspond to intervals when conditions at the site were relatively dry (and hence less ideal for preserving wood). Other organic remains, and the pottery shards, were most concentrated slightly lower in the deposits, between spits 14 (or 15) and 9. These distributional data are consistent with a situation in which the residents lived in timber structures and discarded their waste on to the surrounding terrain. Replenishment of their housing, and abandonment at the end of the occupation period, would accordingly have led to the extensive overlap between wood and other remains, with the wood tending to lie slightly higher in the profile. The phytolith identifications are also consistent with a model in which the sediments accumulated steadily and applied their 'seal of preservation' to the incorporated material.

The accumulation of nearly 2 m of sediment between the base of the excavation and the present-day surface would not appear to correspond in any straightforward way with an increasing elevation above the prevailing water table, despite the very low altitude of the site above sea-level. Spit 18 at the base was identified as drier than spit 5 above the habitation levels, on the basis of phytoliths. Small variations in the local sea-level, secular fluctuations in Luwu's precipitation, and changing proximity of the passing streams and rivers to the site, could all explain this counter-intuitive finding. Almost certainly, the inhabitants of Utti Batue lived in houses erected on wooden piles. This inference is based on the massive posts recorded during survey along the ditch that cuts through the site, the large quantities of excavated timber (making up the second-heaviest category of finds), and all of the relevant historical and ethnographic observations known on the Bugis. Pile housing would have allowed the residents to adjust readily to changes in the degree of waterlogging at the site.

The material record reflects the inhabitants' pottery, both locally produced and imported from overseas, access to textiles, use of stone (with usually minimal preparatory shaping) for a range of purposes, and abundant access to dammar. Glass beads and glass vessels would appear to have served only as minor constituents of the material culture. Iron was not recovered from the excavation, and this suggests that it may not have been particularly common, at least in particular contexts. Certainly, no evidence has been recovered of iron-working at Utti Batue, in contrast to the evident focus on working iron at Pattimang Tua slightly further inland (Bulbeck and Caldwell 2000). The Utti Batue pottery is interpreted to have included abundant storage vessels, and ad hoc 'soft ware' vessels moulded around bundles of goods intended for transport. This soft ware in particular links Utti Batue to the suspected trade route for iron from Lake Matano through Ussu to Malangke. The details of the array of artefacts from Utti Batue, as well as their diversity, accordingly confirm the site's hypothesised role as the maritime node in Luwu's pre-Islamic, export-based economy, where iron and other hinterland produce were centralised for onward distribution to Luwu's trading partners. Full excavation of this unique site would provide unparalleled information on the spatial organisation and material culture of an Indo-Malaysian palace centre during the period when European colonists first moved into the archipelago.

ACKNOWLEDGEMENTS

Robin Westcott of CRCLEME, Geology Department, ANU, executed the clay separation and mineralogical analysis of the bulk samples and the organic lumps, and prepared Figures 4 and 6. John Vickers sectioned the organic lumps for analysis. Vaughn M. Bryant, Jr. and Dawn Marshall of the Palynology Laboratory, Texas A&M University, carried out the organic tests on these lumps that first disproved any identification as coprolites. Karaeng Demmanari of the Makassar Archaeology Office assisted FDB in the identification of the imported ceramics. Tanwir Wolman of the Archaeology Department of Hasanuddin University in Makassar supplied most of the faunal identifications. Geoff Hope, Archaeology and Natural History, ANU, provided advice on the site's likely geomorphological history based on his field inspection, and put FDB into contact with JF. Judith Cameron, in Hope's department, provided advice on textile impressions in Luwu. Peter Bellwood and Campbell Macknight of the ANU, and Truman Simanjuntak of the National Research Centre for Archaeology, also inspected the site and proposed suggestions on how to excavate it.

Moh. Ali Fadillah of the Makassar Archaeology Office organised the permits to excavate in Luwu and facilitated FDB's laboratory work in Makassar. Funding for the entire OXIS project was provided by an Australian Research Council grant to FDB and Ian Caldwell (University of Hull), and by Wenner-Gren Foundation for Anthropological Research International Collaborative Grant No. 19 to FDB and Darmawan Mas'ud Rahman (Universitas Negeri Makassar). Grants from the Australia-Indonesia Institute to FDB, and from the ANU Faculties Research Grant Scheme to FDB and Geoff Hope, respectively defrayed the costs of FDB's laboratory work in Makassar and DB's phytolith analysis.

NOTE

1   Reference to the possibility of a dammar coating on some of the pottery sherds recovered from Utti Batue, inferred from the presence of a dark resinous layer, was not confirmed during recent analysis by Cynthia Lampert. To date, no indication of an organic coating has been forthcoming from her analysis of the pottery, from which confirmation of the initial field observations could be substantiated.

REFERENCES

Ambary, H. M., S. Takashi, H. Michrob, M. T. N. Wibisono & O. Kohji 1993. *Banten, Pelabuhan Keramik Jepang: situs kota pelabuhan Islam di Indonesia* (Banten and the port's Japanese ceramics: an Islamic port city in Indonesia). Jakarta: Pusat Penelitian Arkeologi Nasional.

Bellwood, P. 1997. *Prehistory of Indo-Malaysian Archipelago* (revised edition). Honolulu: University of Hawaii Press.

Bowdery, D. 1998. *Phytolith Analysis applied to Pleistocene-Holocene Archaeological Sites in the Australian Arid Zone*. Oxford: British Archaeological Reports International Series 695.

Bulbeck, D. 1993. New perspectives on early South Sulawesi history. *Baruga* 9, 10–18.

Bulbeck, D. 1996–7. The Bronze-Iron Age of South Sulawesi, Indonesia: mortuary traditions, metallurgy and trade, in F. D. Bulbeck & N. Barnard (eds) *Ancient Chinese and Southeast Asian Bronze Age Cultures*, 1007–76. Taipei: Southern Materials Center Inc.

Bulbeck, D. & I. Caldwell 2000. *The Land of Iron: the historical archaeology of Luwu and the Cenrana Valley. Results of the Origin of Complex Society in South Sulawesi Project (OXIS)*. Hull: University of Hull Centre for South-East Asian Studies.

Bulbeck, D. & B. Prasetyo 1998. Survey of pre-Islamic historical sites in Luwu, South Sulawesi. *Walennae* 1, 29–42.

Bulbeck, F. D. & B. Prasetyo 1999. The Origins of Complex Society in South Sulawesi (OXIS): tentative final report to Lembaga Ilmu Pengetahuan Indonesia. Unpublished Report.

Caldwell, I. 1993. Untitled. *Baruga* 9, 6–8.

Caldwell, I. 1995. Power, state and society among the pre-Islamic Bugis. *Bijdragen tot de Taal-, Land- en Volkenkunde* 151, 396–421.

Caldwell, I. & S. Druce 1998. The tributary and domain lists of Luwuq, Binamu and Bangkala. Unpublished report to the South-East Asian Panel of the British Academy.

Cameron, J. 2000. *Report of Microscopic Analysis of Prehistoric Textile Fragments from Pontanoa Bangka, Sulawesi*. Canberra: Australian National University, Department of Archaeology and Natural History.

Clune, G. & D. Bulbeck 1999. Description and preliminary chronology of Macassar historical earthenware decorations. *Walennae* 3, 39–60.

Dykes, W. R. 1974. *The Genus Iris*. New York: Dover Publications Inc.

Fagan, B. M. 1995. *People of the Earth: an introduction to world prehistory* (eighth edition). New York: Harper Collins College Publishers.

Fahn, A. 1989. *Plant Anatomy*. Oxford: Pergamon Press.

Nik Hassan Shuhaimi b. N. A. R. 1991. Recent research at Kuala Selinsing, Perak. *Bulletin of the Indo-Pacific Prehistory Association* 11, 141–52.

Osazawa, S. 1986. Sago production in kabupaten Luwu, South Sulawesi – a trial for upgrading the economic capability of a traditional sago-producing society, in K. Tanaka, M. Maeda & N. Maeda (eds) *Environment, Landuse and Society in Wallacea*, 51–63. Kyoto: Kyoto University.

Pelras, C. 1996. *The Bugis*. Oxford: Blackwell.

van der Hoop, A. N. J. 1941. *Catalogus der Praehistorische Verzameling*. Bandung: A.C. Nix & Co.

Wheatley, P. 1961. *The Golden Khersonese. Studies in the Historical Geography of the Malay Peninsula Before A.D. 1500*. Kuala Lumpur: University of Malaya Press.

Whitten, A. K., M. Mustafa & G. S. Henderson 1987. *The Ecology of Sulawesi*. Yogyakarta: Gadjah Mada University Press.

PART 3
METHODOLOGY

Chapter 13

Wetland Science

Mike Corfield

The ability of wetlands to preserve the wide spectrum of archaeological evidence described in this volume is due to their complex physical, chemical and microbiological nature, and particularly, but not exclusively, the suppression of the microbiological activity that in non saturated soils results in the decomposition of organic based materials. The chemistry of wetlands derives from the nature of their underlying geology and the dissolved materials in their water inputs; the pH of the groundwater and its redox potential are fundamentally important for the preservation of both organic and inorganic evidence.

An understanding of wetland science is necessary, both so that the significance of the presence or absence of classes of evidence and their condition can be properly interpreted and so that proper strategies can be put in place to preserve wetland sites. A further consideration given by Caple and Dungworth (1998) and Caple (2001) is to assess the practicality of reburial of excavated bulk waterlogged material in wetlands when it is not destined for permanent preservation in museums.

With greater investigation of wetlands it is now possible to refine the list, with priority being given to the maintenance of sustainable wetlands in which archaeological evidence will be preserved *in situ*. Unless we can have a comprehensive understanding of the way that the different categories of evidence are preserved in the ground, and how it may be compromised by changes to wetland burial environments we cannot make proper assessments of the sustainability of the wetland as an archaeological resource. The broad generalisations about what is preserved and under what conditions, needs refining so that the nuances of preservation become clearer. This is a tall order but one that must be addressed if strategies for preservation *in situ* are to be underpinned by sufficiently robust science to persuade the archaeological community that the correct course is being followed.

Secondly, as the papers in this volume suggest, there must be a broader understanding of the way that different classes of evidence survive (*e.g.* Smith 2005). Pollen, plant remains, diatoms and foramenifera tell us much about the past environment and how it was manipulated by the human populations that exploited it. The archaeological record is also increasingly used to inform about past climatic variations and how past human populations reacted to changes in the weather; it is also of increasing importance to climatologists who use it to help build the evidence for present day climate change scenarios (Hulme *et al.* 2002).

The great value of wetlands is that they are neither aquatic nor terrestrial, but lie at the interface between these systems; and because of this, Mitsch and Gosselink (2000) describe them as an enigma to scientists because they possess qualities of each. Gilman (1994) comments that although the value of wetlands has belatedly been appreciated, the hydrological processes defining wetland habitats, and the benefits of wetlands to the hydrological cycle are poorly understood; he calls for more research into the interactions of wetlands with groundwater and surface water which he deems as vital to the assessment of wetland water quality.

WATER BALANCE

The first and obvious requirement of wetlands is that they should be wet. For optimal archaeological preservation ideally, the wetland should be permanently wet and not subject to extreme seasonal fluctuations or management regimes that involve reducing the water table below the level of the archaeological resources for extended periods. Some short term reductions may be permissible provided that the soil has sufficient water retentive characteristics to maintain a preserving environment. The key to keeping wetlands wet is the water balance. Put simply, this means that the sum of all the water inputs to the wetland – ground water, surface water and precipitation – should be equal to or greater than the sum of the outputs – surface water run off, groundwater, evapotranspiration; this has been expressed by Eggelsmann *et al.* (1993) as:

$$P + I = D + E + (R\text{-}C) \text{ mm}$$

Where P = precipitation, I = intrusive water inflows, D = discharge, E = evapotranspiration, R = reserve, C = consumption, and $(R - C)$ = storage.

From this equation it can be seen that if precipitation and any other inflows are less than discharge, evapotranspiration and consumption, then there will be a negative outcome, or a water deficit. In a functioning wetland if a deficit condition continues unchecked the wetland will be drained and its function will be lost. However, archaeological evidence may survive below the ground surface in wetlands that have long since been drained. Such archaic wetlands are dependent on the water table to maintain archaeological remains that are susceptible to desiccation, in good condition, therefore a reduction in the water table will result in a degradation of the evidence. Fig. 13.1 shows the annual variation of the water table at Market Deeping (Lincs), with the rainfall superimposed. It can be seen from this that an evaluation carried out in December would suggest a good waterlogged environment at just over one metre below the ground surface, but in summer the water table falls to 1.5 to 1.6 m for several months and preservation above this level is likely to be poor (Corfield 1998).

The annual cycle of the water table from high in winter to low in summer is a result of losses due to transpiration, an effect that was first noted by Godwin at Wicken Fen (Godwin 1931). It is also important to understand that the water table is not level like surface water, but that it has relief depending on the nature of the soil and piezometric effects. This was graphically demonstrated by Cheetham (2004) at Sutton Common (South Yorkshire) in a monitoring scheme that comprised a 50 m grid of 50 piezometers across the site (Chapman and Cheetham 2002) (Fig. 13.2). Critically this method of monitoring showed that Parker Pearson's earlier test pits had been located at a point where the water table was at its lowest (Parker Pearson & Merrony 1993) and that conversely, the area of greatest interest was where the water table was at its highest. Fig. 13.3 shows two graphic representations of the water table in April and June 2001; it can be seen that the water table is higher in the area of the large enclosure, and that there has been a rapid drop in the water table between the two readings.

In extreme cases water losses from archaeological sites resulting from activities elsewhere can be dramatic, and the cause of de-watering can originate at a considerable distance from the site concerned. At Etton (Northants) during excavation in advance of gravel extraction, the water table fell below the level of the preserved waterlogged deposits in the ring ditches when the water table was impacted upon *c.* 1.5 km

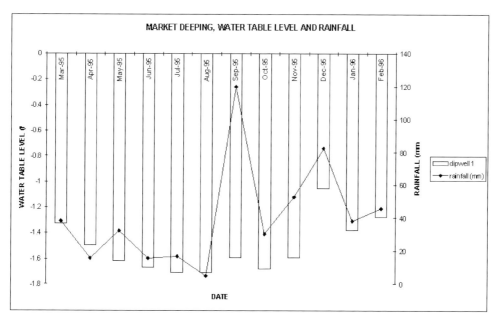

Fig. 13.1 Rainfall and water levels at Market Deeping (Lincs), Mar 1995 – Feb 1996, from Corfield 1996.

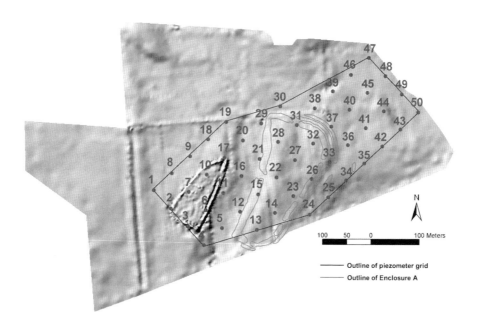

Fig. 13.2 Digital elevation model of Sutton Common showing the piezometric grid. From Cheetham 2004.

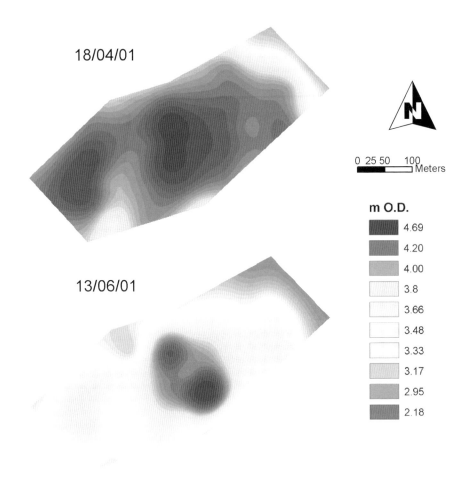

Fig. 13.3 Hydrological model of Sutton Common (S. Yorks) showing the rapid fall of the water table between April and June 2001, from Cheetham 2004 Figs 5.21 A & C.

from the site (French and Taylor 1995). Coles (1995) reported how changes in the water balance resulting from the damming of the Zuider Zee in the Netherlands to create first a freshwater lake and later a polder, compromised an important wetland at De Weeribben in the Overijssel National Park. Prior to the reclamation in 1931 there was a water surplus of 610 mm, with an annual contribution by upwelling from the Zuider Zee of 360 mm. Today, there is a discharge from De Weeribben to the polder that results in a deficit of 250 mm a year, requiring positive management to sustain the wetland.

HYDRAULIC CONDUCTIVITY

Water retention in wetlands will be crucially dependent on the hydraulic conductivity of the soil which is the capacity of the soil to allow the passage of water through it. It is defined by Darcey's Law, which is variously described:

Mitsch and Gosselink (2000) give it as $G = kA_x s$

When G is the flow rate of groundwater (volume per unit time)
k is the hydraulic conductivity
A_x is the groundwater cross-sectional area perpendicular to the direction of flow
s is the hydraulic gradient, or the slope of the water table or piezometric surface.

Eggelsmann *et al.* (1993) give it as $Q = k(\delta h / \delta l)$, where
Q is the discharge across a cross sectional area,
k is the hydraulic conductivity,
h is the hydraulic head,
l is the length of the flow line,
And $\delta h / \delta l$ is the hydraulic gradient.

Both equations are basically the same in that the hydraulic conductivity is the product of the cross section through which the water flow is passing, the gradient of the flow, and the volume of water passing in a given time.

The hydraulic conductivity of the soil provides a measure of the capacity of the soil to retain water, and an awareness of it will help in deciding whether an archaeological site in a wetland can be managed sustainably. The hydraulic conductivity of open soils such as sand will be high, grading through loams and peaty soils to clays that are the least permeable. From the equations above it will be seen that the hydraulic conductivity will also be greater where there is a large head of water and where the gradient is steeper. The hydraulic conductivity of peat soils is complicated by their variability, with low permeability through sapric or well decomposed peats through hemic peats to poorly decomposed fibric peats that can be a thousand times more permeable (Mitsch and Gosselink 2000).

Hydraulic conductivity is a valuable tool for assessing the potential of archaeological sites to preserve wetland evidence, not only will a low hydraulic conductivity give greater water retention, but it will reduce the influx of oxygen rich water to the system. Van de Noort and Davis (1993) used hydraulic conductivity as one of their parameters in the Humber Wetlands archaeological assessment, and in this respect they were the only one of the four English Heritage wetland projects to attempt to provide a measure of scientific evaluation of the potential of the sites they identified to be managed as a wetland resource.

Hydraulic conductivity is also a component of the functional assessment procedures for wetlands, being a generic process used to evaluate the ability of wetlands to support a variety of functions. The basic unit of the functional assessment procedure is the hydrogeomorphic unit or HGMU, which as the name implies is a unit of the landscape having similar hydrology, topography and soil type (Maltby *et al.* 1994). Functional assessment procedures specifically designed for assessing the potential of wetlands to sustain archaeological evidence are being developed (*e.g.* Hogan *et al.* 2002).

THE CHEMISTRY OF WETLANDS

Wetland chemistry is highly complex, being influenced by the inorganic chemistry deriving from the geology and the inflowing water, and the organic chemistry of the peats and the processes of their degradation. There are however two functions that are of considerable significance: pH and redox potential.

pH

pH is a measure of the hydrogen ion concentration of a substance or system. It provides an indication of the degree of acidity or alkalinity, and has been used extensively to explain the quality of preservation of archaeological materials in different soil environments. Fig. 13.4 is from Darvill (1987) as revised by Van de Noort and Davies (1993). It can be seen from this representation that most materials are better preserved in acid environments, but that calcareous based materials require an alkaline environment for their survival. In general upland and ombrigenous bogs in which the water supply is from rainfall rather than from groundwater inputs will be acidic, in the region of pH 4, since their chemistry will be governed primarily by the degradation of plant materials. Lowland fens that receive the bulk of their water from ground and surface water will be closer to neutral (pH 7). This goes some way to explain why the soft tissues of bog bodies such as the Lindow Man, which come from ombrigenous mosses are well preserved, while the bones are not (Stead *et al.* 1986), and why conversely, in the non-acidic wetlands of the Witham Valley, bone implements are well preserved.

Redox potential (Eh)

pH alone gives only a part of the picture. The redox potential is a measure of electron availability in the soil, and hence its oxidation or reduction characteristics. Reduction is said to take place when electrons are gained by ions, giving them a negative charge and oxidation when electrons are lost giving a positive charge. Although the term is derived from reduction and oxidation there is no actual need for oxygen to be involved in the reactions. Oxidised soils are rich in iron III (Fe^{3+}) oxides that give them their colours in the red to yellow range, and oxidised manganese III and IV (Mn^{3+} and Mn^{4+}) minerals that give black colours. In reduced conditions these minerals are reduced to their lower valencies (Fe^{2+} and Mn^{2+}) which are colourless. During periods of temporary dewatering the soluble minerals are oxidised and deposited as gleys

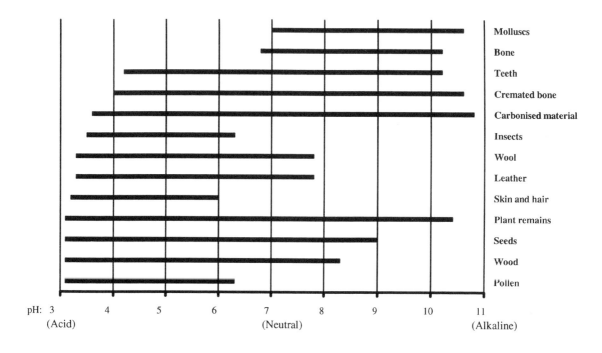

Fig. 13.4 pH and preservation status (after Van de Noort and Davies 1993).

or mottles in the soil. Permanently reduced soils have few of these coloured minerals and are termed as being redox depleted (Vepraskas 1995).

The opening up of a reduced deposit provides a high input of oxygen and precipitation of soluble ferrous iron; at Willingham (Cambs) the initiation of gravel extraction instantly turned the water in the drainage ditches red and the particles of iron oxides stained the feathers of swans (French 2003). The chemistry of these redoximorphic changes is further discussed by Hayes and Vepraskas (2000) in which they describe the changes in soil morphology resulting from ditching and relate it to hydrology and redox dynamics. Basic reduction processes of relevance to archaeological preservation, all at pH 7 can be summarised as follows (*after* Cook 1999):

+400 mV, Manganese IV reduction
$$2MnO_2 + CH_2O + 4H_3O^+ \rightarrow CO_2 + 2Mn^{2+} + 7H_2O$$

about +300 mV: denitrification
$$NO_3^- + CH_2O + H_3O^+ \rightarrow CO_2 + \frac{1}{2}(N_2O) + 2(\frac{1}{2}H_2O)$$

about –200mV: Iron Reduction
$$4Fe(OH)_3 + CH_2O \rightarrow Fe(OH)_2 + CO_2 + 3H_2O$$

about –220mV: Sulphate reduction
$$SO_4^{2-} + 2CH_2O + 2H_3O^+ \rightarrow H_2S + 2CO_2 + 4H_2O$$

≤ -260 mV: methanogenesis
$$2CH_2O \rightarrow CO_2 + CH_4$$

CH_2O is a representation of organic matter.

The reduction processes above are of course reversible should the conditions change. In these equations Cook (1999) indicates the close relationship between pH and redox, a fact often forgotten when constructing preservation charts such as that of Darvill (1987). It is of particular relevance in anaerobic waterlogged environments, and a useful Eh/pH representation of the preservation of environmental evidence has been developed by Retallack (1984) and reproduced in modified form in the English Heritage Guidelines for Environmental Archaeology (English Heritage 2002, Fig. 13.5 below). Caple and Dungworth (1997, 1998) demonstrated the value of Eh/pH diagrams to plot monitoring results; by incorporating the boundaries for the boundaries of water oxidation and reduction it provides a clearer picture of the zones within which there is a likelihood of preservation of organic based materials. Cheetham (2004) has refined this by adding in the boundaries for FeII and FeIII, and the boundaries between sulphate and sulphide (Fig. 13.6). Moderately reduced conditions can be expected below the FeII/FeIII line, while below the sulphate sulphide line conditions will be strongly reducing since this reaction is mediated by anaerobic bacterial activity.

The changes to iron minerals are of particular relevance to archaeologists as they will encounter not only natural iron minerals but also corroded iron artefacts that react in similar ways. The corrosion of metals is usually expressed in Pourbaix diagrams that set out the zones on an Eh/pH chart in which the products of the metal and other reactants will lie. A generic Pourbaix diagram showing the conditions for immunity after general corrosion, and for passivity in the presence and absence of oxygen, is given in Pourbaix (1977). Turgoose (1984) provides a diagram for iron/water/chloride, from which it can be inferred that in a strongly reducing environment iron will be relatively uncorroded, and that in a moderately reducing environment (Eh +200 to –200 mV) magnetite (Fe_3O_4) will be the expected corrosion product.

In the mineral form reduced iron species are mobile in low Eh environments, but as noted above, they are readily precipitated by small changes to the Eh/pH balance for example, at pH 5 when the Eh is greater

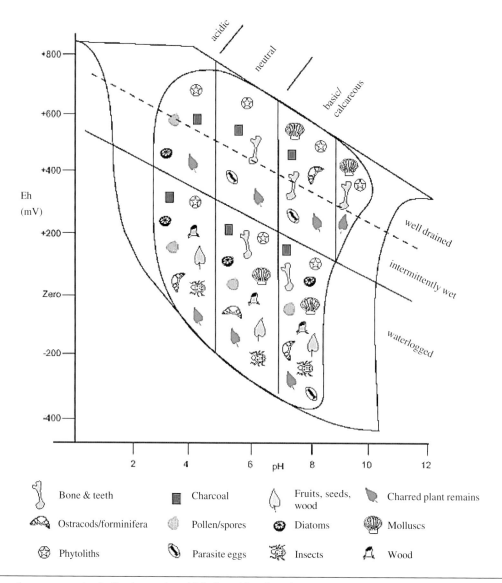

Schematic representation indicating under which depositional environments specific categories of environmental remain can be expected to survive and hence be recovered using appropriate sampling techniques.

Filled area = envelope into which most naturally derived sediments fit. Material outside these limits tends to reflect human activity, e.g. basic slag and other industrial deposits.

Modified from Retallack, 1984.

Fig. 13.5 The combined effects of pH and redox potential on the preservation of archaeological evidence, from English Heritage Centre for Archaeology Guidelines for Environmental Archaeology (after Retallack 1984).

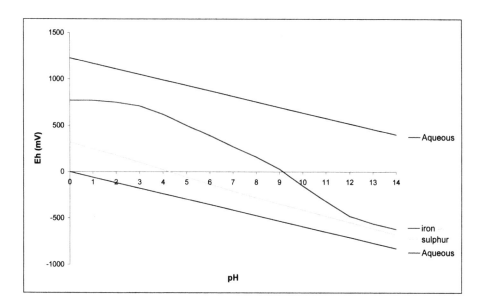

Fig. 13.6 Diagram indicating the boundaries for preservation of organic materials in anaerobic environments (after Cheetham 2004).

than +300 mV or at pH 8 when the Eh is greater than −100 mV. Aged ferric hydrites weather to give limonite ($Fe_2O_3nH_2O$), that in the presence of carbon dioxide may be solubilised to iron 2 bicarbonate ($2Fe(HCO_3)_2$), or may be precipitated as goethite (α FeOOH):

$$2(Fe(HCO_3)_2) + \tfrac{1}{2} O_2 \leftrightarrows 2FeOOH + 4CO_2 = H_2O$$

Limonite has a great affinity for phosphate with which it reacts to form vivianite $Fe_3(PO_4)_2.8H_2O$, the purple blue corrosion seen on archaeological iron which has been buried in, for example, human and animal waste pits. Vivianite corrosion passivates the iron, preventing or limiting further corrosion. However, when excavated iron with this protection dries the corrosion film cracks and makes the iron even more susceptible to corrosion. Limonites, converted by Sulphur compound, are common in fen wetlands, and deposits of hydrogen sulphide (H_2S) and gypsum ($CaSO_4.2H_2O$) are found in the Somerset Levels. With reduced iron, sulphur compounds react to give pyrites (FeS_2), Fell & Ward (1998) used the presence of iron from non saturated soils to demonstrate their previously waterlogged condition, Schweitzer (1993) undertook similar studies on metalwork from the Swiss Lake Villages. Sulphur compounds may also be derived from bacterial activity (see below) when they are found as so called framboidal pyrites in, for example, plant remains and waterlogged wood (Wiltshire *et al.* 1994).

MICROBIOLOGY

Bacteria and actinomycetes, which share characteristics of bacteria and fungi, are the most abundant species in the soil; fungi need aerobic environments and so are only found in the oxygenated levels of wetlands. There are over 400 bacterial genera and estimated to be 10^4 species, but these are likely to be

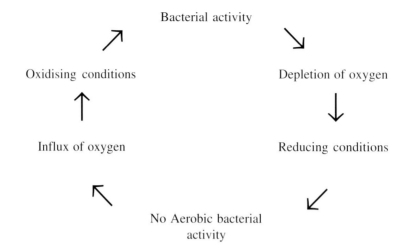

Bacterial activity

Oxidising conditions   Depletion of oxygen

Influx of oxygen   Reducing conditions

No Aerobic bacterial
activity

Fig. 13.7 Diagram to show the relationship between bacteria and redox in wetlands. Aerobic bacteria will function until the available oxygen is depleted, in the reducing conditions there will be anaerobic bacterial activity, but its impact on organic materials will be much less than the aerobic bacteria. Should there be an influx of oxygen oxydising conditions will re-establish and the dormant aerobic bacteria will become active again.

outnumbered by yet to be identified species (Powell 1999). Bacteria are a controlling factor for many of the chemical processes in the soil generally, and in wetlands in particular (see Fig. 13.7).

Broadly, bacteria can be divided into aerobes that require oxygen for their metabolism, and anaerobes that do not; anaerobes can be divided into facultative and obligate anaerobes. Facultative anaerobes can grow either in the presence of oxygen, or without it while obligate anaerobes can only grow in the absence of oxygen. Facultative anaerobes are responsible for depleting wetlands of oxygen, reducing the redox potential and creating the anaerobic environments that enables preservation of organic wetland archaeological evidence. Actinomycetes are important in the breakdown and fermentation of plant remains and the formation of humic acid.

Bacteria mediate important transformations in wetlands, including nitrogen exchange. Ammonia and ammonium compounds are converted to nitrites by bacteria such as *nitrosomonas* sp, and nitrites to nitrates by *nitrobacter* sp, in which form it can be fixed by bacteria in the roots of some wetland plants. Both *Nitrosomonas* and *Nitrobacter* are aerobic and rarely found in anaerobic wetlands, a more common route for nitrate removal is by anaerobic denitrifying bacteria such as *Pseudomonas denitrificans* or *Thiobaccilus denitrificans*that which use nitrates in the oxidation of organic matter, breaking down the nitrates to nitrogen:

$$C_6H_{12}O_6 + 4NO_3^- \rightarrow 6CO_2 + 6H_2O + 2N_2$$

Sulphates released by mineralisation are reduced by bacteria, *Desulphovibrio desulfuricans,* an anaerobic species that is responsible for the formation of hydrogen sulphide. *Thiobacilli thiooxidans* and *ferrooxidans*

System	Redox Potential Range (mV) corrected to pH 7	Microbiology
Oxygen disappearance	+500 to +350	Aerobes
Nitrate disappearance	+350 to +100	Facultative anaerobes
Mn^{2+} formation	Below +400	Facultative anaerobes
Fe^{2+} formation	Below +400	Facultative anaerobes
Sulphide formation	0 to –150	Obligate anaerobes
Hydrogen, methane formation	Below –150	Obligate anaerobes

Table 13.1 Chemistry and microbiology, (Powell 1999, after Sikora and Keeney 1983).

oxidise hydrogen sulphate back to sulphate, while *Thiobacilli thioparus* oxidises it to sulphur. An important reaction is the conversion of sulphate to the amino acids such as cystine and methionine by *Clostridium nigrificans* (Caple and Dungworth 1998).

The role of bacteria in the degradation of archaeological evidence has not generally been studied extensively; an exception is wood, for which there has been a great deal of research that has been reported in the triennial congresses of the International Council of Museums Conservation Committee (ICOM - CC) and the proceedings of its Waterlogged Organic Archaeological Materials Group (WOAM). The proceedings of the triennial meetings of the group include substantial and important reports on wood degradation. More recently the volume by Hoffmann *et al.* (2002) included papers by Björdal and Nilsson on the decomposition of waterlogged wood, and by Helms and Kilstrup on the DNA identification of bacteria in the Nydam bog. Both papers are significant as it is notoriously difficult to identify bacteria to species level, and the use of DNA by the scientists from the National Museum of Denmark is a first for archaeology.

Wood apart, there have been studies of the biological degradation of leather, textiles (eg Peacock 1996), and other artefactual materials which are also reported in the WOAM Series; waterlogged hazelnuts are a rare example of non wood plant material being studied (Pournou 2002).

This and other environmental evidence that can be used to characterise the preservation status of a burial environment; for instance, pollen, as noted above, was used by the Humber Wetlands Project, but the diversity of composition of biological micro and macro material makes it particularly suitable to determine the status of wetland sites. Moreover, as indicators of past environmental changes, diatoms, ostracods and foramenifera can help in the reconstruction of the formation processes and subsequent changes to the wetland. The condition of pollen, along with plant macros and artefactual wood, as has been already observed, can be used as an indicator of the current status of the site.

ACKNOWLEDGEMENTS

I must thank James Cheetham for permission to use the topographical diagrams of the Sutton common water table, and to use his expanded Eh/pH diagram. Jacqui Huntley provided me with the diagram from the English Heritage Environmental Guidelines.

REFERENCES

Bjordal, C. & T. Nilsson. 2002. Decomposition of Archaeological Wood, in P. Hoffmann, J. A. Spriggs, T. Grant, C. Cook & A. Recht (eds) *Proceedings of the 8th ICOM Group on Wet Organic Archaeological Materials Conference*, 235–244. Stockholm 2001. Germany: Bremerhaven.

Caple, C. 2001. Degradation, Investigation and Preservation of Archaeological Evidence in D. R. Brothwell & A. M. Pollard (eds) *Handbook of Archaeological Sciences*, 587–593. Chichester: John Wiley & Sons.

Caple, C. & D. Dungworth. 1997. Investigations into waterlogged burial environments, in A. Sinclair, E. Slater & J. Gowlett. (eds) *Archaeological Sciences: 1995*, 233–40. Oxford: Oxbow.

Caple C. & D. Dungworth. 1998. *Waterlogged Anoxic Archaeological Burial Environments*, English Heritage, London. (Ancient Monuments Laboratory Report 22/98).

Chapman, H. P. & J. L. Cheetham 2002. Monitoring and modelling saturation as a proxy indicator for *in situ* preservation in wetlands: a GIS-based approach. *Journal of Archaeological Science* 29, 277–89.

Cheetham, J. L. 2004. An assessment of the potential for in situ preservation of buried archaeological remains at Sutton common, South Yorkshire. Unpublished PhD Thesis, University of Hull.

Coles, B. 1995. *Wetland Management: a Survey for English Heritage*. Exeter: WARP Occasional Paper 9.

Cook H. 1999. Soil and Water Management: Principles and Purposes, in H. Cook & T. Williamson (eds) *Water Management in the English Landscape, field marsh and meadow,* 15–27. Edinburgh: University Press.

Corfield, M. 1998. The Role of Monitoring in the Assessment and Management of Archaeological Sites, in K. Bernick (ed.) *Hidden Dimensions: The Cultural Significance of Wetland Archaeology,* 302–318. Vancouver: UBC Press.

Darvill, T. 1987. *Ancient Monuments in the Countryside: an archaeological management review.* London: English Heritage Archaeological Report no 5.

Eggelsmann, R., A. L. Heathwaite, G. Grosse-Braukmann, E. Künster, W. Naucke, M. Schuch & V. Schweickle. 1993. Physical Processes and Properties of Mires, in A. L. Heathwaite & Kh Göttlich (eds) *Mires: Process, Exploitation and Conservation,* 171–262. New York: John Wiley and Sons Inc.

English Heritage. 2002. *Environmental Archaeology: a guide to the theory and practice of methods, from sampling and recovery to post excavation,* London: English Heritage.

Fell, V. & M. Ward. 1998. Iron Sulphides: Corrosion Products on artefacts from waterlogged deposits, in W. Mourey & L. Robbiola (eds) *Metals 98, Proceedings of the international conference on metals conservation,* 111–115. London: James and James Scientific.

French, C. 2003. *Geoarchaeology in action: studies in soil micromorphology and landscape evolution,* 163–169. London: Routledge.

French, C. & M. Taylor. 1995. Dessiccation and Destruction: the Immediate Effect of Dewatering at Etton, *Oxford Journal of Archaeology* 4 (2), 139–156.

Gilman, K. 1994. Water Balance of Wetland Areas, Presented at Conference on "The balance of water – present and future", AGMET Gp. (Ireland) & Agric Gp. Of Roy. Meteorol. Soc. (UK), Dublin, 7–9 Sep 1994, 123–142.

Godwin, H. 1931. Studies of the Ecology of Wicken Fen 1: The Groundwater Levels of the Fen. *Journal of Ecology* 19, 139–156.

Hayes, W. A., & M. J. Vepraskas 2000. Morphological Changes in Soils Produced when Hydrology is Altered by Ditching. S*oil Science Society of America Journal,* 64, 1893–1904.

Helms, A. C. & M. Kilstrup 2002. DNA Based Identification of Bacteria inhabiting waterlogged wooden artefacts from the Nydam Bog, in P. Hoffmann, J. A. Spriggs, T. Grant, C. Cook, & A. Recht (eds) *Proceedings of the 8th ICOM Group on Wet Organic Archaeological Materials Conference,* 249–258. Germany: Bremhaven.

Hoffmann, P., J. A. Spriggs, T. Grant, C. Cook, & A. Recht (eds) 2002. *Proceedings of the 8th ICOM Group on Wet Organic Archaeological Materials Conference, Stockholm 2001.* Germany: Bremerhaven.

Hogan, D. V., P. Simpson, A. M. Jones & E. Maltby 2002. Development of a protocol for the reburial of organic archaeological remains, in P. Hoffmann, J. A. Spriggs, T. Grant, C. Cook, & A. Recht (eds) *Proceedings of the 8th ICOM Group on Wet Organic Archaeological Materials Conference,* 187–212. Germany: Bremhaven.

Hulme M., G. J. Jenkins, X. Lu, J. R. Turnpenny, T. D. Mitchell, R. G. Jones, J. Lowe, J. M. Murphy, D. Hassell,

P. Boorman, R. McDonald & S. Hill 2002. *Climate Change Scenarios for the United Kingdom: the UKCIP02 Report,* Norwich: Tyndall Centre for Climate Change Research.

Maltby, E., D. V. Hogan, C. P. Immirzi, J. H. Tellman & M. J. Van der Peijl 1994. Building a New Approach to the Investigation of Wetland Ecosystem Functioning, in W. J. Mitsch (ed) *Global Wetlands Old and New*, 637–658, London: Elsevier Science.

Mitsch, W. J. and J. G. Gosselink 2000. *Wetlands*, New York: John Wiley & Sons.

Parker Pearson, M. & C. Merrony 1993. Sutton Common: desiccation assessment 1993 – interim report. Sheffield: Department of Archaeology and Prehistory, University of Sheffield, South Yorkshire Archaeological Services, and Doncaster Museum.

Peacock, E. E. 1996. Biodegradation and characterization of water-degraded archaeological textiles for conservation research. *Biodeterioration and Biodegradation* 38(1), 49–60.

Pourbaix, M. 1977. Electrochemical Corrosion and Reduction, in Brown, B.F. (ed.) *Corrosion and Metal Artefacts – a Dialogue Between Conservators and Archaeologists and Corrosion Scientists,* 1–16. Washington D.C.: National Bureau of Standards special publication 479.

Pournou, A. 2002. A comparative Study of Conservation Methods for Waterlogged Hazelnuts, in P. Hoffmann, J. A. Spriggs, T. Grant, C. Cook, & A. Recht (eds) (2002), *Proceedings of the 8th ICOM Group on Wet Organic Archaeological Materials Conference*, 493–499. Germany: Bremerhaven.

Powell, K. L. 1999. The impact of recent changes within the subsurface environment upon the integrity of buried wood: implications for the in situ preservation of archaeological timbers Unpublished PhD thesis, The Robens Centre for Public and Environmental Health, University of Surrey

Retallack, G. J. 1984. Completeness of the rock and fossil record: some estimates using fossil soils. *Palaeobiology* 10, 59–78.

Schweizer, F. 1993. Bronze objects from lake sites: from patina to 'biography'. *Ancient and historic metals: conservation and scientific research: proceedings of a symposium organized by the J. Paul Getty Museum and the Getty Conservation Institute, November 1991. 1993*, 33–50.

Sikora, L. J. and D. R. Keeney 1983. Further aspects of soil chemistry under anaerobic conditions, in A. J. P. Gore (ed) *Ecosystems of the World 4A, Mires: Swamp, Bog, Fen and Moor.* London: Elsevier Scientific.

Smith, R. J. 2005. *The Preservation and Degradation of Wood in Wetland Archaeological and Landfill Sites.* Unpublished PhD thesis: University of Hull.

Stead, I. M., J. B. Bourke & D. Brothwell 1986. *Lindow Man: the body in the bog.* London: British Museum.

Vepraskas, M. J. 1995.) Redoximorphic Features for Identifying Aquic Conditions. Unpublisjhed Technical Bulletin 301, North Carolina State University, Raleigh, NC.

Van de Noort, R. & P. Davies 1993. *Wetlands Heritage: An Archaeological Assessment of the Humber Wetlands,* Humber Wetlands Project, University of Hull.

Wiltshire, P. E. J., K. J. Edwards & S. Bond 1994. Microbially derived metallic sulphide spherules, pollen and the waterlogging of archaeological sites, in *American Association of Stratigraphic Palynology Contribution Series* No. 29, 207–221.

Chapter 14

In situ Preservation: Geo-Archaeological Perspectives on an Archaeological Nirvana

Malcolm Lillie

INTRODUCTION

The ever present threat to waterlogged archaeological sites has been highlighted by numerous studies. These include those of the major wetlands projects that have been funded by English Heritage in the Somerset Levels, The Fens, The North West Wetlands and the Humber Wetlands (*e.g.* Coles and Coles 1986, 1989, Hall and Coles 1994, North West Wetlands Survey 1990–3, Van de Noort and Davies 1993), amongst others.

Coles (1995a,145) has noted that 'once exposed, preservation *in situ* is extremely difficult, if not impossible'. This observation reflects the fact that the waterlogged burial environment is extremely sensitive to any shift in the fragile balance that results in a 'stable', preserving, context. Numerous threats to this resource have been shown to exist, from the more obvious threats of aggregates extraction (whether minerals or peats) and coastal erosion, to less 'visible' and anthropogenically induced impacts. The latter include de-watering for agricultural and domestic purposes through abstraction, aquifer draw-down and a general reduction in water tables, as highlighted by Van de Noort and Davies for the Humberhead Levels, England (1993, 109–10). Indeed, the removal of saturation as a result of lowering of water tables has been inferred as one of the major causes of degradation of the waterlogged resource (*e.g.* Chapman and Cheetham 2002), and has been intimately associated with aggregates extraction (French 2003). The waterlogged resource is significant not only for its exceptional preservation of the organic part of the archaeological record, but also for the palaeoenvironmental archive contained within the matrix surrounding the archaeological remains (*e.g.* Coles and Coles 1986).

In the present study an overview of recent work at the Knights Hospitaller's Preceptory, Beverley, East Yorkshire is presented, and comparison is made between this and two other key waterlogged sites (Fig. 14.1). These sites are Sutton Common, which is located in the Humber Wetlands region as defined by the Humber Wetland Survey (Van de Noort and Davies 1993) and Flag Fen, located in the Cambridgeshire Fens (Pryor 1992). The sites comprise differing landscape contexts, with waterlogged archaeological remains of differing character and date making up the archaeological record. The sedimentological characteristics and general burial conditions of these sites are considered in relation to the potential for the preservation of the contained archaeological remains *in situ*.

MEASURING *IN SITU* PRESERVATION

Despite nearly two decades of concerted effort in the management of wetland archaeological sites (*cf.* Coles 1995a, 147), it is apparent that no clear consensus has been achieved with regard to the ultimate goal of preservation *in situ*. Despite this, a range of fundamental parameters have been suggested as being necessary in achieving this aim.

Corfield (1996, 32) noted that: 'effective preservation *in situ* requires a thorough understanding of the factors that might bring about the deterioration of the archaeological evidence'. Whilst it is generally accepted that preservation of an archaeological site relies on a thorough understanding of the soils, hydraulic conductivity and site specific and landscape wide hydrological regimes, such an holistic understanding requires multi-disciplinary studies of a broad range of parameters. In addition to these, others such as redox potential, pH and microbial activity need to be assessed (*cf.* Caple and Dungworth 1997), and Welch and Thomas (1996, 16) have observed that 'good preservation of organic material depends on the maintenance of a stable chemical environment [which is] characterised by anoxic and reducing conditions of near neutral pH'.

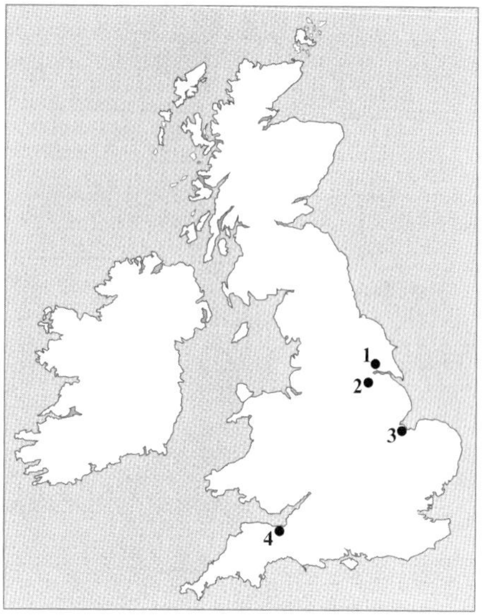

Fig. 14.1 Location of key sites discussed in the text 1 = Beverley, 2 = Sutton Common, 3 = Flag Fen, 4 = Sweet track, Somerset levels.

The above parameters indicate the sort of data that needs to be generated if we are to approach the 'ideal' of *in situ* preservation. However, it should be noted that, *contra* Van de Noort *et al.* (2001a, 96) 'the creation and maintenance of a static environment that is sustainable in the long term' is unachievable as stasis does not exist in such waterlogged contexts, although the vectors of degradation can indeed be inhibited.

Fundamentally, the principal factor influencing preservation within wet environments is considered to be the presence of saturation, which reduces the levels of oxygen accessing sediments, thereby restricting the activities of aerobic bacteria (see Corfield 1996, Caple 1996, Chapman and Cheetham 2002). This element has been a principal factor in the development of management strategies for preserving sites in both the UK and abroad (*cf.* Coles 1995b).

Unfortunately, the measurement of water tables is not as straightforward as it at first appears, as it is possible to have more than one level of water table, due to the specific characteristics of the burial environment, and groundwater flows are influenced by head potential and the permeability of sediments (Welch and Thomas 1996, 17–18). Due to these dynamic characteristics, recent archaeological studies of groundwater character have employed piezometers in order to obtain insights into the overall shape of the water table and highlight any complicating factors occurring at different stratigraphic levels in the site and its immediate environs (*cf.* Welch and Thomas 1996, Chapman and Cheetham 2002).

It has been suggested that redox potential and pH are the most useful measures in characterising an anoxic environment, with pH being directly related to the location of the site (Corfield 1996, 35). In terms of the redox potential of the burial environment, recent research by Caple and Dungworth (1997) has suggested that in order to maintain optimum burial conditions, redox levels need to be maintained at between -100 and -400 mV. Corfield (1996, 35) reported that redox potential and pH were the most useful measures in characterising an anoxic environment, although levels of +200 to -400mV were quoted for anaerobic sediments. Despite these differing redox level recommendations, redox potential provides a quantifiable measure of the rate of flow of electrons in the burial environment, thereby enabling the generation of 'boundary conditions' which are useful in distinguishing anoxic from aerobic environments (*cf.* Caple and Dungworth 1997, 234).

The values obtained from redox probes *in situ* require adjustment in order to enable direct comparison between different locations and data sets. Adjustment of recorded values to the Standard Hydrogen Electrode (SHE) is undertaken by convention. Although there are many suitable electrodes that can be used in obtaining redox and other data, in order to reduce ambiguity a single reference electrode has been chosen for reporting electrode potentials from the sites studied below (Compton and Sanders 1996, Cheetham 2004). Consequently, the monitoring of redox potential outlined here has been carried out using a silver/ silver chloride reference cell, which requires an adjustment of +222 mV (*cf.* Stumm and Morgan, 1981) from the raw value obtained after direct measurement of the redox probe.

When considering the potential ranges from which conclusions can be drawn relating to the burial environment, the work of Patrick and Mahaptra (1968) is informative. The following figures are used to denote soil status:

Oxidised soils	=	>+400 mV
Moderately reduced	=	+200 to +400 mV
Reduced	=	-100 to +200 mV
Highly reduced	=	-300 to -100 mV

In attempting to develop a more holistic understanding of archaeological burial environments, the departments of Geography and Biology at the University of Hull have recently had three doctoral students undertaking studies of these environments. The research has included investigation of site hydrology (Cheetham 2004), the decay of woody tissue in differing burial environments (Smith 2005) and soil microbiology (on-going at the time of writing), with a combination of parameters being used to characterise the burial environments. These studies have been undertaken under the supervision of the author and Dr Steve Ellis (Geography) and Dr Ray Goulder (Biology). Funding for each of these students was obtained from a number of sources, including English Heritage, the Countryside Agency and the University of Hull.

For the purposes of this study, the following discussion outlines the results of monitoring at one specific site and contrasts this against others that have a proven potential to preserve waterlogged deposits, using techniques that have been developed and/or applied at varying levels of resolution, since 1999.

CASE STUDY – KNIGHTS HOSPITALLER'S PRECEPTORY, BEVERLEY

The case study site considered here is that of the Medieval moated site of the Knights Hospitaller's Preceptory of Holy Trinity (inner trinities), Beverley, East Yorkshire (NGR TA 03853967). This site was originally investigated as part of a mitigation strategy (Lillie and Cheetham 2002a and b), and regularly monitored by the author between 10th September 2001 and 12th August 2002, with intermittent follow-up visits occurring until 5th January 2003. Analysis of hydrology, redox potential and pH was undertaken at two locations during the monitoring programme. This study was undertaken as a condition of the scheduled monument consent for the construction of a car park, and was initiated in order to assess any impacts that the development might have on the contained ditch deposits and associated hydrology of the moated site (Lillie and Cheetham 2002a and b).

According to Evans (1997, 67), the site was known locally as the Trinities, and consisted of an inner (moated) site and a larger enclosure known as the Outer Trinities. At present the Inner Trinities appears to occupy a relatively level terrace (Evans 1997), located on the extensive till deposits that flank the eastern edge of the Yorkshire Wolds, England (BGS Sheet 72: Beverley [Solid and Drift edition]). The tills of this region are characterised as being impermeable, compact fine clays and silts, with local variations. In effect, where the moat ditch is excavated through this material, the sediments within the ditch are relatively isolated from adjacent groundwater tables, thereby presenting an 'ideal' context for controlled *in situ* preservation. Previous archaeological investigations at this site had indicated that the archaeology within the ditch was recorded at heights of between *c.* 7.0 m OD and 4.75 m OD (York Archaeological Trust 1991, 5).

Due to the nature of the stratigraphy encountered within the ditches, and the limited potential for studying the sites hydrology within the confines of the development, two monitoring locations were established in the western ditch (Fig. 14.2). The ditch sequences consisted of brick and rubble fill in the upper *c.* 1.3m of the profile, with clays and organic-rich silts with occasional sandstones gravel inclusions (< 30mm diameter) to the base. At the monitoring locations monitoring point 1 comprised five piezometers, which were sunk to depths of 3.0 m, 2.17 m (due to presence of stone slabs at depth), 1.75 m, 1.13 m, and 0.5 m depth respectively, and a dip well. At monitoring point 2 the equipment installed comprised a cluster of five piezometers which were sunk to depths of 3.0 m, 2.28 m, 1.75 m, 1.13 m, and 0.5 m respectively, and a dip well, which were located adjacent to a cluster of twelve redox probes. The redox probes comprise four 3 mm gauge insulated copper wire probes with platinum tips inserted to 0.5 m depth, four probes inserted to 1.5 m depth and four probes to 3.0 m depth. At both monitoring locations the equipment was housed in pre-formed concrete sectional manholes which facilitate ease of access and serve to protect the equipment from the elements and any inquisitive members of the general public.

The monitoring at Beverley was undertaken over a *c.* 17 month period, thereby ensuring that any seasonal variability was accounted for during this study. The deepest piezometers at both monitoring points in the ditch sequences exhibited a slow rate of recharge, with both locations com-

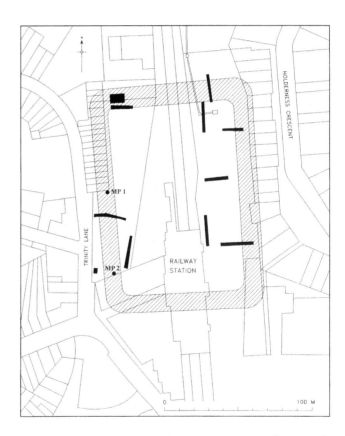

Fig. 14.2 The location of the Inner Trinities showing the position of the moat shown in tone) and evaluation trenches (shown in black) (after Evans 1997: 68), and the location of the monitoring points used in the current study (MP 1 and MP 2) 14.3a.

pleting the recharge on 19.03.02, suggesting that the recharge rates reflect the situation prevailing within the sediments of the ditch on its western side. A series of peaks in the data (Figs 14.3a and b) probably reflect individual rainfall events, or a combination of these, over a restricted time period. These data will be compared against precipitation data after collation of the results obtained during the follow-up monitoring visits undertaken between December 2003 and December 2004.

When studying the results of the water table monitoring (Figs 14.3a and b), it is apparent that the deeper piezometers, number 1 at monitoring point 1 and piezometers 1 and 2 at monitoring point 2, do not mirror the water table activity evidenced by the uppermost piezometers. This suggests that whilst the uppermost piezometers may be reflecting short term fluctuations, as noted above, the water tables indicated by the lower piezometers remain isolated from these influences. As such, the water table activity reflected by the lower piezometers should be reflecting water table trends within the ditch, trends that are effectively isolated from surface water inputs.

It is clear from the above observations that the low hydraulic conductivity of these deposits indicates that there is a very good potential for the long-term *in situ* preservation of the organic archaeological remains that are still preserved within the ditch feature. The heights of the water tables within the ditch at both monitoring locations suggest saturation occurring between 6.61 and 6.88 m OD at monitoring point 1 and 7.40–8.00 m OD at monitoring point 2. Given that the archaeology within the ditch sequences is recorded at between 7.0 and 4.75 m OD it is apparent that there is a high water table occurring in relation to the organic-rich silts in the ditches at both of these locations.

The redox monitoring at the southern end of the ditch (Fig. 14.4a) has shown that between 11.12.01 and 05.01.02, a consistent set of readings between *c.* -80 to -125 mV were recorded for cluster 1 at 1.5 m depth, and *c.* 0 to -200 mV for cluster 2, at 3.0 m depth. Whilst the latter readings were generally more positive than those in evidence at cluster 1, they were within the limits of reducing conditions as outlined by Caple and Dungworth (1997, 235), but they were in fact close to the sulphur/sulphide boundary. The readings obtained occasionally reach levels that would suggest a potentially compromised burial environment between 29.04.02 and 19.06.02. This period corresponds to the lowest water level point in the water table monitoring data obtained at the adjacent piezometers at 7.429 m OD on 29.04.02 and a similar trough at 7.551 m OD on 19.06.02 (Fig. 14.3b). Subsequently, between 22.01.01 and 13.08.02, the redox data records a number of 'peaks' towards more positive readings up until 16.05.02, and a return to more stable reducing conditions by 13.08.02. At 0.5 m depth the redox probes exhibit Eh (mV) values of +375 to +700, indicating oxidising conditions in the upper, post-Medieval, deposits within the ditch at this location.

In general, whilst some fluctuations in redox potential are occurring at the 1.5m and 3.0m depths within the ditch sequences, these events appear to mirror the water table data in that the lowest recorded water tables correspond to elevated (more positive) redox potentials in the deposits to 1.5m depth (compare Figs 14.3a and b with Fig. 14.4a).

It is concluded that, as no data on the redox potentials of the burial environment were available prior to the current study, the levels recorded, when considered alongside the clear evidence for saturation of these deposits, may well reflect those occurring in the burial environment prior to the initiation of the monitoring programme. As such, any longer term fluctuations in these redox potentials should prove informative in terms of the preservation potential of the ditch sequences. This observation is of particular significance to future monitoring projects. In a situation where a development is known to be scheduled it is essential that baseline data on the archaeological deposits associated with the development be generated before work is undertaken at the site in question. Ideally baseline data would be collected over a 12 month monitoring period before development occurs in order to ensure that seasonal variability is accounted for.

The pH data (Fig. 14.4b) indicate that levels between pH 6.9 and 7.5 occurred in the early stages of the monitoring. Site 1 was recorded as being more alkaline than site 2, but the overall pH levels were broadly

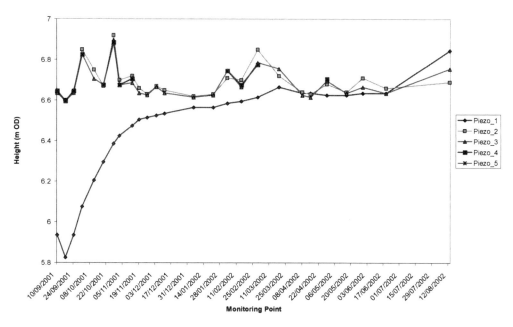

Fig. 14.3a Water table heights across the survey period 10.09.01 to 12.08.02 for monitoring point 1. Car park surface height at this location is 8.55 m OD.

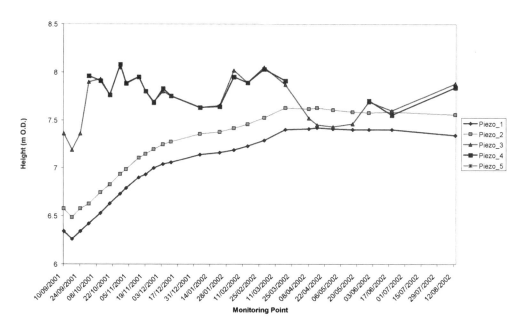

Fig. 14.3b Water table heights across the survey period 10.09.01 to 12.08.02 for monitoring point 2. Car park surface height at this location is 8.60 m OD.

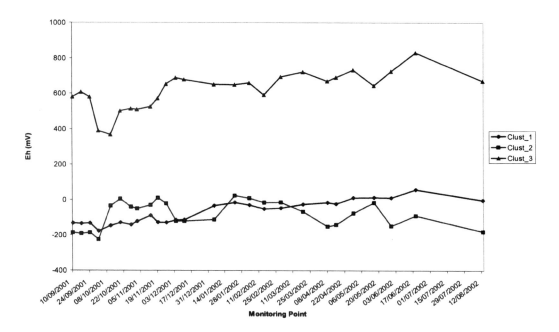

Fig. 14.4a Corrected Eh (mV) readings for redox probes, clusters 1–3 at Beverley.

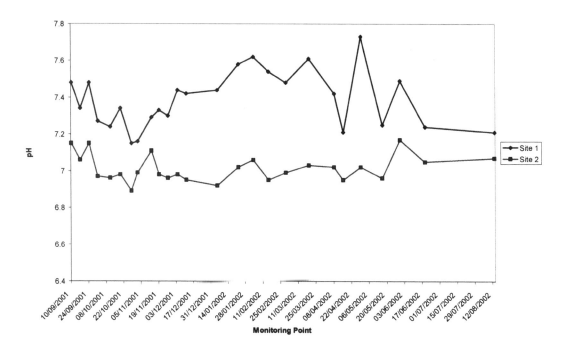

Fig. 14.4b pH levels recorded from the dip wells at monitoring points 1 and 2.

neutral to alkaline across the site. The pH level recorded between 01.01.02 and 13.08.02 mirror each other between both monitoring points. Site 1 remains the more alkaline across the monitoring period, and this location also exhibits a more pronounced reaction in pH levels. This is highlighted by comparison of the data recorded between 15.04.02 and 19.06.02, where the piezometer levels at site 1 and the pH levels match exactly, with troughs in water table activity being accompanied by lowered pH levels at *c.* 7.21–7.25, and peaks of 7.73 (29.04.02) and 7.49 (30.05.02). On the basis of these observations it is clear that pH is influenced by surface water inputs at this location. At site 2 the peaks and troughs in water table activity recorded across the period 15.04.02 to 19.06.02 are again broadly mirrored by the pH levels in evidence. Generally, the period of the water table trough recorded at monitoring point 2 between 08.04.02 and 16.05.02 matches a low period of pH between 6.95 and 7.02. The subsequent peaks in the data directly reflect the peaks and troughs occurring at both locations between 30.05.02 and 19.06.02. The pH levels remain more stable across the final two monitoring dates between 19.06.02 and 13.08.02.

On the basis of the monitoring undertaken at the site of the Knights Hospitaller's Preceptory, Beverley, it was concluded that, in general, the ditch sequences reflect good conditions for the continued preservation of the contained organic materials. Difficulties arise due to the lack of baseline data against which the monitoring data could be compared; however, in the absence of this data the results are sufficient to suggest that the development, a car park, has had a minimal impact on the hydrology and burial environment in the moat ditch sequences. This observation is supported by the data, which indicates that the parameters studied may be showing reactions to short-term rainfall events despite the fact that an artificial surface had been laid over the ditch at this location. Finally, it has been suggested that the low hydraulic conductivity of the ditch silts and clays, attested by the prolonged re-charge of the lower piezometers, is significant in ensuring the maintenance of saturation within the ditch, a situation that is complimented by the impervious nature of the tills into which the ditch has been excavated in antiquity.

GEOMORPHIC CONTEXT AND THE POTENTIAL FOR *IN SITU* PRESERVATION

The case study presented above has highlighted the value of a context, such as the ditch of a moated site, which has defined and confined parameters against which a study of the potential for *in situ* preservation can be assessed. A combination of factors such as the sedimentary characteristics of the dominant sub-surface geology (in this case impermeable till), the low hydraulic conductivity of the silts and clays in the lower levels of the ditch, and the clearly defined boundaries of the archaeological context (the moat ditch itself) has enabled viable conclusions to be developed from the monitoring programme to date. However, while such contexts exist, and are often in need of considered management strategies, *in situ* preservation will be less straightforward in a wider, landscape setting.

Despite this observation, recent conservation efforts at sites such as the Windover burial pool in Florida (Coles 1995a) and the Sweet Track (Brunning 1999), have benefited from attempts aimed at managing the burial environment. The specific context of both of these sites may well reflect their value in terms of conservation efforts. When considering the Sweet Track, an active management system has been in place since 1983 (Brunning 1999, 33). Initial study of the water levels at the Shapwick Heath nature reserve indicated that fluctuating conditions existed, with periodic exposure of the upper parts of the trackway in evidence. Since 1993 pumping into the area of the reserve and the creation of a bund has been used to maintain high water levels around the trackway at this location.

Brunning (1999, 35–8) has noted that the greatest decay appeared to have resulted from the long-term activity of bacteria under anaerobic or micro-aerobic conditions and to a lesser degree fungal decay. Other factors such as seasonal fluctuations in water levels, outflow of ground water and the topography of the peat surface are clearly potential vectors of a compromised burial environment at this location. Further complicating factors, even in a relatively 'simple' conservation environment such as the Sweet Track have been identified. These include the fact that the trackway pegs and planks will have been exposed to differing micro-environmental perturbations during the active 'life' of the structure and differing rates of peat accumulation will have influenced the potential for decay/preservation vectors. In addition, across the entire length of the Sweet Track variations in water levels will also have served either to promote or inhibit fungal/bacterial activity (*cf.* Boddy and Ainsworth 1984, 95–6).

The above example serves to highlight the inherent complexities of any attempts at the conservation and management of waterlogged archaeological sites in unconstrained settings. In order to provide a more holistic overview of this area of conservation practice, two examples – Flag Fen, Peterborough (Pryor 1992) and Sutton Common, South Yorkshire (Van de Noort *et al.* 2001a and b) – are briefly considered below, in an attempt to further highlight the difficulties inherent in managing the archaeological resource of wetlands.

Flag Fen

At Flag Fen (Fig. 14.5), the author undertook a limited period of groundwater monitoring between 1st February 2002 and 8th April 2002 with colleagues from the Wetland Archaeology and Environments Research Centre, University of Hull (Lillie and Cheetham 2002c). This monitoring was undertaken during the 'wettest' part of the year in order to assess the nature of water table activity at the interface between the wetland/dryland edge of the site. The site comprises a kilometre long post alignment constructed between the gently sloping dryland edge to the west, southeastwards to the platform and on to the slightly steeper dryland of Northey 'island' (Pryor 1992, 439). The archaeological resource is dated to the period *c.* 1350–950 BC, and almost all of the human activity recorded at this location occurs between -0.5m and +4.5m OD (Pryor 1992).

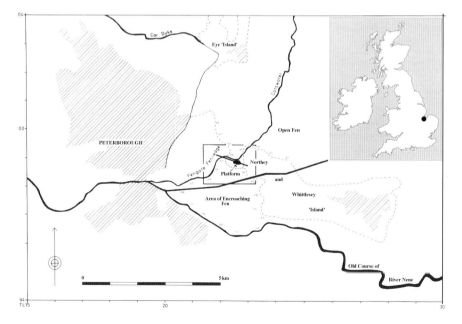

Fig. 14.5 Location map for Flag Fen, with topography of the Fen basin in the vicinity of the platform during the Bronze Age (after Pryor 1992, 441 [Fig. 14.2]).

The research undertaken at this location (Lillie and Cheetham 2002c) focussed on the post alignment, which is defined by the dryland margin at the Fengate Fen-edge to the northwest, and the timber platform which is located to the southeast, close to the margins of the Northey and Whittlesey 'island' (Pryor 1992). For monitoring purposes five peizometer clusters were positioned central to a borehole survey which was undertaken in December 2001, and which followed the post alignment from the northwest, in a southeasterly direction towards the platform (and artificial lake). This lake was designed to ensure continued saturation, through seepage, of the deposits underlying and adjacent to it (Pryor pers. comm. 2001).

However, drainage activities reported by Pryor (1992), included the enlarging and deepening of the Mustdyke in 1972, and a second phase of deepening in 1982. This excavation work had taken the base of the Mustdyke to a level that was *c.* >1m below the top of the underlying Pleistocene clay-capped gravels. This level is *c.* 2m below the height of the highest timbers of the Bronze Age platform (Pryor 1992: 442).

The height of the known archaeological resource in the immediate vicinity of the piezometer clusters considered here is placed at *c.* 0.7m OD to -0.13m OD at the platform, *i.e.* towards the southeastern end of the piezometer line, and 1.6m OD at the power station end of the post alignment in the northwest. Four trial excavations undertaken in 2000 by Soke Archaeological Services indicated that the top of the posts in the alignment occurred at between 0.6 and 0.93m OD (Britchfield *pers. comm.* 2002).

The threats of de-watering associated with drainage were originally highlighted by borehole surveys undertaken in 1983 and 1984, which indicated that the water tables were either at, or just below, the levels of the highest horizontal timbers at Flag Fen, and well below the tops of the surviving posts (Pryor 1992, 442). Excavations along the line of the post alignment indicate that the timbers in the vicinity of piezometer cluster 2 were at 0.81m OD in 1999–2000, those in the vicinity of cluster 3 were at 0.60m OD, and those to the northwest of cluster 4 were at 0.93m OD.

Soil and sediment analysis of eight profiles obtained from the Fengate fen-edge at the power station sub-site and the Cat's Water sub-site are summarised in French (1992). In brief, an argillic brown earth (or brown forest soil) was overlain by a thin peat and alluvial deposits, which thicken in a southeastwards direction (French 1992, 458). The gross stratigraphy as described by French (1992, 458–61) is mirrored in the stratigraphic survey that was undertaken in advance of the positioning of the piezometer clusters used in the current analysis (Fig. 14.6).

The monitoring points were positioned on the basis of the stratigrahic/sedimentological survey and the associated GPS survey (undertaken by Henry Chapman) which also enabled the GIS development of models showing the surface topography, buried landsurface topography, and a model of the depths of the Holocene peats and alluvial deposits across the study area (Fig. 14.6).

On the basis of the monitoring that was undertaken it is apparent that the piezometer clusters all produced readings which appear to be a genuine reflection of the prevailing water tables at this location. Despite this observation, it should be noted that only limited evidence for water table activity was actually recorded within the peizometers. At the dryland margins (monitoring point 1), it would appear that limited water table movement is occurring within the lower peats. However, the water level does not exceed 0.61m OD, indicating that on the four occasions where water levels were recorded (Fig. 14.7a), the depth never exceeded a height of 0.122m above the base of the peats and their contact with the underlying sandy clays with gravels.

Away from the dryland margins, at monitoring point 4 (Fig. 14.7b), the readings obtained indicate that at the lowest monitoring point in P4a (at 2.30m depth [1.707 to -0.593m OD]), we are seeing a consistent trend of water table activity between -0.2 and +0.1m OD. These results mirror those obtained from location 3, and as such would appear to indicate that at this location water through-flow is occurring both within the lower sandy clay unit (unit 1), and within the basal *c.* 0.30m of the overlying detrital 'mud-like' material.

The Fen peats and alluvial deposits that have been identified throughout the survey area have exhibited

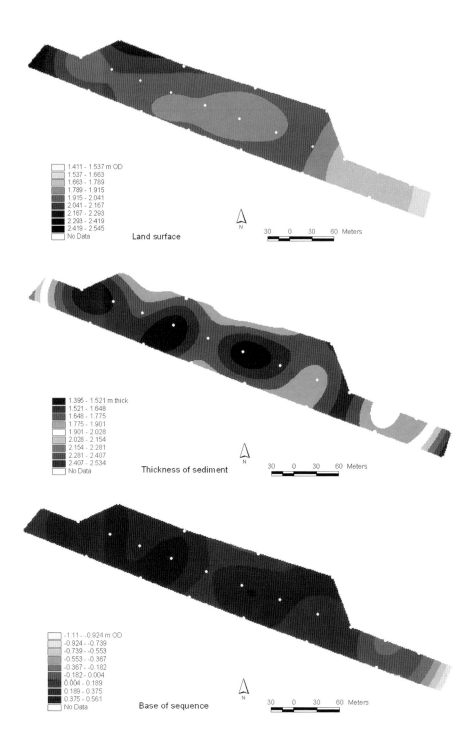

Fig. 14.6 Stratigraphy at Flag Fen in relation to Piezometer locations.

water table activity in only the very basal part of the sequence. Towards the dryland margins, the basal 0.129m of the Fen peats are intermittently saturated by water table activity (Fig. 14.7a). Moving away from the fen edge, monitoring point 2 indicated that saturation occurred in the basal 90mm of the Fen sequence, but that saturation is not constant. Monitoring point 4 (Fig. 14.7b) exhibited similar water table activity to monitoring point 3, with heights of between 0.237 and 0.347m OD being recorded, indicating saturation across the monitoring period of between 144 to 254mm in the Fen sequence.

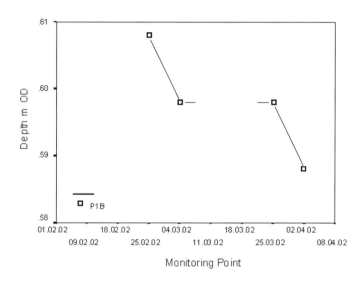

Fig. 14.7a Water levels recorded in Piezometer cluster 1 at Flag Fen.

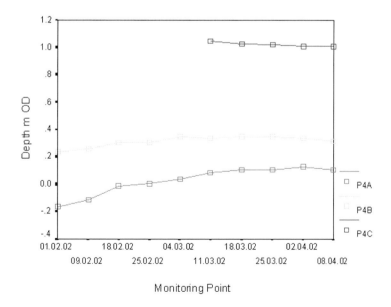

Fig. 14.7b Water levels recorded in Piezometer cluster 4 at Flag Fen.

Overall, the two monitoring points nearest to the dryland margins have shown that a severely compromised situation exists in relation to the buried archaeological resource. Where the base of the sequence falls away from the dryland margins, in a southeasterly direction, water table activity is evidently more constant, but the saturation levels within the organic-rich fen sequences never exceed 0.557m OD. Pryor (1992) reports that the archaeological resource is recorded as being between -0.5m and +4.5m OD in this area. As none of the piezometer clusters indicates water table activity above 0.557m OD, it is apparent that the upper *c*. 4.0m of the waterlogged archaeological resource in the area of the post alignment should be considered to be compromised to the point where destruction of the resource can be anticipated.

The main conclusion is that, given the recorded water table activity, all waterlogged archaeology located above *c*. 0.5m OD within the survey area should be considered to be drying out and under serious threat. We can anticipate that the upper part of the posts along the alignment will be within Zone 1 of preservation as defined by Chapman and Cheetham (2002, 278). This zone, the dry zone, has been highlighted as the key factor in the deterioration of wood structure and loss of archaeological information. At the dryland margins the situation is apparently exacerbated due to the elevated topography at this location.

Sutton Common

Recent sedimentological investigations at Sutton Common (Lillie and Schofield 2002, Lillie 1997) have highlighted the complex nature of the stratigraphy associated with the palaeochannel and areas adjacent to the Iron Age enclosures at this location. This particular site has been the subject of detailed water table monitoring by Cheetham (Van de Noort *et al*. 2001a and b, Chapman and Cheetham 2002), within a programme designated as the '3 M's' approach by the then Centre for Wetland Archaeology at the University of Hull. This approach combined monitoring, modelling and management in an attempt to 'achieve sustainability of the wet-preserved archaeological resource, or "near-zero" change' (Van de Noort *et al*. 2001b, 279).

As a result of the modelling of water table activity at Sutton Common it was suggested that a maximum height of 4.1m OD of permanent water table height could be achieved by manipulation of the site's hydrology, and that certain areas of the site could not be preserved *in situ* (Van de Noort *et al*. 2001b, 285). On the basis of these observations, changes to the drainage regimes were instigated in the autumn of 1999. One significant problem highlighted at Sutton Common is the variability in the deposits in and around the site (Van de Noort *et al*. 2001a, 97). This variability is particularly marked in the palaeochannel sequences, which are not as uniform as might be suggested by their description as 'water retentive peats' (Van de Noort *et al*. 2001a, 97).

Lillie and Schofield (2002) excavated sixty-six boreholes across the palaeochannel and adjacent areas of superficial peat formation in the vicinity of Sutton Common (Fig. 14.8). This analysis indicated that the channel infill sequences comprise low-energy silts and clays intercalated with periodic 'higher' energy flow being reflected by discrete sand and clay-silt horizons with woody detritus throughout. There is a general trend towards the formation of more organic-rich and peaty deposits towards the upper part of the infill sequences throughout the study area. Variability is evidenced within the channel deposits between the two enclosures at Sutton Common (and to either side of the causeway).

To the south of the causeway between the enclosures the sequences recovered from transect 5 comprise intercalated peats, silty peats and clays, with erosive contacts between sediment units and inconsistencies between the sequences identified within the palaeochannel reflecting variability in flow regimes and the periodic absence of flow from the area delimited by the palaeochannel. To the north of the causeway the sequences identified in transect 7 exhibit subtle variations towards the top of the infill sequence that may well reflect differing sediment accumulation regimes occurring under the influence of the causeway. The sediments on the northern side of the causeway do not exhibit the same degree of clay inclusion in the upper

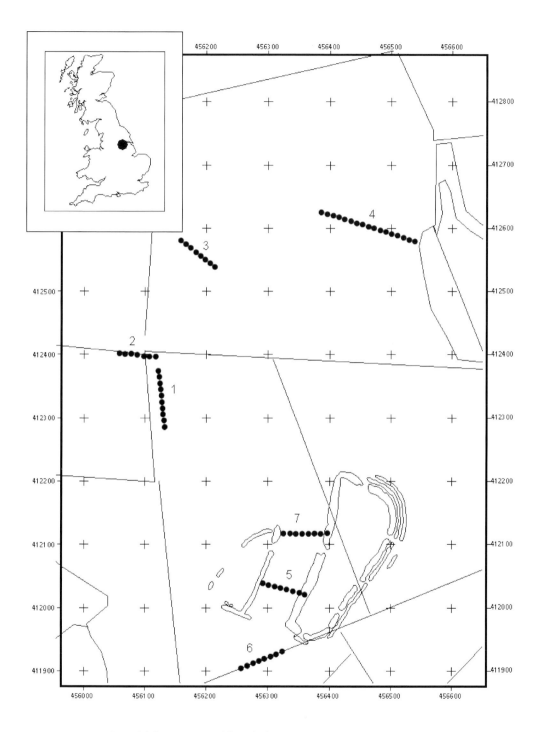

Fig. 14.8 Location of borehole transects at Sutton Common.

part of the sequence as evidenced to the south. This variability results in surface water ponding on the southern side of the causeway between the enclosures.

Elsewhere in the survey area considerable difference in the nature of the sedimentary sequences is in evidence, with shallow superficial peats over sands occurring to the north of the enclosures, away from the palaeochannel, and deeper sediment sequences in evidence to the east of the larger enclosure associated with Shirley Wood as outlined by Lillie (1997) and Schofield (2001). Further exacerbation of the complexities inherent in the Sutton Common landscape setting occurs within the limits of the monument itself, where the ditches (even where ploughed-out) have a proven, and occasionally deeply stratified, mixed organic infill sequence.

Sedimentologically, the landscape context at Sutton Common is complex when compared to locations such as the Knights Hospitaller's moated site in Beverley, the Sweet Track and Flag Fen, as considered above, although the latter examples are clearly limited parts of wider landscape settings. The characterisation of the Sutton Common landscape, and any mitigation strategies associated with it, will clearly require an approach that has seldom been applied in archaeological resource management to date. Whilst English Heritage and the various agencies involved in the management at Sutton Common have, in effect, made this site a case study in approaches to *in situ* preservation, it is apparent that the considerable complexities in the depositional sequences in and around the enclosures will influence and indeed impact upon future mitigation strategies, including [and obviously not limited to] the raising of water tables around the site.

CONCLUSIONS

It has been suggested that 'the ultimate aim of *in situ* preservation is the creation and maintenance of a static burial environment that is sustainable in the long term' (Van de Noort *et al.* 2001a, 96). However, as the above discussion has highlighted, the precise geomorphic context of the waterlogged sequences that are to be preserved in this way must be studied in detail. In addition, whilst monitoring of parameters such as water tables and the chemistry of the burial environment are fundamental to understanding the site context, a more holistic landscape-wide approach needs to be employed if effective management strategies are to be implemented. Similarly, an effective understanding of the data generated through monitoring of redox, pH and water tables (amongst others) will be more easily developed if baseline information on these parameters is generated *before* developments that may further impact on the burial environment are undertaken (including excavation).

Progress has clearly been made since the early 1990s, but the majority of mitigation continues to be reactive as opposed to pro-active. Access to wetland archaeological sites is limited (Van de Noort *et al.* 2002), though the threats imposed by drainage, water abstraction and aggregates extraction continue to increase. Whilst it has been suggested that the principal factor influencing preservation within wet environments is the presence of saturation, *e.g.* Corfield (1996), Caple (1996), Chapman and Cheetham (2002), achieving elevated and stable levels of saturation in an unconstrained landscape context is, at best, extremely difficult, and at worst, unachievable.

ACKNOWLEDGEMENTS

The investigation and analysis of the Knights Hospitaller's Preceptory and Flag Fen was undertaken with James Cheetham, who calculated redox potentials and assisted with fieldwork and report writing. The work at Beverley was undertaken for CgMs Consultants on behalf of Tesco Stores Ltd. Henry Chapman undertook the modelling of the Flag Fen data. The borehole data from Sutton Common was obtained by the author and Dr Ed Schofield whilst the latter was employed at the Wetland Archaeology & Environments Research

Centre (WAERC), University of Hull. The observations presented in this paper represent those of the author, and were developed during discussions with colleagues, between July 2000 and September 2003, whilst the author was the director of WAERC.

REFERENCES

Boddy, L. & A. M. Ainsworth. 1984. Decomposition of Neolithic wood from the Sweet Track. *Somerset Levels Papers* 10, 93–96.

Brunning, R. 1999. The *in situ* preservation of the Sweet Track, in B. Coles, J. Coles, & M. Schou Jørgensen (eds) *Bog bodies, Sacred Sites and Wetland Archaeology,* 33–38. Exeter: WARP.

Caple, C. 1996. Parameters for monitoring anoxic environments, in M. Corfield, M. P. Hinton, T. Nixon, and M. Pollard (eds). *Preserving archaeological remains in situ,* 113–123. London: Museum of London Archaeological Service.

Caple, C. & D. Dungworth 1997. Investigations into waterlogged burial environments, in A. Sinclair, E. Slater & J. Gowlett (eds) *Archaeological Sciences 1995: Proceedings of a conference on the application of scientific techniques to the study of archaeology,* 233–40. Oxford: Oxbow Monograph 64.

Chapman, H. P. and J. L. Cheetham 2002. Monitoring and modelling saturation as a proxy indicator for *in situ* preservation in wetlands – a GIS-based approach. *Journal of Archaeological Science* 29, 277–89.

Cheetham, J. L. 2004. An assessment of the potential for in situ preservation of buried archaeological remains at Sutton common, South Yorkshire. Unpublished PhD Thesis, University of Hull.

Coles, B. 1995a. Paradox and Protection: The significance, vulnerability and preservation of wetland archaeology, in M. Cox, V. Straker & D. Taylor (eds) *Wetlands: Archaeology and Nature Conservation,* 144–55. London: HMSO.

Coles, B. 1995b. *Wetland management – a survey for English Heritage.* Exeter: Wetland Archaeology Research Project.

Coles, B. & J. Coles 1986. *Sweet Track to Glastonbury.* London: Thames and Hudson.

Coles, B. & J. Coles 1989. *People of the Wetlands.* London: Thames and Hudson.

Compton, R. G. & G. H. W. Sanders 1996. *Electrode Potentials.* Oxford: Oxford University Press.

Corfield, M. 1996. Preventive conservation for archaeological sites, in A. Roy & P. Smith (eds) *Archaeological conservation and its consequences,* 32–37. London: International Institute for Conservation of Historic and Artistic Works.

Evans, D. H. 1997. Excavations and watching briefs on the site of the Knights Hospitaller's Preceptory, Beverley, 1991–94. *East Riding Archaeologist* 9, 66–115.

French, C. 1992. Fengate and Flag Fen: summary of the soil and sediment analyses. *Antiquity* 66, 458–61.

French, C. 2003. *Geoarchaeology in Action: Studies in soil micromorphology and landscape evolution.* London: Routledge

Hall, D. & B. Coles 1994. *Fenland Survey: an essay in landscape and persistence.* English Heritage Monograph.

Lillie, M. C. 1997. The Palaeoenvironmental Survey of the Rivers Aire, former Turnbridge Dike (Don north branch), and the Hampole Beck, in R. Van de Noort and S. Ellis (eds) *Wetland Heritage of the Humberhead Levels: An Archaeological Survey,* 47–78. Hull: Humber Wetlands Survey.

Lillie, M. C. and J. L. Cheetham 2002a. Monitoring of moat ditch deposits at Knights Hospitaller's Preceptory, Beverley (TA 03853967). Unpublished WAERC Report No. KHP-BEV/02-2, University of Hull.

Lillie, M. C. and J. L. Cheetham 2002b. Monitoring of moat ditch deposits at Knights Hospitaller's Preceptory, Beverley (September 2001–August 2002) (TA 03853967). Unpublished WAERC Report No. KHP-BEV/02-1, University of Hull.

Lillie, M. C. and J. L. Cheetham 2002c. Water Table Monitoring at Flag Fen, Peterborough (TL 227989). Unpublished WAERC Report No. SAS-FF/02-01, University of Hull.

Lillie, M. C. and J. E. Schofield 2002. Sutton Common, Askern: borehole survey. Unpublished WAERC Report No. SCOM/02-01, University of Hull.

North West Wetlands Survey. 1990–3. *Annual Reports.* Lancaster: Lancaster University Archaeology Unit.

Patrick, Wm. H. & I. C. Mahaptra 1968. Transformation and availability to rice of nitrogen and phosphorous in waterlogged soils. *Advances in Agronomy* V20, 323–59.

Pryor, F. 1992. Current Research at Flag Fen, Peterborough. *Antiquity* 66, 439–57.

Schofield, J. E. 2001. Vegetation Succession in the Humber Wetlands. Unpublished PhD Thesis, University of Hull

Smith, R. J. 2005. *The Preservation and Degradation of Wood in Wetland Archaeological and Landfill Sites.* Unpublished PhD thesis: University of Hull.

Stumm, W. & J. J. Morgan 1981. *Aquatic Chemistry.* New York: John Wiley & Sons.

Van de Noort, R. & P. Davies. 1993. *Wetland Heritage: An archaeological assessment of the Humber Wetlands.* University of Hull: Humber Wetlands Project.

Van de Noort, R., H. Chapman and J. Cheetham 2001a. *In situ* preservation as a dynamic process: the example of Sutton Common, UK. *Antiquity* 75, 94–100.

Van de Noort, R., H. Chapman and J. Cheetham. 2001b. Science-based conservation and management in wetland archaeology: the example of Sutton Common, UK. in B. A. Purdy (ed) *Enduring Records: the environmental and cultural heritage of wetlands*, 277–286. Oxford: Oxbow Books.

Van de Noort, R., W. Fletcher, G. Thomas, I. Carstairs & D. Patrick 2002. *Monuments at Risk in England's Wetlands.* Exeter: University of Exeter Report for English Heritage.

Welch, J. & S. Thomas 1996. 'Groundwater modelling of waterlogged archaeological deposits' in M. Corfield, M. P. Hinton, T. Nixon, and M. Pollard (eds). *Preserving archaeological remains in situ*, 16–20. London: Museum of London Archaeological Service.

York Archaeological Trust 1991. *Report on an Archaeological Evaluation at Station Yard, Beverley.* York: Y.A.T.

Chapter 15

The Present and the Past: The Interpretation of Sub-Fossil Molluscan Assemblages and the Relevance of Modern Studies, with Specific Reference to Wet-Ground Contexts in the UK

Paul Davies

INTRODUCTION

The interpretation of sub-fossil Mollusca from archaeological contexts and Holocene soils and sediments has always relied upon there being a relationship between the present and the past. It is possible to reconstruct past environments and landscapes, since virtually all of the species found as Holocene sub-fossils are still to be found in the British Isles. A standard uniformitarianist approach can readily be adopted, where the modern ecologies (or aspects of the modern ecologies) of species (or groups of species) can be applied back through time. While this is not without its difficulties, many of which are outlined below, Thomas (1985, 149) correctly noted that sub-fossil analysis has worked 'because the data generally make ecological sense'.

This paper will consider some of the general theoretical difficulties of a uniformitarian approach to the interpretation of sub-fossil Mollusca, in particular concentrating on the issues surrounding the theoretical problem of non-analogue communities (Gee and Giller 1991). Attention will then turn to wetlands, with a discussion of assemblages recovered from Holocene overbank alluvial deposits and modern floodplain faunas, although the implications of the demonstrated relationships between present and past are wider. This will be followed by discussion of the interpretation of sub-fossil molluscan assemblages recovered from early-mid Holocene tufa deposits, and how modern ecological work might help improve the level at which we can reconstruct the environments in which such deposits formed.

MODERN ECOLOGIES AND ECOLOGICAL GROUPS

Although snail shells were recognised as being present in archaeological contexts and Holocene sediments over a century ago (Evans 1972), the development of sub-fossil molluscan analysis as a powerful tool for environmental reconstruction can in essence be traced back to two seminal papers. These, by A. E. Boycott

(1934, 1936), synthesising knowledge of modern ecologies of land snails and freshwater snails, respectively (freshwater snails are not further considered here). Both papers made a number of important observations. First, it was recognised that while many species of Mollusca had very broad ecologies and could tolerate a wide variety of conditions, a number were rather more exacting in their requirements. For land snails, Boycott (1934) proposed that species could be grouped into several major categories:

> *Obligate hygrophiles* – species restricted to damp places
> *Facultative hygrophiles* – other species found in damp places but not restricted to them
> *Xerophytes* – species restricted to dry places
> *Facultative xerophytes* – other species found in dry places but not restricted to them
> *Synanthropes* – species showing a liking for sites associated with human activity
> *Anthropophobes* – species avoiding sites associated with human activity
> *Rupestrals* – species particularly associated with stone walls
> *Woodland* – species found in woodlands
> *Woodland 'restricted'* – species usually found only in woodland
> *Non-woodland* – species not normally found in woodlands

While the groups were not necessarily mutually exclusive (*i.e.* one particular species of snail might belong to more than one group), some of the groupings were much better defined than others – the 'woodland restricted' group, for example. The value of this to sub-fossil studies was quickly recognised, and Sparks (1961) was able to apply a modified version of these groupings to quantitatively-based fossil studies, recognising four ecological groups (marsh, dry land, woodland and *Vallonia* sp.) as being useful for interpreting past environments. The difficulties, however, were still recognised, with Sparks (1961, 77) reporting that 'these groups must not in any way be considered final, as the variations in the groups have so far been little studied'.

The following ten years saw a tremendous increase in the application of molluscan analysis to archaeological contexts and Holocene sedimentary sequences, particularly by M. P. Kerney and J. G. Evans. Evans (1972) was able to provide a major synthesis on the use and interpretation of Mollusca recovered from archaeological sites, and to propose the following ecological groupings as useful in interpretation:

> *Woodland species*
> *Pomatias elegans*
> *Intermediate (or catholic) species*
> *Open-country species*
> *Marsh species* –amphibious
> obligatory marsh
> characteristic but not confined
> others with no special affinity
> *Freshwater slum*
> *Alien species* – those introduced to the British Isles in the historic period
> *Burrowing species*
> *Anthropophobic species*
> *Synanthropic species*

The synthesis described which species should go into which group, and also summarised the literature on species ecology and their sub-fossil occurrences. The link between known ecology and palaeoecology was made explicitly apparent for each species present in British Holocene deposits, and limitations (as well as potential) were recognised. Subsequently, other ecological groupings have been used. Preece (1980), for example, has used 'swamp' (obligate hygrophiles), 'terrestrial A' (catholic/generalist) and 'terrestrial B' (more shade-demanding) groupings in interpreting Mollusca recovered from tufa deposits (see below), but to some extent these are variants on the themes already established by Sparks and Evans.

Given the above developments it can be seen that interpretation of sub-fossil assemblages can potentially proceed in two ways, either through an autecological approach (based upon individual species ecology), or a synecological approach (where present and past communities can be compared). In reality it is somewhat artificial to separate these fully, since, as will become apparent, there is overlap. To do so, however, is useful heuristically. In relation to Mollusca, the approaches and their difficulties have previously been outlined by Evans (1972, 1991), Cameron (1978a) and Thomas (1985), upon whose work much of the following two sections is based.

Interpretation difficulties – autecology

At its most basic level an autecological approach can assign each recovered species to an ecological group based upon its present day habitat preferences, *i.e.* the classic uniformitarianist approach. The changing percentages of these groups through a sequence can then be used to indicate environmental change, for example from shaded to open-country. While Harris and Thomas (1991) point out that this tends to reduce interpretation to a summation of micro-habitats, they also recognise the value in that it avoids matching past and present *communities* (see below). Nevertheless, as for other macro- or micro-fossil analyses where there is a continuation into the present, autecological uniformitarianism leads to a number of difficulties which can be summarised as follows:

Identicality of ecology between past and present is often assumed rather than demonstrated. Indeed, identicality of *species* is also assumed, not proven, based upon identicality of shell morphology, which conceivably may not be a reliable guide.

Specificity of ecology is also problematic. In reality, very little detail is known about molluscan species ecology. Distributions are controlled by a large number of interacting variables, both biotic and abiotic, most of which are unknown or unpredictable for most species. The modern work that has been done seems to make it increasingly difficult even to assign species to an appropriate habitat group. Basically, poor knowledge of modern ecologies leads to weak and generalised interpretation of sub-fossil material, and this can be particularly true when it comes to wetlands (see below). Even if we had very detailed modern ecological data, for example on tolerances to environmental variables, this does not get around the potential problem of identicality of species (see above), nor serve to demonstrate that the same variables acted in the same way upon the species, either singly or in concert, in the past. The species that seem to be very restricted in their distributions (stenotopic species), and may therefore be good candidates for status as indicator species, may, to some extent, have become more restricted in their distributions through the Holocene, not least because of the increasing effects of human modification of the landscape.

Finally, the degree to which we can accept identicality of environments is also problematic. Modern grasslands are not, in detail, like Neolithic or Bronze Age grasslands, modern woodlands are not, in detail, like prehistoric 'wildwoods', and modern, managed wetlands are not, in detail, like unmanaged wetlands. The extent to which we can use individual species' present habitat requirements and project this back in time is potentially uncertain. As well as relating to the above, this also relates to the problems of non-analogue communities (see below).

Interpretation difficulties – synecology

A synecological approach to interpretation may in part be considered as the summation and synthesis of the autecology of each species within the assemblage – in effect treating it as akin to an ecological community. An extra dimension can be given, however, by the use of analogy at the community level. Autecology uses *individual* ecologies analogously between present and past, while synecology matches, or attempts to match, present communities with past *representations* of communities (the fossil assemblages recovered). Sometimes the assemblages recovered may represent an amalgamation of several past communities, both spatially, for example when post-mortem transport brings shells from a wide area together at one point, or temporally, where the recovered assemblage represents an accumulation of several successive communities that occupied a given spot through time. In such circumstances a synecological approach may be problematic. At other times, however, for example from the surfaces of buried soils, or within overbank alluvial or tufa deposits, the recovered assemblage may be very closely representative of the once living community (Evans 1991), and the principle of one-to-one community matching between past and present becomes theoretically possible. An example is the work of Bush (1988), which utilised numerical analysis techniques in matching past and present wetland molluscan samples and used this as a basis for legitimising the interpretations offered for the fossil samples and for tracking the direction of environmental change. Nevertheless, this analogue community approach is open to criticism, summarised briefly as follows (note again that there is overlap between some of the arguments and with some of the autecological arguments):

Paucity of applicable data is one of the major problems. Although the number of modern studies, and thus our knowledge of modern communities, continues to increase, the applicability of the data to the past is usually somewhat limited. Many modern studies have been conducted on woodland faunas from a variety of woodland types (*e.g.* Wardhaugh 1997), but interpretation of fossil assemblages still tends not to go beyond 'woodland' as a generic habitat classification for an environment dominated (at least in terms of the most obvious vegetation) by trees. Even then (see below) there can still be an element of doubt in some cases. Part of the problem is that modern ecological studies are not asking questions of relevance to palaeoecological interpretations (Davies 2003a). Another problem is that many of the modern studies are from geographical areas different to those of most of the fossil material they could potentially help to interpret. This is problematic since molluscan faunas are geographically dependent (see, for example, Cameron and Redfern 1972, Cameron 1978b). A further potential problem is that modern studies have tended to concentrate on established, stable habitats, whereas many of the fossil assemblages may derive from transient, unstable ones (see below).

Similarity of past and present faunas may, of course, also be purely coincidental, *i.e.* the match does not necessarily imply similarity of habitat (in detail) between past and present. This has already surfaced as a problem above, and serves to compound the problem of ending up with 'limp' environmental reconstructions (grassland, woodland, marsh *etc.*), rather than detailed discussion of the habitat structure and other characteristics.

The treatment of *past assemblages as communities* can also be questioned. As mentioned above, except in defined circumstances (Evans 1991), two of which are expanded upon below, recovered assemblages are often not representative of past communities – they have suffered additions, removals and amalgamation. The use of an analogous community approach is therefore best restricted to those exceptional circumstances where autochthony is more certain.

In short, most of the problems relate to non-analogue communities. Some molluscan communities today may not be represented in the past, and some in the past may not be represented in the present. Even if there are real similarities between past and present, they may not be recognisable due to one or more of the factors detailed above.

Interpretation difficulties – summary and caution

It is time, perhaps, to remember the earlier comment of Thomas (1985), that notwithstanding its difficulties, sub-fossil molluscan analysis does work. It is vital to appreciate that some of the above difficulties are theoretical, which is not the same thing as saying they are actually real and apply at all times. Indeed, much of the 'art' of interpretation involves assessing the difficulties and making informed decisions as to which do or do not apply. This is particularly true of the autecological difficulties; just because one cannot prove identicality of ecology through time does not mean that one cannot assume it. When assumptions are made and they seem to be consistent with other evidence (either molluscan or not) then the assumptions are strengthened. Proof is an unobtainable goal in such analyses. Similarly, while it may not be possible to prove identicality of environment through time, that does not prevent a consideration of general environmental parameters such as vegetation characteristics, for example shaded or open, structured or unstructured, grazed or ungrazed, or hydrological characteristics, for example wet versus dry, or seasonally flooded as compared to catastrophically flooded. The general features of the environment will have an effect on the molluscan species present irrespective of the actual habitat-type as we would classify it. Of course, it helps to know how terms like wooded, open, wet, dry *etc.* relate to molluscan faunas; that, above all, is where appropriate modern ecological work has a real value.

The synecological difficulties are more 'real'. This has led, with exceptions (*e.g.* Bush 1988) to something approaching reluctance in wholeheartedly adopting such an approach and seeing what happens. Nevertheless, it is clear that sometimes the sub-fossil record can preserve molluscan assemblages that ought to be fairly close (in terms of species and abundance) to the molluscan communities from which they were derived (Evans 1991 and above). In such cases, at least some of the difficulties are removed. Furthermore, if in such cases there are closely comparable modern communities from narrowly defined environments similar to that which one would expect to be represented in the fossil sequence, the comparability must mean something, particularly if the relationship is replicated elsewhere. Such closely related sub-fossil assemblages and modern communities have been demonstrated recently with respect to the Mollusca of floodplains. In central southern England the assemblages recovered from overbank alluvial deposits within Holocene floodplain fills show a high degree of comparability with faunas from modern floodplain environments. Some of the similarities have been detailed elsewhere (Davies *et al.* 1996, Davies and Grimes 1999, Davies 2003a), and recently synthesised (Davies 2003b). To date, the implications of the work for other appropriate contexts have not been articulated. Here I wish to outline briefly the major points and then discuss their relevance to assemblages recovered from another type of wet-ground deposit – tufa.

HOLOCENE OVERBANK ALLUVIAL ASSEMBLAGES AND MODERN FAUNAS

Evans (1991) and Evans *et al.* (1992) recognised that a number of distinct species groupings could be identified as recurrent within assemblages recovered from Holocene overbank alluvium within river valleys of the central southern chalklands of England. Davies (1992, 1998) also demonstrated that these could be defined numerically using multivariate techniques. The recurrent groups (termed taxocenes; WGT = wet-ground taxocene) and their environmental interpretations are as follows:

WGT-1 – A low diversity assemblage with mainly open-country species, *Vallonia pulchella, Trichia hispida, Cochlicopa lubrica* and Limacidae being the predominant species. Amphibious and marsh species at low abundance or absent. This taxocene most likely represents an open, structurally simple, low diversity, relatively dry-ground environment, akin to grazed pasture or meadow which is only occasionally flooded.

WGT-2 – As WGT-1 but with marsh/swamp species well represented, principally *Lymnaea truncatula, Carychium minimum* and Succineidae. This taxocene represents an open but wetter and more structurally diverse environment than taxocene 1, such as lightly grazed pasture or winter flooded meadow.

WGT-3 – Marsh and/or woodland species predominant, often high diversity. The marsh species *Zonitoides nitidus* and *Carychium minimum* are usually characteristic, as is a paucity of *Vallonia pulchella* and Succineidae. This most probably represents a winter flooded environment of high structural diversity such as fen/marsh or, when woodland species are predominant, woodland.

WGT-4 – A low diversity taxocene, principally *Lymnaea truncatula* although Succineidae may also be well represented. Likely to represent a low diversity and perhaps transient environment such as a mudflat.

WGT-5 – The amphibious *Anisus leucostoma*, *Lymnaea truncatula* and *Pisidium personatum* are distinctive, although land species (either open-country or woodland) may also be abundant. This represents an environment akin to those outlined as WGT- 2 or 3, but flooded for longer periods and possibly with permanent pools of water.

WGT-6 – Taxocene 1 plus abundant *Pupilla muscorum*. It is difficult to know what environment this represents, since *Pupilla* is typically xerophilic although it can presently live in marshes and was a marsh species during the Late-glacial (Kerney *et al.* 1964). Either this represents an environment in which alluvial material has rapidly accreted and dried, providing a baked fissured surface (which the species favours as a xerophyte), or it is some sort of long, rank grassland with a thick thatch. A recent study undertaken at Kingsmead Bridge in Wiltshire suggested that the former was more likely at that site (Davies 1996, but see below also).

WGT-7 – *Lymnaea peregra* distinctive, with other amphibious species and a land component that can either be open-country or wooded. This taxocene is as yet geographically restricted to the upper Kennet valley (Evans *et al.* 1988). At the moment, it remains possible that *Lymnaea peregra* is allochthonous, brought in by flooding, and environmental interpretation is therefore uncertain.

WGT-8 – Freshwater species, particularly *Valvata cristata* and *Pisidium casertanum*. This taxocene represents the transition from wetland to aquatic habitats or *vice versa*. The environment indicated is permanently wet and subject to stagnation.

It should be noted that the term taxocene is in itself interpretative. It implies an ecological relationship between the species, in other words it accepts them as being representative of a past community (it does not imply interaction between the species). This does not preclude the fact that other species might be represented in the assemblage too; taxocenes and assemblages are not the same. Further details of the geographical occurrences of the taxocenes can be found in Davies (2003b).

The environmental interpretations offered for the taxocenes were based, in effect, upon general faunal characteristics and the autecology of some of the species or groups of species involved. One of the key faunal characteristics is diversity. Modern studies (and fossil ones where there is corroborating evidence) suggest that low diversity taxocenes are likely to be characteristic of simple, relatively unstructured environments, such as pasture or mudflat. High diversity taxocenes are much more likely to represent environments with more structural diversity, for example tussocky fen, marsh or woodland. Further discussion of interpretative issues relating to taxocenes can be found in Evans *et al.* (1992).

Although many of the taxocenes have been recognised in overbank alluvial deposits dating to the Neolithic and Bronze Age (Davies 2003b), some of the deposits discussed by Evans *et al.* (1992) and Davies (1998) consisted of relatively recent overbank alluvium, certainly Roman or post-Roman. In addition, the same recurrent assemblages have been found in very recent (either late-Medieval or post-Medieval) overbank alluvial deposits within the Bristol Avon and Wellow Brook, Somerset valleys (Davies unpublished). Logically, there-fore, it is reasonable to enquire as to whether such recurrent groupings have

Fig. 15.1 Fen at Bossington, Hampshire.

any modern analogues. Several modern studies have been undertaken recently within the same geographical area (the central southern chalklands), and the details can be found in the published papers, but one or two points are worth restating here as an indication of the value of comparative modern studies.

Davies *et al.* (1996) discussed wet-ground molluscan faunas recovered from reedswamp pasture, rank grassland, fen and carr. They found, unsurprisingly perhaps, that the less complex the vegetation structure, the less was the faunal diversity. In addition however, while in the context of their particular study the carr fauna could be separated from those of fen and rank grassland (both tussocky), due to several species which only occurred in the carr, none of those species could be considered as indicator species of carr more generally, all having been found in other wetland habitats in other published studies. The implication, not stated at the time, is that it is most likely impossible to differentiate between highly structured non-wooded and wooded wet-ground environments in the sub-fossil record. Further work where large tussocks within the fen were actually dissected and sampled (Davies unpublished) confirmed that even the strongest of the so-called 'woodland' species (*e.g. Acanthinula aculeata* and *Clausilia bidentata*) were present. As far as the Mollusca were concerned, fen and carr offer more or less identical conditions (Figs 15.1 and 15.2), and WGT-3 as detailed above ought just to be considered representative of highly-structured environments which may or may not be wooded.

Davies and Grimes (1999) demonstrated that small-scale spatial variation in hydrological and vegetation characteristics had a significant effect on molluscan fauna. By assessing the fauna of sheep-grazed relic carrier and drain features of an abandoned floodplain water-meadow system on the River Wylye, Wiltshire, they showed that each had unique faunal characteristics. Furthermore, the fauna of the carriers (*Vallonia-Trichia-Cochlicopa*) were identical to those of WGT-1 (above), and those of the drains (carrier fauna plus *C. minimum, L. truncatula*) were identical to those of WGT-2. Clearly, modern analogue faunas could be found.

Further work (Davies 2003b) found that relic water-meadow drains within cattle-grazed pasture on the floodplain of the River Test, Hampshire (Fig. 15.3), contained a fauna (*Vallonia-Trichia-Cochlicopa-Pupilla*)

Fig. 15.2 (left) Carr at Bossington, Hampshire. The molluscan fauna is similar to that of the fen in Fig. 15.1. Fig. 15.3 (above) Relic water meadow system, River Test, Hampshire.

identical to that of WGT-6 (above). This contrasted with the sheep-grazed pasture of the River Wylye (above) and suggested that it might be possible to distinguish between cattle-grazed and sheep-grazed floodplain pasture in the sub-fossil record.

In summary, therefore, some basic ecological surveys have shown that there are modern analogues for WGT-1, 2, 3 and 6. This is somewhat ironic, since the interpretive approach used in designating taxocenes was intended from the outset not to be reliant on species-based analogy (Evans *et al.* 1992). Nevertheless, of crucial importance here is the fact that all of these modern studies were in part designed in order to demonstrate whether the wet-ground taxocenes previously proposed had any modern analogues, though that was sometimes subordinate to other research questions. Geographic areas and precise sites were targeted because they were most likely to have analogous faunas. There was a clear link between the modern and sub-fossil work, and there is a clear benefit in designing modern ecological work explicitly (though not necessarily wholly) to consider palaeoecological questions. The other taxocenes may also have modern parallels, but finding them will depend upon surveying carefully chosen locations.

TUFA DEPOSITS – THE SUB-FOSSIL RECORD AND POTENTIAL OF MODERN WORK

Late-Glacial and Holocene tufa deposits are widespread in the chalk and limestone dominated areas of Britain, principally in the south of England but also in areas adjacent to the Yorkshire and Lincolnshire Wolds, in South Wales and elsewhere (Evans 1972, Pentecost 1993). Paludal tufas, forming as a result of calcium carbonate precipitation from spring-flushes or groundwaters rising to surface level (Ford and Pedley 1996), have been particularly useful. These tufas have often formed over long periods of time (thousands of years) and contain well preserved and abundant molluscan remains that can furnish detailed terrestrial environmental sequences (*e.g.* Kerney *et al.* 1980, Preece 1980, Evans and Smith 1983, Preece and Robinson 1984). They are of particular value since many have been obtained from parts of the UK where pollen is not readily preserved and vegetation history is otherwise largely unknown. Detailed molluscan bio-chronozones are now well established for southern England (see Kerney 1977, Kerney *et al.* 1980 and recent re-assessments by Preece and Bridgland 1998, 1999). Furthermore, the tufa deposits often seal, or are interstratified with, palaeosols representing early to mid-Holocene land surfaces. It is particularly noticeable that such British tufa deposits often have associated *in-situ* Mesolithic archaeology, for example at Blashenwell, Dorset (Preece 1980), Cherhill, Wiltshire (Evans and Smith 1983), Cwm Nash (Evans *et al.* 1978), Bossington (Davies 1992, Davies and Griffiths 2005). Other such sites are listed in Evans (1972).

Associated Mesolithic material is also evident at several Irish sites (Preece and Robinson 1982, Preece *et al.* 1986). In a number of studies some of the molluscan evidence obtained has been interpreted as suggestive of Mesolithic interference with woodland vegetation (*e.g.* Evans *et al.* 1978, Preece 1980, Preece *et al.* 1986, Davies 1992; Davies and Griffiths 2005), although whether such clearings were in fact purposefully created rather than naturally created and opportunistically utilised (Brown 1997) is impossible to assess. Nevertheless, the importance of tufa deposits as repositories of both environmental and cultural material is clear.

Generally, the environmental histories that emerge from tufa deposits are in good agreement, both with each other and with respect to what is known about the Late-Glacial and Holocene periods from other environmental evidence. Broadly they show cold, open Late-Glacial conditions giving way as the Holocene progresses to warmer and eventually fully wooded habitat. This is very clearly shown at Sidlings Copse, Oxfordshire, where the molluscan evidence was complemented by pollen data (Preece and Day 1994). Extensive tufa deposition seems to have ceased by around 4000 BP, the cessation most often being explained as a result of changing hydrological conditions consequent upon increasing human impact and landscape openness (*e.g.* Preece and Robinson 1982). Further discussion of the 'tufa decline' can be found in Griffiths and Pedley (1995) and Goudie *et al.* (1993).

The woodland fauna generally found in mid-Holocene tufa deposits are extremely species-rich in comparison with fauna documented from modern woodland habitat (Meyrick and Preece 2001), reflecting the ideal conditions offered to Mollusca in warm, damp, shaded and undisturbed habitat. This difference is attributable in part to the lack of modern analogue habitat (warm, damp, shaded, undisturbed *and* calcium carbonate forming conditions), and perhaps also in part to the lack of quantitative studies from wet woodland. Most modern studies have been of relatively dry woodlands (*e.g.* Bishop 1976, Wardhaugh 1997) although of course there is usually some inherent spatial variability in the dampness of the sites studied. There have been few studies of wet-woodlands (carrs, or more particularly marsh woodland – based on mineragenic rather than organic substrates), and even fewer within the same general geographic area as substantial tufa deposits; in short, there are limited appropriate modern data. It is worthy of note that in terms of numbers of species some of the modern samples from fen and carr at Bossington, Hampshire (Davies *et al.* 1996), an area where there are also substantial tufa deposits (Pentecost 1993), approached the richness found by Meyrick and Preece (2001) in fossil tufa deposits, although the species composition was different.

Recently, it has been noted that modern tufa deposition is most likely severely underestimated, at least in some areas (Baker and Simms 1998). Although deposition of large quantities of tufa in any one location is not likely to approach that normal in early-mid Holocene deposits, the actual number of tufa deposition points within suitable geological areas (the limestone of the Mendips and the Wessex area, Somerset) was demonstrated to be potentially much higher than previously estimated. Subsequent visits by the author and colleagues to some of the sites mentioned by Baker and Simms (1998) have confirmed tufa deposition. Additionally, survey of other likely sites has added to the list of both modern and fossil deposits. Crucially, some of the more substantial modern deposition is occurring in woodland with the type of ground conditions often inferred in sub-fossil interpretation, principally wet, tussocky, and herb- and calcium carbonate-rich (Fig. 15.4). The undisturbed analogy is less secure, in that the woodlands do not necessarily have a demonstrably long history, although they are certainly not heavily impacted upon by human activity. Deposition at other sites is in open conditions with vegetation ranging from short to long and tussocky. Relevant too is the fact that substantial early-mid Holocene deposition in the Test Valley, Hampshire, has recently been documented as having formed largely in open-conditions, although a woodland 'phase' is apparent (Davies and Griffiths 2005).

Given that there are now potentially some useful modern analogue sites within the same geographical area (southern Britain), it would seem an ideal opportunity to test sub-fossil inferences which have often been internally-derived, *i.e.* with relation to the sub-fossil Mollusca alone. It has already been outlined how the fauna of fen-type habitat closely resemble those from carr or wet-woodland-type habitat, and it would seem that the possibility that some of the 'woodland' fauna of fossil tufa deposits might not in fact represent woodland, but

Fig. 15.4 Modern tufa deposition site within woodland, Mendip.

other highly structured wet-ground environments, ought to be tested. The interpretations
offered may well be correct, but will
obviously be strengthened by being 'tested'
via analogous comparison. Certainly, species
composition will vary between past and
present, some species common to Holocene
tufa deposits now being either absent from
the UK or national rarities (Meyrick and
Preece 2001). Nonetheless, it is true to say
that the degree to which analogous comparisons can be made can only be established
after the fact.

WETLANDS AND MOLLUSCA – FUTURE DIRECTIONS

The lack of modern faunal studies that are
appropriate for analogous comparison with
sub-fossil assemblages has already been
highlighted, and this clearly needs address-
ing. The fact that neither the studied habitats
nor the species involved may be identical
with fossil counterparts is to some extent
irrelevant or, as in the case of alluvial fauna
above, needs demonstrating in any case.
Similar habitat (and this means similar as
far as the Mollusca are concerned) will result
in similar general faunal characteristics (see
above). More narrowly, the following two
aspects would seem worthy of study.

First, seasonal variation of molluscan
fauna has been outlined as a possibility on

Fig. 15.5 River Wylye, overbank flood conditions.

the basis of sub-fossil studies in English river valleys (*e.g.* Robinson 1988, Davies 1996) and recently
demonstrated in modern studies elsewhere (Valero *et al.* 1998). Research into the temporal variation of
active molluscan communities on floodplains subject to seasonal inundation (Fig. 15.5) would perhaps help
further in the interpretation of sub-fossil wet-ground taxocenes (see above), although finding floodplains
that flood in a manner similar to that which would have occurred prior to recent flood management regimes
may be difficult.

Second, the spatial variation of molluscan fauna in spatially complex habitat needs to be further
addressed. As mentioned above, one of the reasons why there may not be any published modern studies
demonstrating the same level of species richness as some mid-Holocene tufa assemblages may relate to the
fact that similar habitats (wooded, tussocky, wet, carbonate-rich) have yet to be adequately studied. The
modern floodplain study of Davies and Grimes (1999) has demonstrated a high degree of spatial separation
of species over relatively small distances due to minor variations in hydrology and vegetation structure, a
feature that has also been demonstrated elsewhere (Vareille-Morel *et al.* 1999). Separation is likely to be

much more marked in spatially complex habitats (Fig. 15.4), particularly if they are also subject to seasonal change. The death assemblages, and future sub-fossil assemblages, will reflect this complexity.

Finally, and by way of summary, it is worthwhile reiterating that improving the interpretation of sub-fossil molluscan sequences is most likely to be dependent upon more modern studies, a point also made by Evans and O'Connor (1999). Further, it is important that future modern studies are designed specifically with the aim of furthering palaeoecological interpretation.

ACKNOWLEDGEMENTS

The basis of this work relied very heavily on the advice and guidance of Professor J. G. Evans. Lisa Thomas is thanked for introducing me to some of the modern tufas forming on the Mendips, Somerset.

REFERENCES

Baker, A. & M. J. Simms 1998. Active deposition of calcareous tufa in Wessex, UK, and its implications for the late-Holocene tufa decline. *The Holocene* 8, 359–65.

Boycott, A. E. 1934. The habitats of land Mollusca in Britain. *Journal of Ecology* 22, 1–38.

Boycott, A. E. 1936. The habitats of fresh-water Mollusca in Britain. *Journal of Animal Ecology* 5, 116–86.

Bishop, M. J. 1976. Woodland Mollusca around Nettlecombe, Somerset. *Field Studies* 4, 457–64.

Brown, T. 1997. Clearances and clearings: deforestation in Mesolithic/Neolithic Britain. *Oxford Journal of Archaeology* 16, 133–46.

Bush, M. B. 1988. The use of multivariate analysis and modern analogue sites as an aid to the interpretation of data from fossil mollusca assemblages. *Journal of Biogeography* 15, 849–61.

Cameron, R. A. D. 1978a. Interpreting buried land snail assemblages from archaeological sites – problems and progress, in D. R. Brothwell, K. D. Thomas & J. Clutton-Brock (eds) *Research Problems in Zooarchaeology*, 19–23. London: Institute of Archaeology.

Cameron, R. A. D. 1978b. Terrestrial snail faunas of the Malham area. *Field Studies* 4, 715–28.

Cameron, R. A. D. & M. Redfern 1972. The terrestrial Mollusca of the Malham area. *Field Studies* 4, 589–602.

Davies, P. 1992. Sub-fossil Mollusca from Holocene overbank alluvium and other wet-ground contexts in Wessex. Unpublished Ph.D thesis, Department of Archaeology, University of Wales, College of Cardiff.

Davies, P. 1996. The ecological status of *Pupilla muscorum* (Linné) in Holocene overbank alluvium at Kingsmead Bridge, Wiltshire. *Journal of Conchology* 35, 467–71.

Davies, P. 1998. Numerical analysis of subfossil wet-ground molluscan taxocenes from overbank alluvium at Kingsmead Bridge, Wiltshire. *Journal of Archaeological Science* 25, 39–52.

Davies, P. 2003a. Chaos and patterns: reconstructing past environments using modern data. The molluscan experience, in V. Straker & K. Robson-Brown (eds) *Proceedings of the Archaeological Sciences 1999 Conference, University of Bristol*, 36–41. Oxford: British Archaeological Reports International Series 1111.

Davies, P. 2003b. The interpretation of Mollusca from Holocene overbank alluvial deposits: progress and potential, in A. J. Howard, M. G. Macklin & D. G. Passmore (eds) *The Alluvial Archaeology of North West Europe and the Mediterranean*, 291–304. Rotterdam: Balkema.

Davies, P. & H. I. Griffiths 2005. Molluscan and ostracod biostratigraphy in the River Test at Bossington, Hampshire. *The Holocene* 15(1), 97–110.

Davies, P. & C. J. Grimes 1999. Small scale spatial variation of pasture molluscan faunas within a relic watermeadow system at Wylye, Wiltshire, UK. *Journal of Biogeography* 26, 1057–63.

Davies, P., C. H. Gale & M. Lees 1996. Quantitative studies of modern wet-ground molluscan faunas from Bossington, Hampshire. *Journal of Biogeography* 23, 371–7.

Evans, J. G. 1972. *Land Snails in Archaeology*. London: Seminar Press.

Evans, J. G. 1991. An approach to the interpretation of dry-ground and wet-ground molluscan taxocenes from central-southern England, in D. R. Harris & K. D. Thomas (eds) *Modelling Ecological Change*, 75–89. London: Institute of Archaeology.

Evans, J. G. & T. O'Connor 1999. *Environmental Archaeology: principles and methods.* Stroud: Sutton.

Evans, J. G. & I. F. Smith 1983. Excavations at Cherhill, north Wiltshire 1967. *Proceedings of the Prehistoric Society* 49, 43–117.

Evans, J. G., C. French & D. Leighton 1978. Habitat change in two Late-glacial and Post-glacial sites in southern Britain: the molluscan evidence, in S. Limbrey & J. G. Evans (eds) *The Effect of Man on the Landscape: the lowland zone*, 63–75. London: Council for British Archaeology.

Evans, J. G., P. Davies, R. Mount & D. Williams 1992. Molluscan taxocenes from Holocene overbank alluvium in central southern England, in S. Needham & M. G. Macklin (eds) *Alluvial Archaeology in Britain*, 65–74. Oxford: Oxbow Monograph No. 27.

Evans, J. G., S. Limbrey, I. Máté & R. Mount 1988. Environmental change and land-use history in a Wiltshire river valley in the last 14,000 years, in J. C. Barrett & I. A. Kinnes (eds) *The Archaeology of Context in the Neolithic and Bronze Age: recent trends*, 97–103. Sheffield: Department of Archaeology and Prehistory, University of Sheffield.

Ford, T. D. & H. M. Pedley 1996. A review of tufa and travertine deposits of the world. *Earth-Science Reviews* 41, 117–75.

Gee, J. H. R. & P. S. Giller 1991. Contemporary community ecology and environmental archaeology, in D. R. Harris & K. D. Thomas (eds) *Modelling Ecological Change*, 1–12. London: Institute of Archaeology.

Griffiths, H. I. & H. M. Pedley 1995. Did changes in the late Last Glacial and early Holocene atmospheric CO_2 concentrations control rates of precipitation? *The Holocene* 5, 238–42.

Goudie, A. S., H. A. Viles, & A. Pentecost 1993. The Holocene tufa decline in Europe. *The Holocene* 3, 181–186.

Harris, D. R. & K. D. Thomas 1991. Modelling ecological change in environmental archaeology, in D. R. Harris & K. D. Thomas (eds) *Modelling Ecological Change*, 91–102. London: Institute of Archaeology.

Kerney, M. P. 1977. A proposed zonation scheme for Late-glacial and Postglacial deposits using land Mollusca. *Journal of Archaeological Science* 4, 387–90.

Kerney, M. P., E. H. Brown & T. J. Chandler 1964. The Late-glacial and Post-glacial history of the chalk escarpment near Brook, Kent. *Philosophical Transactions of the Royal Society of London B* 248, 135–204.

Kerney, M. P., R. C. Preece & C. Turner 1980. Molluscan and plant biostratigraphy of some Late-Devensian and Flandrian deposits in Kent. *Philosophical Transactions of the Royal Society of London B* 291, 1–43.

Meyrick, R. A. & R. C. Preece 2001. Molluscan successions from two Holocene tufas near Northampton, English Midlands. *Journal of Biogeography* 28, 77–93.

Pentecost, A. 1993. British travertines: a review. *Proceedings of the Geologists Association* 104, 23–39.

Preece, R. C. 1980. The biostratigraphy and dating of the tufa deposit at the Mesolithic site at Blashenwell, Dorset, England. *Journal of Archaeological Science* 7, 345–62.

Preece, R. C. & D. R. Bridgland (eds) 1998. *Late-Quaternary Environmental Change in North-west Europe: excavations at Holywell Coombe, south-east England.* London: Chapman & Hall.

Preece, R. C. & D. R. Bridgland 1999. Holywell Coombe, Folkestone: a 13,000 year history of an English chalkland valley. *Quaternary Science Reviews* 18, 1075–125.

Preece, R. C. & S. P. Day 1994. Comparison of Post-glacial molluscan and vegetational sequences from a radiocarbon-dated tufa sequence in Oxfordshire. *Journal of Biogeography* 21, 463–78.

Preece, R. C. & J. E. Robinson 1982. Molluscan and Ostracod faunas from Post-glacial tufaceous deposits in County Offaly. *Proceedings of the Royal Irish Academy 82B*, 115–31.

Preece, R. C. & J. E. Robinson 1984. Late Devensian and Flandrian environmental history of the Ancholme valley, Lincolnshire: molluscan and ostracod evidence. *Journal of Biogeography* 11, 319–52.

Preece, R. C., P. Coxon & J. E. Robinson 1986. New biostratigraphic evidence of the Post-glacial colonization of Ireland and for Mesolithic forest disturbance. *Journal of Biogeography* 13, 487–509.

Robinson, M. A. 1988. Molluscan evidence for pasture and meadowland on the floodplain of the upper Thames basin, in P. Murphy & C. French (eds) *The Exploitation of Wetlands*, 101–12. Oxford: British Archaeological Reports British Series 186.

Sparks, B. W. 1961. The ecological interpretation of Quaternary non-marine Mollusca. *Proceedings of the Linnean Society of London* 172, 71–81.

Thomas, K. D. 1985. Land snail analysis in archaeology: theory and practice, in N. R. J. Fieller, D. D. Gilbertson & N. G. A. Ralph (eds) *Palaeobiological Investigations: research design, methods and data analysis*, 131–56. Oxford: British Archaeological Reports International Series 266.

Valero, M. A., R. Marti, M. D. Marcos, F. Robles & S. Mascoma 1998. *Lymnaea truncatula* (Mollusca: Lymnaeidae) in the rice fields of eastern Spain. *Vie et Milieu-Life and Environment* 48, 73–8.

Vareille-Morel, C., G. Dreyfuss & D. Rondeland 1999. The characteristics of habitats colonized by three species of *Lymnaea* (Mollusca) in swampy meadows on acid soil: their interest for control of fasciolosis. *Annales de Limnologie* 35, 173–8.

Wardhaugh, A. A. 1997. The terrestrial molluscan fauna of some woodlands in north-east Yorkshire: a framework for quality scoring and association with old woodland flora. *Journal of Conchology* 36, 19–30.

Chapter 16

Non-Marine Ostracods as Indicator Taxa in Ancient Wetlands

Huw I. Griffiths

INTRODUCTION

Freshwater ostracods are now one of the standard tools in aquatic palaeobiology, and a large literature has developed detailing their use in biostratigraphy and, most particularly, in palaeolimnology, palaeoecology and palaeoclimatology. Despite this, ostracods have been almost universally ignored by environmental archaeologists, possibly because of their small size, and also because it is commonly believed that their taxonomy is difficult. In fact, the taxonomy of freshwater ostracods is comparatively easy. Global (and continental) species richnesses are relatively low when compared to other environmental proxy groups (*e.g.* diatoms, forams, Mollusca, insects) and as such, learning the taxonomy of the suites of species found within a particular region does not represent a major challenge. This is particularly true since continental-level faunal listings have become available for Europe (Meisch 2000), North America (Delorme 1991), Africa (Martens 1984) and South America (Martens and Behen 1994). Moreover, national checklists are available for a large number of countries not covered by these works (*e.g.* New Zealand – Chapman 1963, Barclay 1968; Turkey – Altınsaçı and Griffiths 2002; Iran – Griffiths *et al.* 2001; Israel – Martens and Ortal 1999). Since Griffiths *et al.*'s (1993) introduction to the basic techniques for the use of ostracods in archaeological work, the accessibility of relevant literature has expanded considerably. Notable advances are Meisch's (2000) excellent review of the taxonomy and ecology of freshwater ostracods from central and western Europe, and Griffiths and Holmes' (2000) monograph on ostracod palaeoecology, which includes a valve-based key to the British Quaternary species.

Ostracods offer several methodological and interpretative advantages to environmental archaeologists:

– Ostracod shells ('valves') are readily and abundantly preserved in calcareous, aerobic sediments (sometimes several thousands per gram). Thus their remains are often present where pollen is not.

– Ostracods, living within a variety of habitats within a water body, provide more information about lake environments than does pollen – palynological signals largely reflect lake-edge vegetation and the fall-out from a broader spatial catchment.

– At a broader geographic scale, ostracods provide data on temperatures, vegetation, permanence/ephemerality, salinity and solute chemistry, and thus on climatic change.

– Ostracods are often present with other groups of invertebrates (Mollusca, chironomids, Cladocera and diatoms) and their analysis can be undertaken in conjunction with these (*e.g.* Griffiths *et al.* 1994, 2002).

– Assemblages are frequently autochthonous, and the presence or absence of appropriate ranges of juvenile moult stages means that allochthony can be tested.

– Ostracod species-level taxonomy is comparatively well understood and stable, and individual species' ecologies are relatively well described.

In addition (and in contrast to most other taxa preserved in Quaternary sediments, *e.g.* Cladocera, diatoms, insects and mites), freshwater ostracods are evolutionarily dynamic within the Quaternary, and show short time-scale morphological evolution. This allows certain forms to be used for broad-scale site dating, especially in the case of the more ancient sedimentary sequences (see Griffiths 2001). Finally, a more recent development is the application of geochemical techniques to ostracod shell calcite. Trace element ratios have been examined, but the most notable of these methodologies is the analysis of stable isotope ratios – notably $^{18}O/^{16}O$ and $^{13}C/^{12}C$. This powerful tool allows the ecological and geochemical investigation of material originating from the same subfossil samples. In this way investigators may not only make palaeoecological inferences, but also examine quantitative changes in temperature and nutrient status within sediment successions (*e.g.* von Grafenstein *et al.* 1994, 1999). Thus, given the advantages associated with ostracod work, the lack of interest in ostracods from archaeological sites is surprising; to date only one German diploma and two British PhD theses have been produced either entirely or focussed largely on this subject (Mount 1991, Nöthlings 1992, Griffiths 1995).

Ostracods, like all other organisms, respond to complex and varying suites of environmental variables, and species responses in space and time may sometimes vary (Marmonier and Danielopol 1988 provide an interesting example). Thus, even though the ecologies of all species are known in broad terms, and those of some have been described in detail, the use of single 'indicator species' for palaeoenvironmental determinations is unwise. A more robust approach is the use of 'hard wired' Key Biological Factors (KBFs) for reconstructing environments. Useful characteristics may include reproductive or feeding mode, or tolerances for seasonal desiccation, salinity or anoxia. Another crude but very effective interpretational method is that of Absolon (1973). From examination of a large number of Last Glaciation late-glacial and Holocene calcareous deposits in central Europe, Absolon was able to propose distinct ostracod species associations that correlated with distinct depositional regimes. He was also the first to notice different faunal associations in late-glacial and Holocene lake deposits – the late-glacial '*candida*-Fauna' (which is cold, oligotrophic and essentially bottom-dwelling) and an early Holocene '*cordata*-Fauna' composed of more eurythermal taxa, and also free-swimming and plant associated forms. Absolon's categories remain sound, although in the case of the '*candida*-Fauna' and '*cordata*-Fauna' it is now clear that there is some overlap between the two, and some slight regional differences also exist, notably for very large water bodies such as the Alpine lakes.

OSTRACODS FROM ARCHAEOLOGICAL CONTEXTS

Middle Pleistocene (Marine Oxygen Isotope Stage [MIS] >5) and older sites

In the UK, formal ostracod reports exist for Middle Pleistocene sites at Purfleet (MIS 9) (Schreve *et al.* 2001), Beeches Pit (MIS 11) (see Fig. 16.1) (Preece *et al.* 2000, Griffiths unpubl.), Clacton-on-Sea (MIS 11) (Holmes 1996) and Little Oakley (possibly MIS 15 or earlier) (Robinson 1990). In addition, several of the more important European Palaeolithic sites have yielded rich faunal assemblages, which have included

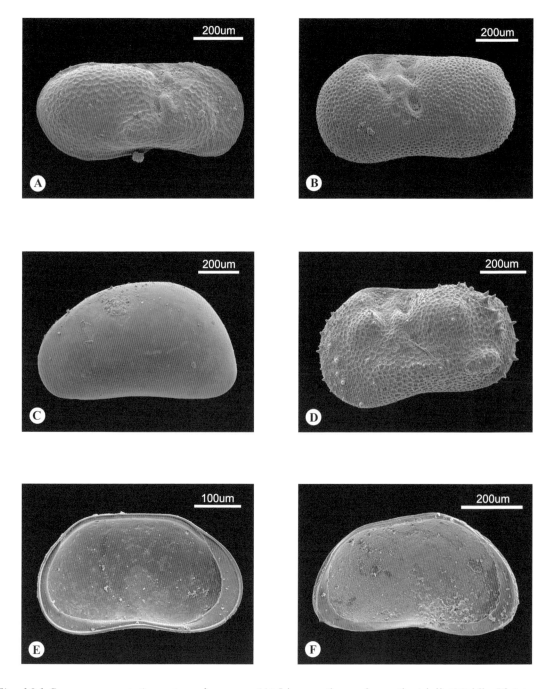

Fig. 16.1 Some representative ostracod genera: (A) Limnocythere *cf.* sanctipatricii *(Middle Pleistocene, Beeches Pit, UK), (B)* Ilyocypris bradyi *(Middle Pleistocene, Beeches Pit, UK), (C)* Candona candida *(Middle Pleistocene, Beeches Pit, UK), (D)* Ilyocypris *sp. indet. (Middle Pleistocene, Purfleet, UK), (E)* Nannocandona faba *(Holocene, Alport tufa, UK), (F)* Potamocypris fallax *(Holocene, Alport tufa, UK).*

ostracods. Unfortunately, some sites have poor stratigraphic controls and their dates are uncertain, particularly within the context of the MIS chronology. European Middle Pleistocene sites include the German travertine at Ehringsdorf (MIS 7), which contains human remains, artefacts and hearths (Diebel and Wolfschläger 1975) and the *Homo erectus* site at Bilzingsleben (possibly MIS 11) (Diebel and Pietrzeniuk 1990). Two further *Homo erectus* sites with ostracod faunas are Vértesszőlős in Hungary and Stránská Skála in the Czech Republic. Vértesszőlős (which may date to MIS 13) includes human remains, artefacts, footprints, hearths, and diverse mollusc and mammal faunas. The latter include many small mammals (rodents and insectivores) and also 24 larger mammal species (primarily ungulates) recovered from a 'kitchen midden' deposit (Kretzoi and Dobosi 1990). At Stránská Skála (possibly MIS 15 or earlier) ostracods were found only in non-archaeological deposits at Černovice (Molčíková in Musil 1995), and so were not associated directly with the artefacts and burnt and butchered vertebrate remains found in the talus cones (Musil 1995). The only sites known to the author to pre-date these are the Plio-Pleistocene ostracod assemblages from *Homo erectus* sites in Israel's Arava Valley and reported on briefly by Rosenfeld *et al.* (1997).

The ostracod faunal assemblages of some of the sites (*e.g.* Stránská Skála) were studied by inexperienced or non-specialist workers. Of the remainder, chronoendemic forms are known from Purfleet (Schreve *et al.* 2001), Beeches Pit (Preece *et al.* 2000, Griffiths unpubl.), Little Oakley (Robinson 1990), Ehringsdorf (Diebel and Wolfschläger 1975) and Vértesszőlős (Diebel and Pietrzeniuk 1990). All include varied faunas that can be tied to broad climatic and ecological regimes, and chronospecies that can assist in dating. The most disappointing in faunistic terms is the Bilzingsleben tufa, where not only are no extinct species known, but the ecological signal itself is poor.

Last Interglacial (MIS 5e) sites

Although several ostracod sequences are known from sites attributed to the Last Interglacial (Griffiths 1996), very few of these have associated archaeology. Humans are believed to have been absent from the British Isles in the Ipswichian, and archaeology is also unknown from the Low Countries (Speleers 2000). To date, the only European sites with both archaeology and ostracod records are the German localities of Taubach (Diebel and Pietrzeniuk 1977), Burgtonna (Diebel and Pietrzeniuk 1978) and Gröbern, Grabschütz and Neumark Nord (Fuhrmann and Pietrzeniuk 1990a, b, c). These localities all feature vertebrate remains, molluscs, pollen and Palaeolithic artefacts. The sites at Burgtonna and Taubach are a series of travertines (= tufa *sensu* Pedley 1990), but the three remaining sites are extensive lake deposit sequences in depressions in Saalian till (see von Kolfschoten 2000) that were exposed during the former DDR's widespread open cast mining of lignite (brown coal). The ostracod record from each of these sites is stratigraphically relatively complete (especially so in the case of the three lake sites), and the faunas are diverse and well preserved. The signals from the tufa sites are difficult to interpret, and this is further hampered by poor stratigraphic control, so that formal correlation is difficult for much of both successions (von Kolfschoten 2000). This is not the case, however, at the palaeolakes of Gröbern, Grabschütz and Neumark Nord, and the results of ostracod analysis have provided detailed records of changes in temperature, hydrochemistry and water depth (*i.e.* climatic proxy data), which were particularly useful for reconstruction of long timescale site palaeoecology and palaeoclimatology. It is notable that the modern North American and Siberian species *Fabaeformiscandona rawsoni* appears almost exclusively in Last Interglacial sites (including at several of these localities), suggesting that the species may be a useful biostratigraphic marker of MIS 5e in Europe.

Last Glaciation (MIS 4–2) sites

Last Glacial archaeological sites with ostracods are very poorly documented. One of the few exceptions is the Kastritsa rock shelter in northwest Greece (Galanidou *et al.* 2000). This site, situated on a limestone hill near Lake Pamvotis, features lithic (and other) artefacts and an extensive bone assemblage dominated by deer. Dating of the site has been problematic, however. A series of raised beach deposits had previously been related to palaeolake levels, but this evidence has now been re-evaluated on the basis of sediment sequences from Lake Pamvotis which contain pollen, molluscs and ostracods, coupled with data from ostracod ecology and the stable isotope analysis of ostracod calcite (Frogley *et al.* 1999, 2001). These not only reconstruct climate and lake levels, but also show the site to be rather older than originally thought (Galanidou *et al.* 2000). Within northern Europe, the only site that seems to fall within the Last Glaciation is Königsaue in Sachsen-Anhalt, Germany (see Mania and Toepfer 1973). However, stratigraphic control over this site is again poor, and provenance within either MIS 2 or MIS 3 seems plausible.

Holocene (MIS 1) sites

Not surprisingly, the majority of ostracod records from archaeological sites come from Holocene deposits, and much of the published work is from either lake deposits or tufas. To date, most work has been undertaken on sites in Britain (Preece 1980, Robinson 1986, 1991, Griffiths and Mount 1993, Griffiths 1995, Kontrovitz *et al.* 1995, Holmes and Griffiths 1999) or Germany (Mania 1967a, b, Gunther 1986, Nöthlings 1992, Griffiths *et al.* 1994, Keding *et al.* 1995, Janz and Matzke Karasz 2001). However, there are also publications that refer to investigations undertaken in Italy (Taliana *et al.* 1996), Ireland (Griffiths 1999), Poland (Bilan 1988), Switzerland (Oertli 1967), the United Arab Emirates (Gebel *et al.* 1989), Greece (Finke and Malz 1988, Zangger 1993, Zangger and Malz 1989), Belize (Bradbury *et al.* 1990, Alcala-Herera *et al.* 1994) and the southern United States (Palacios-Fest 1994).

Almost all of the most valuable studies have been based on extended lacustrine sequences, detailing time series data that allow sequentially-based interpretation of environmental change over time, albeit at a range of sampling resolutions (the Italian cave site of 'Grotta del lago' is unusual, although the cave infill was largely of lacustrine deposits: Taliana *et al.* 1996.) Some of the longest and best resolved records studied to date are those from the German sites of Duvensee and Stellmoor (Gunther 1986, Griffiths *et al.* 1994), both near Hamburg in northern Germany, and both with abundant archaeology (Bokelmann 1971, 1991). The Duvensee and Stellmoor sequences cover the period from the Last Glaciation late-glacial (Bölling/ Alleröd) through to the early or mid Holocene.

The Stellmoor site is particularly well known because of the massive assemblages of weapons and reindeer bones derived from the seasonal activities of Late Palaeolithic (and later) hunters, and described in detail by Rust (1937, 1943) (see also palaeoeconomic details in Bokelman 1991). Both Stellmoor and Duvensee record changes in lake level and trophic status over time, and these directly reflect on site palaeoeconomy, whilst providing palaeoclimatic inferences that augment data obtained from palynological analyses (*e.g.* Averdieck 1986, Usinger 1981). More significantly, they also record the history of lake evolution, which directly reflects upon local and regional environmental change, and on resource utilisation by hunter-gatherer societies. Although not as detailed, the record from Star Carr in Yorkshire, England, records much the same suite of processes (Holmes and Griffiths 1999). Many of these relate to the natural 'ageing' of the lake basin (*i.e.* eutrophication, humification, infilling and encroachment) and also, at a regional scale, the evidence reflects climate changes, most notably those occurring across the Late-glacial/ Holocene boundary.

One of the most detailed successions studied so far (albeit covering a shorter time-span) comes from Lough Boora in the Irish Midlands. Lough Boora, an extensive lake marl deposit overlain by peat, has one of the earliest Mesolithic sites in Ireland (O'Sullivan 1998). The Lough Boora assemblage consists of a hearth, artefacts and bone (some burnt) from a hunting camp on a peat-buried promontory (see Ryan 1980, 1984, van Wijngaarden-Bakker 1989). A pollen record from a peat monolith overlying the site and taken during excavation was studied by O'Connell (1980), but this only detailed environmental change at the site subsequent to its burial. In contrast, the underlying lake marl features very large numbers of mollusc and ostracod remains. The latter, when sampled in high resolution time series, revealed a detailed record of repeated fluctuations in lake level and nutrient status throughout the early Holocene (Griffiths 1999), with comparatively rapid trophic overturn and infilling. These acted to remove the lake's value as an economic resource (*e.g.* as a source of fish and waterfowl) for Mesolithic people.

Tufa successions containing good ostracod sequences are not uncommon, although to date, only a few Holocene sites are of archaeological interest (*e.g.* Preece 1980, Preece *et al.* 1986, Griffiths and Mount 1993, Davies and Griffiths 2005). Tufas are potentially a good source of palaeoenvironmental, palaeoclimatic and palaeohydrological data. However, in practice, they are complicated but highly constrained environments and, in ecological terms, are relatively stable; tufa-depositing waters tend to be cold, oligotrophic and calcium rich, and are often groundwater fed (Pedley 1990, Ford and Pedley 1996). As a result, many tufa deposits contain little information on environmental changes beyond the immediate depositional environment, and this may be difficult to relate to the broader catchment. Moreover, as tufas usually derive from flowing water systems, the taphonomic processes within them can be complex and include both extensive spatial and temporal averaging (Griffiths *et al.* 1996).

It is clear that much of the work on tufa-ostracod associations to date has been over simplistic. Despite this, some tufa sites do feature ostracod signals that reflect events within the broader environment, and this is particularly true when molluscan evidence is also available. For example, effects on local environment by forest disturbance and/or clearance have been picked up at sites in both Ireland (Preece *et al.* 1986) and southern England (Davies and Griffiths 2005). In the latter case (at Bossington, Hampshire) this was identified by changes within both the mollusc and ostracod faunas. Unusually, the latter also features groundwater ostracod species (*e.g. Pseudocandona breuili* and *P. eremita*), the influence of which decreases throughout the sequence. The succession from Bossington (which contains Mesolithic artefacts and beaver remains) is unusual amongst tufa studies as several profiles were examined from different areas adjacent to the palaeo-River Test, and interpretation is enhanced by the incorporation of elements from a complex ecological mosaic.

The well-known sensitivity of ostracod faunas to changes in salinity/solute chemistry has been harnessed in several archaeological contexts. At Ain al-Faidha / Al-'Ain in Abu Dhabi (United Arab Emirates), Gebel *et al.* (1989) were able to reconstruct salinity regimes that are believed to have had broad-scale relevance to Neolithic site utilisation, notably by progressive climatically-driven evaporative concentration of lake waters. Further afield, in Arizona, Palacios-Fest (1994) reconstructed the salinities of Hohokam agricultural drainage ditches using ostracod faunal and shell geochemical data (Mg/Ca and Sr/Ca ratios), and was able to identify both climate-related flood phases and progressive soil salinisation between AD 1025 and 1425. Similarly, Bradbury *et al.* (1990) used pollen, mollusc and ostracod data to examine Mayan agriculture in Belize, deriving a record of progressive salinisation believed to correlate with the eventual collapse of Mayan civilisation locally.

OPPORTUNISTIC SAMPLING

Thus far, two other approaches have been used with regard to ostracods from archaeological sites – spot sampling or identification of opportunistically retrieved material for ecological determination, and offsite studies. Many of the latter have involved multidisciplinary studies on large-scale research projects, some of which have been primarily archaeological, whilst others have revealed archaeology incidentally.

Many of the ostracod samples submitted for analysis from archaeological contexts simply represent ostracods that have been spotted in sieve residues during, for example, molluscan work. Samples such as these are usually highly taphonomically biased towards larger-sized adult and subadult valves, and also to the large-bodied species (*Herpetocypris, Candona* etc.) retained in larger mesh-size sieve residues. This will markedly distort population profiles (which can provide information on the degree of autochthony or otherwise of the assemblage) as well as providing an unrepresentative view of the original fauna (biased towards benthic species at the expense of smaller free-swimming taxa). Despite this, even this evidence can sometimes be useful, as for example in the case of a two species assemblage recovered from Roman sewers at York (Meyrick 1976), two samples submitted from the Turkish Neolithic site of Çatalhöyük in the Konya Basin (Griffiths unpubl.) and a small Middle Pleistocene assemblage (with accompanying Foraminiferida) from Boxgrove (Whatley and Haynes 1986). Most notable amongst these is a single sample from Turkish Lake Kuş which provided evidence for a tsunami event in the Byzantine Period (Leroy *et al.* 2001).

Increasingly, ostracods have been used during programmes to collect essentially off-site, broad-scale palaeoecological and palaeoclimatic data that can be referred back to particular sites. Examples of work of this kind are de la Vega Leinert *et al.*'s (2000) work in the Bay of Skaill, Orkney, Scotland, which reflects broad-scale environmental trends affecting and associated with human activity at the Neolithic site of Skara Brae. A further example comes from work in the Konya Basin in Turkey (Roberts *et al.* 1999), in which multiple proxies (ostracods, pollen, diatoms) have been used to reconstruct broad-scale environments and climatic contexts for sites such as Çatalhöyük, and also for processes such as the establishment of Neolithic farming and plant and animal domestication.

WAYS FORWARD

At present ostracod analysis from wetland archaeological contexts remains under-exploited, largely because of a shortage of specialist workers and also because of a widespread (and mistaken) belief that the taxonomy of the group is difficult. It is clear, however, that the best results are obtained by involvement of the ostracod worker at an early stage in the study and, if possible, during field sampling, as ostracod analysis works best with a full data set! Moreover, the answers obtained from ostracod analysis largely depend on the questions being asked – for example, whether detailed palaeosite reconstructions are required and at what spatial scale or, as is often the case for Middle Pleistocene Palaeolithic sites, whether ostracods can also assist in dating.

Clear developments for the near future will involve:

– Strategic integration of ostracod specialists into archaeological work in wetland environments, preferably during the project planning phase. This will ensure that meaningful field samples are taken, and that these are treated appropriately post-excavation.

– Development of ostracod transfer functions and combined transfer functions (see Reed *this volume*). Already an ostracod transfer function has being developed for Spain (Mezquita *et al.* 2005), and a combined ostracod and diatom transfer function for Turkey is nearing completion (Reed and Griffiths unpubl.). These will allow more robust site reconstructions, based on quantitative rather than qualitative (and semi-intuitive) inferences.

– Further development of stable isotope ostracod shell geochemical techniques. These have already been shown to be of value in a range of palaeolimnological and environmental impact studies (see Griffiths and Holmes 2000 for a brief review) but still remain to be applied to ostracods from archaeological sites.

– Application of new technologies, particularly at sites where replicated spatio-temporal sampling is possible. The use of techniques such as Ground Penetrating Radar (GPR) (*e.g.* Pedley *et al.* 2000) can provide data that not only allow more powerful analysis of within-succession biofacies, but also can be integrated into a Geographical Information System (GIS) and modelled (*e.g.* Ferrier and Wadge 1997).

As yet the only attempt at placing archaeological ostracod data within a GIS is Finke and Malz's (1988) reconstruction of the palaeoenvironments of Greek Lake Lerna, but developments such as these offer exciting possibilities for the future.

ACKNOWLEDGEMENTS

Horst Janz (Tübingen) and Renate Matzke Karasz (München) are thanked for help with literature, and Paul Davies (Bath Spa) for helpful discussions. Dan Schreve (Royal Holloway College, Egham) provided advice on MIS dating, although any mistakes are the author's own. Rachel Fazakerly, Tony Sinclair and John Garner (Hull) provided the SEM work and plate, and Jane Reed (Hull) made critical comments on the manuscript. John G. Evans (Cardiff) is thanked for inspiring the author's interest in ostracods.

REFERENCES

Absolon, A. 1973. Ostracoden aus einigen Profilen spät- und postglazialer Karbonatablagerungen in Mitteleuropa. *Mitteilungen der Bayerischen Staatssammlung für Paläontologie und historische Geologie* 13, 47–94.

Alcala-Herera, J. A., J. S. Jacob, M. L. M. Castillo & R. W. Neck 1994. Holocene palaeosalinity in a Maya wetland, Belize, inferred from the microfaunal assemblage. *Quaternary Research* 41, 121–30.

Altınsaçlı, S. & H. I. Griffiths 2002. A review of the occurrence and distribution of recent non-marine Ostracoda in Turkey. *Zoology in the Middle East* 27, 61–76.

Averdieck, F. R. 1986. Palynological investigations in sediments of ancient lake Duvensee, Schleswig-Holstein (north Germany). *Hydrobiologia* 143, 407–10.

Barclay, M. H. 1968. Additions to the freshwater ostracod fauna of New Zealand. *New Zealand Journal of Marine and Freshwater Research* 2, 67–80.

Bilan, W. 1988. Late Quaternary Ostracoda from lacustrine sediments of the Orle Basin. *Folia Quaternaria* 58, 69–74.

Bokelmann, K. 1971. Duvensee, ein Wohnplatz des Mesolithikums in Schleswig-Holstein, und die Duvensee-gruppe. *Offa* 28, 5–26.

Bokelmann, K. 1991. Some new thoughts on old data on humans and reindeer in the Ahrensburgian Tunnel Valley in Schleswig-Holstein, Germany, in N. Barton, A.J. Roberts & D. A. Roe (eds) *The late glacial in north-west Europe: human adaptation and environmental* change, 72–81. London: Council for British Archaeology Research Report No. 77

Bradbury, J. P., R. M. Forester, W. A. Bryant & A. P. Covich 1990. Paleolimnology of Laguna de Cocos, Albion Island, Rio Hondo, Belize, in M. D. Pohl (ed.) *Ancient Maya Wetland Agriculture: excavations on Albion Island, northern Belize*, 119–54. Boulder (Co): Westview Press.

Chapman, M. A. 1963. A review of the freshwater ostracods of New Zealand. *Hydrobiologia* 22, 1–40.

Davies, P. & H. I. Griffiths 2005. Molluscan and ostracod biostratigraphy in the River Test at Bossington, Hampshire. *The Holocene* 15(1), 97–110.

de la Vega Leinert, A. C., D. H. Keen, R. J. Jones, J. M. Wells & D. E. Smith 2000. Mid-Holocene environmental changes in the Bay of Skaill, Mainland Orkney, Scotland: an integrated geomorphological, sedimentological and stratigraphical study. *Journal of Quaternary Science* 15, 509–28.

Delorme, L. D. 1991. Ostracoda, in J. H. Thorp & A. P. Covich (eds) *Ecology and Classification of North American Freshwater Invertebrates*, 691–722. San Diego: Academic Press.

Diebel, K. & E. Pietzeniuk 1977. Ostracoden aus dem Travertin von Taubach bei Weimar. *Quartärpaläontologie* 2, 119–37.

Diebel, K. & E. Pietzeniuk 1978. Die Ostrakodenfauna aus dem jungpleistozänen (weichselkaltzeitlichen) Deckschichten von Burgtonna in Thüringen. *Quartärpaläontologie* 3, 207–221.

Diebel, K. & E. Pietzeniuk 1980. Pleistozäne Ostracoden vom Fundort des *Homo erectus* bie Bilzingsleben. *Ethnographisch-Archäologische Zeitschrift* 21, 26–35 (pls 15–16 in *Ethnographisch-Archäologische Zeitschrift* 20: 655–6).

Diebel, K. & H. Wolfschläger 1975. Ostracoden aus dem jungpleistozänen Travertin von Ehringsdorf bei Weimar. *Abhandlungen des Zentralen Geologischen Instituts, (Paläontologische Abhandlungen)* 23, 91–136.

Ferrier, G. & G. Wadge 1997. An integrated GIS and knowledge-based system as an aid for the geological analysis of sedimentary basins. *International Journal of Geographical Information Science* 11, 282–97.

Finke, E. A. W. & H. Malz 1988. Der Lernäische See. Auswertung von Satellitenbildern und Ostracodenfaunen zur Rekonstruktion eines vergangenen Lebensraumes. *Natur und Museum* 118, 213–22.

Ford, T. D. & H. M. Pedley 1996. A World review of tufa and travertine deposits. *Earth Science Reviews* 41, 117–75.

Frogley, M. R., P. C. Tzedakis & T. H. E. Heaton 1999. Climate variability in northwest Greece during the last interglacial. *Science* 285, 1886–9.

Frogley, M. R., H. I. Griffiths & T. H. E. Heaton 2001. Historical biogeography and Late Quaternary environmental change at Lake Pamvotis, Ioannina (NW Greece); evidence from ostracods. *Journal of Biogeography* 28, 745–756.

Fuhrmann, R. & E. Pietrzeniuk 1990a. Die Ostrakoden fauna des Intergalzials von Gröbern (Kreis Gräfenhainichen), in L. Eissmann (ed.) *Die Eemwarmziet und die frühe Weichselzeit im Saale-Elbe-Gebiet: Geologie, Paläontologie, Palökologie. Altenberger Naturwissenschaftliche Forschungen* 5, 168–93.

Fuhrmann, R. & E. Pietrzeniuk 1990b. Die Ostrakodenfauna des Interglazials Grabschütz (Kreis Delitzsch), in L. Eissmann (ed.) *Die Eemwarmziet und die frühe Weichselzeit im Saale-Elbe-Gebiet: Geologie, Paläontologie, Palökologie. Altenberger Naturwissenschaftliche Forschungen* 5, 202–27.

Fuhrmann, R. & E. Pietrzeniuk 1990c. Die Aussage der Ostrakodenfauna zum Sedimentationsablauf im Interglazialbecken, zur klimatischen Entwicklung und zur stratigraphischen Stellung des Interglazials von Neumark-Nord (Geiseltal), in D. Mania, M. Thomae, T. Litt & D. Weber (eds) *Neumark-Gröbern. Beiträge zur Jagd des mittlepaläolithischen Menschen. Veröffenentlichungen der Landesmuseum Vorgeschichtlich Halle* 43, 161–6.

Galanidou, N., P. C. Tzedakis, I. T. Lawson & M. R. Frogley 2000. A revised chronological and paleoenvironmental framework for the Kastritsa rockshelter, northwest Greece. *Antiquity* 74, 349–55.

Gebel, H. G., C. Hanns, A. Liebau & W. Raehle 1989. The late Quaternary environments of 'Ain al-Faidha / Al-'Ain, Abu Dhabi Emirate. *Archaeology in the United Arab Emirates* 5, 9–48.

Griffiths, H. I. 1995. The application of freshwater ostracods to the study of late Quaternary palaeoenvironments in North-western Europe. Unpublished PhD thesis, University of Wales, Cardiff.

Griffiths, H. I. 1996. European Quaternary freshwater Ostracoda; a biostratigraphic and palaeobiogeographic primer. *Scopolia* 34, 1–168.

Griffiths, H. I. 1999. Freshwater Ostracoda from the Mesolithic lake site at Lough Boora (Co. Offaly, Ireland). *Irish Journal of Earth Sciences* 17, 39–49.

Griffiths, H. I. 2001. Ostracod evolution and extinction – its biostratigraphic value in the European Quaternary. *Quaternary Science Reviews* 20, 1743–1751.

Griffiths, H. I. & J. A. Holmes 2000. *Non-marine Ostracods and Quaternary Palaeoenvironments*. London: Quaternary Research Association Technical Guide 8.

Griffiths, H. I. & R. Mount 1993. Ostracods, in J. G. Evans, S. Limbrey, R. Mount & I. Máté (eds) An environmental history of the Upper Kennet Valley, Wiltshire, for the last 10,000 years, 181–4. *Proceedings of the Prehistoric Society* 59: 139–95.

Griffiths, H. I., A. J. Rouse & J. G. Evans 1993. Processing freshwater ostracods from archaeological deposits, with a key to the valves of the major British genera. *Circaea – the Journal of the Association for Environmental Archaeology* 10, 53–62.

Griffiths, H. I., V. Ringwood & J. G. Evans 1994. Weichselian Late-glacial and early Holocene molluscan and ostracod sequences from lake sediments at Stellmoor, north Germany. *Archiv für Hydrobiologie* 99 (S), 357–80.

Griffiths, H. I., K. E. Pillidge, C. J. Hill, J. G. Evans & M. A. Learner 1996. Ostracod gradients in a calcareous stream: implications for the palaeoecological interpretation of tufas and travertines. *Limnologica* 26, 49–61.

Griffiths, H. I., A. Schwalb & L. R. Stevens 2001. Evironmental change in NE Iran: the Holocene ostracod fauna of Lake Mirabad. *The Holocene* 11, 757–764.

Griffiths, H. I., J. M. Reed, M. J. Leng, S. Ryan & S. Petkovski 2002. The recent palaeoecology and conservation status of Balkan Lake Dojran. *Biological Conservation* 104, 35–49.

Gunther, J. 1986. Ostracod fauna of Duvensee, an ancient lake in northern Germany. *Hydrobiologia* 143, 411–16.

Holmes, J. A. 1996. Ostracod faunal and microchemical evidence for Middle Pleistocene sea-level change at Clacton-on-Sea (Essex, UK), in M. C. Keen (ed) *Proceedings of the Second European Ostracodologists' Meeting, Glasgow, 23–27 July, 1993*, 135–40. London: British Micropalaeontological Society.

Holmes, J. A. & H. I. Griffiths 1999. Ostracoda from Star Carr, in P. Mellars & P. Dark (eds) *Star Carr in Context: new archaeological and palaeoenvironmental investigations at the Early Mesolithic site of Star Carr, North Yorkshire*, 175–8. Cambridge: McDonald Institute for Archaeology.

Janz, H. & R. Matzke Karasz 2001. Holozäne Ostrakoden aus Karbonatablagerungen im Bereich der neolithischen Feuchtbodensiedlung Unfriedshause. *Mitteilungen der Bayerischen Staatssammlung für Paläontologie und historische Geologie* 41, 33–63.

Keding, E., P. Frenzel & S. Dušek 1995. Mollusken und Ostrakoden aus der arcahäologischen Grabung Harrhausen – paläoökologische Aussagemöglichkeiten. *Alt-Thüringen* 29, 95–107.

Kontrovitz, M., J. Slack & H. I. Griffiths 1995. Ostracoda from the moats of a Medieval Castle, Scotland. *Geological Society of America Abstracts* 27, A-415.

Kretzoi, M. & V. T. Dobosi 1990. *Vértesszőlős: site, man and culture*. Budapest: Akadémiai Kiadó.

Leroy, S., N. Kazancı, Ö. Ileri, M. Kibar, E. McGee & H. I. Griffiths 2001. An abrupt (seismic) event recorded in Lake Manyas (N-W Turkey) sediment dating from the Byzantine period. *Terra Nostra* 3, 125–31.

Mania, D. 1967a. Pleistozäne und holozäne Ostracodgesellschaften aus dem ehemaligen Ascherslebener See. *Wissenschaftliche Zeitschrift der Universität Halle* 16: 501–50.

Mania, D. 1967b. Der ehemalige Ascherslebener See (Nordharzvorland) in spät- und postglazialer Zeit. *Hercynia* 4, 199–260.

Mania, D. & V. Toepfer 1973. Königsaue. Gliederung, Ökologie und mittelpaläolithische Funde der letzten Eiszeit. *Veröffentlichungen des Landesmuseums für Vorgeschichte in Halle* 26,: 9–164.

Marmonier, P. & D. L. Danielopol 1988. Découverte de *Nannocandona faba* Eckman (Ostracoda, Candoninae) en basse Autriche. Son origine et son adaptation au milieu interstitiel. *Vie et Milieu* 38, 35–48.

Martens, K. 1984. *Annotated Checklist of Non-marine Ostracods (Crustacea, Ostracoda) from African Inland Waters*. Tervuren: Koninklijk Museum voor Midden-Afrika. Zoologisches Dokumentatie N. 20.

Martens, K. & F. Behen 1994. A checklist of the Recent non-marine Ostracoda (Crustacea, Ostracoda) from the inland waters of South America and adjacent islands. *Travaus scientifiques du Musée national d'histoire naturelle de Luxembourg* 22, 1–81.

Martens, K. & R. Ortal 1999. Diversity and zoogeography of inland-water Ostracoda (Crustacea) in Israel (Levant). *Israel Journal of Zoology* 45, 159–73.

Meisch, C. 2000. *Freshwater Ostracoda of Western and Central Europe (Süßwasserfauna von Mitteleuropa 8/3)*. Heidelberg: Spektrum Akademischer Verlag.

Meyrick, R. W. 1976. Ostracods, in P. C. Buckland (ed.) *The Environmental Evidence from the Church Street Roman Sewer System*, 31–2. London: Council for British Archaeology/York Archaeological Trust. The Archaeology of York.

Mezquita, F., J. R. Roca, J. M. Reed & G. Wansard 2005. Quantifying species-environment relationships in non-marine Ostracoda for ecological and palaeoecological studies: examples using Iberian data. *Palaeogeography, Palaeoclimatology, Palaeoecology* 225, 93–117.

Mount, R. 1991. An environmental history of the upper Kennet River Valley and some implications for human communities. Unpublished PhD thesis, University of Wales, Cardiff.

Musil, R. 1995. Stránská Skála Hill. Excavations of open-air sediments 1964–1972. *Anthropos (Brno)* 26, 1–213.

Nöthlings, F. 1992. *Süßwasser-Ostrakoden aus holozänen Sedimenten im Bereich der neolithischen Siedlung Pestenacker*. Köln: Geologischen Institut der Universität zu Köln.

O'Connell, M. 1980. Pollen analysis of fen peat from a Mesolithic site at Lough Boora, County Offaly, Ireland. *Journal of Life Sciences of the Royal Dublin Society* 2, 45–9.

Oertli, H. J. 1967. Ostrakoden aus der subrezenten Seekreide des Burgäschisees, in *Seeberg Burgäschisee-Süd, Teil 4. Chronologie und Umwelt. Acta Bernensia* 2, 129–33.

O'Sullivan, A. 1998. *The archaeology of lake settlement in Ireland*. Dublin: Discovery Programme/Royal Irish Academy. Discovery Programme Monographs 4.

Palacios-Fest, M. R. 1994. Nonmarine ostracode shell chemistry from ancient Hohokam irrigation canals in central Arizona: a paleohydrochemical tool for the interpretation of prehistoric human occupation in the North American Southwest. *Geoarchaeology* 9, 1–29.

Pedley, H. M. 1990. Classification and environmental models for cool freshwater tufas. *Sedimentary Geology* 68, 143–54.

Pedley, H. M., I. Hill, P. Denton & J. Brassington 2000. Three-dimensional modelling of a Holocene tufa system in the Lathkill Valley, north Derbyshire, using ground-penetrating radar. *Sedimentology* 47, 721–37.

Preece, R. C. 1980. The biostratigraphy and dating of the tufa deposit at the Mesolithic site at Blashenwell, Dorset, England. *Journal of Archaeological Science* 7, 345–62.

Preece, R. C., P. Coxon & J. E. Robinson 1986. New biostratigraphic evidence of the Post-glacial colonisation of Ireland and for Mesolithic forest disturbance. *Journal of Biogeography* 13, 487–509.

Preece, R. C., D. R. Bridgland, S. G. Lewis, S. A. Parfitt & H. I. Griffiths 2000. Beeches Pit, West Stowe, Suffok (TL 789719), in S. G. Lewis, C. A. Whiteman & R. C. Preece (eds) *The Quaternary of Norfolk and Suffolk: field guide*, 185–95. London: Quaternary Research Association.

Roberts, N., S. Black, P. Boyer, W. J. Eastwood, H. I. Griffiths, M. Leng, R. Parish, J. M. Reed, D. Twigg & H. Yiğitbasioğlu 1999. Chronology and stratigraphy of Late Quaternary sediments in the Konya Basin, Turkey: results from the KOPAL project. *Quaternary Science Reviews* 18, 611–30.

Robinson, E. 1986. Ostracods from Meare East. *Somerset Levels Papers* 12, 102–5.

Robinson, J. E. 1990. The ostracod fauna of the Middle Pleistocene interglacial deposits at Little Oakley, Essex. *Philosophical Transactions of the Royal Society of London* 328 (B): 409–23.

Robinson, E. 1991. Ostracods from the Late Bronze Age river channel, in S. Needham (ed.) *Excavations at Runnymede Bridge, 1978: the late Bronze Age waterfront site*: 275–6. London: British Museum Press / English Heritage.

Rosenfeld, A., G. Avni, H. Ginat, A. Honigstein & I. Saragusti 1997. Pliocene-Pleistocene palaeo-lakes in the Arava Valley, southern Israel – ostracods and *Homo erectus*, in *Abstracts – Thirteenth International Symposium on Ostracoda (ISO 97), University of Greenwich, July, 1997*: unpaginated.

Rust, A. 1937. *Das Altsteinzeitliche Rentierjägerlager Meiendorf*. Neumünster: Karl Wachholtz Verlag.

Rust, A. 1943. *Die Alt- und Mittelsteinzeitlichen Funde von Stellmoor*. Neumünster: Karl Wachholtz Verlag.

Ryan, M. 1980. An early Mesolithic site in the Irish Midlands. *Antiquity* 54, 46–7.

Ryan, M. 1984. Archaeological excavations at Lough Boora, Boughal Townland, Co. Offaly, 1977, in J. Cooke (ed.) *Proceedings of the 7th International Peat Congress, Dublin (Volume 1)*, 407–13. Dublin: Office of Public Works.

Schreve, D. C., D. R. Bridgland, P. Allen, J. J. Blackford, C. P. Gleed-Owen, H. I. Griffiths, D. H. Keen & M. J. White 2001. Sedimentology, Pleistocene river palaeontology and archaeology of Middle Thames terrace deposits at Purfleet, Essex, UK. *Quaternary Science Reviews* 21, 1423–1464.

Speleers, A. 2000. The relevance of the Eemian for the study of the Palaeolithic occupation of Europe. *Geologie en Mijnbouw* 79, 283–91.

Taliana, D., M. Alessio, L. Allegri, L. Capasso Barbato, C. de Angelis, D. Esu, O. Girotti, E. Gliozzi, S. Improta, I. Mazzini & R. Sardella 1996. Preliminary results on the 'Grotta del Lago' Holocene deposits (Triponzo, Nera River Valley, Umbria, Central Italy). *Il Quaternario – Italian Journal of Quaternary Sciences* 9, 745–52.

Usinger, H. 1981. Ein weit verbreiteter Hiatus in spät-glazialen Seesedimenten: Mögliche Ursache für Fehlinterpretation von Pollendiagrammen und Hinweis auf klimatisch verursachte Seespiegelbewegungen. *Eiszeitalter und Gegenwart* 31, 91–107.

van Wijngaarden-Bakker, L. H. 1989. Faunal remains and the Irish Mesolithic, in C. Bonsall (ed) *The Mesolithic in Europe: papers presented at the Third International Symposium, Edinburgh, 1985*, 125–33. Edinburgh: John Donald Publishers Ltd.

von Grafenstein, U., H. Erlenkeuser, A. Kleinmann, J. Müller & P. Trimborn 1994. High-frequency climatic oscillations during the last deglaciaition as revealed by oxygen-isotope records of benthic organisms (Ammersee, southern Germany). *Journal of Paleolimnology* 11, 349–57.

von Grafenstein, U., H. Erlernkeuser & P. Trimborn 1999. Oxygen and carbon isotopes in modern fresh-water ostracod valves: assessing vital offsets and autecological effects of interest for palaeoclimate studies. *Palaeogeography, Palaeoclimatology, Palaeoecology* 148, 133–52.

von Kolfschoten, T. 2000. The Eemian mammal fauna of central Europe. *Geologie en Mijnbouw* 79, 269–81.

Whatley, R. C. & J. R. Haynes 1986. Foraminifera and Ostracoda, in M. B. Roberts (ed) Excavation of the Lower Palaeolithic site at Amey's Eartham Pits, Boxgrove, West Sussex: a preliminary report, 232–4. *Proceedings of the Prehistoric Society* 52, 215–45.

Zangger, A. 1993. *The Geoarchaeology of the Argolid*. Berlin: Gerb. Mann Verlag / Deutches Archäologisches Institut Athen.

Zangger, A. & H. Malz 1989. Late Pleistocene, Holocene, and recent ostracods from the Gulf of Argos, Corinth. *Courier Forschungs-Institut Senckenberg* 113, 159–75.

Chapter 17

The Role of Quantitative Diatom-Based Palaeolimnology in Environmental Archaeology

Jane M. Reed

INTRODUCTION

Diatoms (single-celled algae) are abundant in virtually all aquatic environments. Through a combination of their abundance, diversity and sensitivity to a wide range of environmental variables, and their often high quality preservation (Battarbee 1986), their analysis is at the forefront of research into past environmental change in a wide range of lacustrine, estuarine and marine environments. In recent decades, the value of diatoms as environmental indicators has been enhanced greatly by the development of *transfer functions* based on large modern regional data-sets of diatom and water quality data, from which we are able to model diatom response to a range of environmental variables such as salinity, nutrient content or pH (Birks *et al.* 1990). Application of a transfer function to fossil diatom data preserved in lake sediment cores allows past change to be reconstructed quantitatively and in a reproducible manner (Stoermer and Smol 1999). With the publication of many transfer functions now achieved, the research has reached the point where other diatom workers may take advantage of this powerful tool for environmental reconstruction.

Recent reviews of diatom analysis in archaeology (Battarbee 1988, Juggins and Cameron 1999) have highlighted its potential value at all levels in archaeological endeavour, including artefact provenancing, the study of site formation processes and local or regional off-site environmental reconstruction, but the authors have also underlined the continuing paucity of diatom-based studies incorporated into archaeological research design. Equally, there has been limited application of the quantitative transfer function technique. In estuarine research, for example, where diatoms offer great potential for elucidating site location patterns (Juggins and Cameron 1999), the pioneering work of Juggins (1992), aimed at improving salinity estimates when compared to those provided by qualitative or semi-quantitative ecological classification schemes (reviewed in Denys and de Wolf 1999), has not yet been built upon in other coastal zone research. In contrast, quantification of the diatom record is now a key focus in most fields of palaeolimnological research in inland waters, but, with a few exceptions, most archaeologically-orientated studies still rely on classification schemes (*e.g.* Whitmore *et al.* 1996 or examples cited in Juggins and Cameron 1999).

This paper discusses the application of transfer functions to reconstruct environmental change in two spheres of research of relevance to archaeology in wetlands – those of salinity, lake-level and climate reconstruction in semi-arid, Mediterranean regions, and of total phosphorus reconstruction related to anthropogenic eutrophication of lakes during the Holocene. As noted above, comprehensive reviews are published elsewhere of the scope for diatom research in archaeology (Battarbee 1988, Juggins and Cameron 1999 and papers cited therein), and the recently-published volume of which Juggins and Cameron (1999) forms a part provides a detailed overview of the wider applications of diatoms in the environmental and earth sciences. The aim of this paper is to facilitate the critical use of published transfer functions by other users. Following brief reviews of the potential value to archaeological research of salinity and nutrient reconstructions, and a summary of the transfer function technique, the emphasis is on the degree to which published transfer functions may provide a reliable environmental reconstruction when applied by other users.

DIATOM-INFERRED SALINITY AS A CLIMATE INDICATOR

Closed-basin, saline lakes lack a surface outflow and are found in semi-arid climate zones where precipitation exceeds evaporation; they are widespread in Africa, Spain, the Near and Middle East, the continental interior of the former Soviet Union, western North America, Latin America, Canada and Australia (Hammer 1986). In saline lakes, lake levels are often highly responsive to changes in effective moisture (precipitation minus evaporation), and minor climatic fluctuations can cause major change in water level compared to freshwater systems (Harrison *et al.* 1991). Salts tend to be conserved in the system and these seasonal and inter-annual fluctuations are accompanied by significant changes in salinity and ionic composition with dilution and concentration of the lake waters (Hardie and Eugster 1970). In essence, an inverse relationship often exists between salinity and lake level (Street-Perrott and Harrison 1985). The sensitivity of diatoms to salinity is well known, and diatom-inferred salinity (or the related variable, conductivity) is an important proxy for changes in effective moisture over time (Fritz *et al.* 1999).

In semi-arid regions, other sources of palaeoclimate data are relatively rare and palaeolimnological data provide a key source of data on past climate change for regions where patterns of climate change are often still poorly understood (*e.g.* Roberts and Wright 1993, Roberts *et al.* 1999, Reed *et al.* 2001). Thus, while local, site-specific palaeoenvironmental studies are the ideal in archaeology (Juggins and Cameron 1999), the derivation of a strong regional palaeoclimate framework has still not been achieved in many semi-arid regions, including parts of the Mediterranean such as Turkey.

As an apposite example, in an excellent study of the last 1100 years of lake sediment history of Lake Naivasha, a crater lake in the Eastern Rift Valley of Kenya, Verschuren *et al.* (2000) showed clear evidence for alternating drought and humid phases from inferences of lake-level change based on mutually supportive conductivity estimates, derived from diatom and chironomid (midge larvae) transfer functions. In this case the area lacked archaeological evidence, but had a strong pre-colonial oral history of periods of abandonment, famine and unrest alternating with ages of prosperity and political stability, which had been genealogically dated. The results showed a strong correlation between periods of unrest and inferred phases of lowstand (*i.e.* high salinity, drought episodes) in the lake, thus reinforcing the oral chronology of events and indicating a strong link in this case between cultural development and climate change.

DIATOM-INFERRED TOTAL PHOSPHORUS AS AN INDICATOR OF TROPHIC STATUS

Eutrophication (nutrient enrichment) arises most often as a result of artificial input of nitrates and phosphates from anthropogenic sources such as sewage and agricultural activity or, less directly, from accelerated input related to the effects of activities such as forest clearance on soil erosion (Mason 1997). Diatom-eutrophication studies are based on their response to total phosphorus, which is present in very small quantities in the natural environment and is demonstrably the most important ecological limiting factor in eutrophication (Harris 1995). Equally, it is the most significant variable driving changes in diatom species assemblage composition associated with nutrient enrichment (*e.g.* Bennion 1995).

Diatom-based eutrophication studies have been confined mainly to freshwater lakes in temperate regions, and are most often associated with conservation issues surrounding accelerated nutrient enrichment over the last *c.* 150 years (*cf.* Battarbee 1999) rather than with longer-term Holocene changes in water quality. A classic exception is a study of Holocene changes in trophic status of Diss Mere, Norfolk (Fritz 1989), the later quantification of total phosphorus for which (Birks *et al.* 1995) is summarised in Hall and Smol (1999). The eutrophication history correlates closely throughout with archaeological and documentary evidence for human settlement patterns, ranging from the effects of forest clearance in the Neolithic to the subsequent expansion of the nearby town and associated agricultural activities during the Roman and Medieval periods, and demonstrates a long history of human impact on water quality.

There is also potential for archaeological eutrophication studies in saline lake regions. In this case, the link between nutrient availability and human activities may become more difficult to define, since salinity-related change in ionic composition of the lake waters can itself lead to enhanced nutrient availability in more saline waters (Saros and Fritz 2000). However, based on salinity and nutrient classification schemes, Whitmore *et al.* (1996) successfully managed to disentangle the effects of climate change and anthropogenic eutrophication in a Holocene study of saline lakes on the Yucatan Peninsula, Mexico. An inferred phase of maximum nutrient levels correlated clearly with intense local occupation of an urban Maya site during a deep water (low salinity) phase of stable occupation around 2000 BP. The study was again important in demonstrating a link between phases of prosperity and favourable climate, coupled with eutrophication as an indicator of the intensity of urban settlement.

THE TRANSFER FUNCTION TECHNIQUE

The development of transfer functions for diatom-based palaeolimnological research was initiated during the 1980s in studies aimed at quantifying the effects of acid rain on lake-water pH (ter Braak and van Dam 1989, Birks *et al.* 1990). Weighted averaging is a simple statistical technique which has proved to date to be the most effective means of modelling species distributions (Birks *et al.* 1990).

Two stages are involved, comprising regression of a modern data-set and subsequent calibration against fossil data. In regression, modern diatom surface sediment species distribution is modelled in relation to water quality using weighted averaging techniques on a regional data-set of samples from sites which span the ecological gradient of interest (*e.g.* for salinity, fresh to hypersaline, <0.5 gl^{-1} to >40 gl^{-1}). The results provide an estimate of the ecological optimum and tolerance range for each taxon (species) present.

Weighted averaging is based on the ecologically valid assumption that species follow a normal, or unimodal, distribution, such that they tend to reach peak abundance at a preferred value of the environmental variable (the optimum, calculated as the centroid of the range), and tail off to either side; each site will be dominated by species whose optima are close to the value of the environmental variable at that site (Juggins 1992). The tolerance range is calculated as one abundance-weighted standard deviation either side of the optimum; those species with narrow tolerance ranges are effectively "indicator species" with highly specific requirements. The optimum is calculated for each species as follows (Juggins 1992):

$$U_{wa} = \sum\nolimits_{i=1}^{n} y_i x_i / \sum\nolimits_{i=1}^{n} y_i \qquad (1)$$

where U_{wa} = the optimum; n = number of sites; y_i = species abundance at site i; x_i = measured environmental value at site i.

To establish the reliability of the resultant transfer function, this step is preceded by rigorous exploratory statistical analysis of the data-set to ensure the significance on diatom distribution of the variable to be reconstructed; its predictive ability can additionally be assessed using techniques described clearly in Fritz *et al.* (1999). Once the optima are calculated, subsequent calibration against fossil assemblages then provides a quantitative estimate of the past environmental variable, based on those fossil taxa which also occur in the modern data-set (*i.e.* for which an optimum exists), as follows:

$$x_{wa} = \sum_{k=1}^{m} y_k U_k / \sum_{k=1}^{m} y_k \qquad (2)$$

where X_{wa} = estimate of the environmental variable for a fossil sample; m = number of species; y_k = abundance of species k; U_k = optimum of species k.

Thus, weighted averaging simply models the modern distribution of each species along an environmental gradient as an average of its occurrence at different values of the environmental variable, weighted by its abundance. Calibration of fossil assemblages is a similar step but with fossil chemistry, rather than species optimum, as an unknown.

Some of the published transfer functions (below) provide comprehensive lists of species optima and tolerance ranges. For those in possession of the modern data-set, calibration may be performed using purpose-built programs such as CALIBRATE (Juggins and ter Braak 1992). Although more time consuming, it can be seen from the above that the calibration step is also easy to perform manually, being achieved by simple arithmetic calculation, using published optima, on a spreadsheet of fossil species percentage abundance data against depth.

PUBLISHED TRANSFER FUNCTIONS

There is a growing body of published transfer functions for salinity (expressed in g l[-1]) or the related variable conductivity (expressed as µS cm[-1]) (S/C) and for total phosphorus (TP, expressed as µg l[-1]), especially from North America and the Europe-Africa regions. Those which provide comprehensive lists of optima and tolerance ranges are listed in Table 17.1. Others are either based on small data-sets and without rigorous exploratory statistical analyses (Yang and Dickman 1993, Kashima 1995), are awaiting publication (*e.g.* for salinity: Juggins unpublished, Kazakhstan; Reed unpublished, Turkey), or have been published or applied to palaeolimnological sequences but without inclusion of the necessary details on species optima (*e.g.* Fritz *et al.* 1993, Anderson and Rippey 1994, Bennion *et al.* 1996, Lotter *et al.* 1998).

Table 17.1 shows that all the published transfer functions are statistically reliable predictors of the environmental variable concerned, with high apparent r^2 values (>0.5) indicating a strong linear relationship between observed (*i.e.* measured) and diatom-inferred values of S/C or TP in the sites making up the regional data-set. In comparing the two groups of summary statistics, the stronger significance of salinity as an ecological limiting factor is suggested by the consistently higher r^2 values of the S/C transfer functions (0.79–0.95) compared to the nutrient transfer functions (0.57–0.79), perhaps reflecting the more complex ecological relationships between diatom distribution and nutrient availability in fresh waters (*e.g.* Yang and Dickman 1993).

APPLICABILITY OF PUBLISHED TRANSFER FUNCTIONS

A danger in applying transfer functions is that they will produce a quantitative estimate regardless of its

Transfer function	Authors	No. sites	Range	r^{2*}
Salinity				
British Columbia, Canada	Cumming & Smol 1993	59	fresh-hypersaline	0.95
British Columbia, Canada (extended)	Wilson *et al.* 1994	111	0.04-369 g l⁻¹	0.89
Northern Great Plains, N. America	Fritz *et al.* 1993a	66	0.65-270 g l⁻¹	not cited
Western Victoria, Australia	Gell 1997	62	0.5-133 g l⁻¹	0.79
Conductivity				
Africa	Gasse *et al.* 1995	282	0.4-99 mS cm⁻¹	0.87
Spain	Reed 1998a	74	0.2-338 mS cm⁻¹	0.91
Total phosphorus				
British Columbia, Canada	Hall & Smol 1992	46	5-28 µg l⁻¹	0.73
British Columbia, Canada (extended)	Reavie *et al.* 1995	59	6-42 µg l⁻¹	0.75
Southern England, UK	Bennion 1994	31	7-1123 µg l⁻¹	0.79
Central Europe, Alpine and pre-Alpine	Wunsam & Schmidt 1995	86	2-266 µg l⁻¹	0.57

*the r^2 correlation coefficient quoted is the 'apparent' r^2 value (Birks *et al.* 1990), based on the entire data-set. This may hide weaknesses in parts of the environmental gradient which are poorly-covered, which may be revealed by additional bootstrapping or jackknifing techniques of predictive ability; most of the above transfer functions have also been tested in this manner.

Table 17.1 Description of published transfer functions for salinity, conductivity and total phosphorus.

reliability. While the internal statistical robustness of the transfer function itself is important, other factors come into play when applying it to fossil assemblages. One of the most easy to identify is a poor match between fossil and modern species assemblage composition, such that the reconstruction is in effect based on an unreliably small percentage of the fossil diatom species represented. Application of a transfer function may also be invalid if the modern data-set is not relevant to a study in another region due to regional variation of species preferences. As discussed below, this may be due either to real ecological differences in niche space occupied, or in some cases may be an artefact of the restricted range of sites covered in a transfer function. The final section of this paper is dedicated to illustrating how reliability may be assessed and to recommending how far transfer functions are likely to be applicable to core sequences from other geographic regions, using examples taken primarily from the author's work in Europe.

Assessing the match between fossil and modern samples

The strongest measures of goodness of fit between fossil and modern samples are modern analogue matching techniques, and calculation of the root mean squared error of prediction for fossil samples (Birks *et al.* 1990), but both techniques require access to the modern data-set. Although a weaker means of assessment, the sum of percentages upon which the reconstruction was based (the denominator of the calibration equation (2), above) can provide a simple means of assessing species presence in the fossil flora compared to that of the modern.

Fig. 17.1a shows the percentage of fossil diatom species represented in three Late Quaternary conductivity reconstructions from lakes in southern central Turkey (Reed *et al.* 1999, Roberts *et al.* 2001) and one from the Laguna de Medina, southwestern Spain (Reed 1995). Diatoms were clearly well represented in the modern flora of the Turkish sites; apart from a small number of subsamples in Eski Acıgöl and one subsample in Pınarbası, 80–100% of the fossil taxa have a modern analogue. This is in contrast to the Medina sequence, where in many levels only 0–50% of fossil taxa are represented. This may be clarified by reference to the results of modern analogue matching on this sequence using MAT (S. Juggins unpublished) (Fig. 17.1b), by reference to the data-set for the Spanish transfer function on which it was based (Reed 1998a). Details are given in Reed (1995); essentially the results show that only one sample of the entire sequence has a statistically 'good' match in the modern flora. In addition to the low percentage abundance on which many of the reconstructions were based, reflecting the complete absence of some fossil taxa in the modern data-set, this indicates the additionally low abundance of many of the taxa which are represented. A graphic illustration is provided in Reed (1995), where detrended correspondence analysis of the modern and fossil data-sets showed virtually no overlap between the two. Thus, while there are more rigorous techniques for assessing the goodness of fit, the percentage data alone provide a good general indicator of representativity.

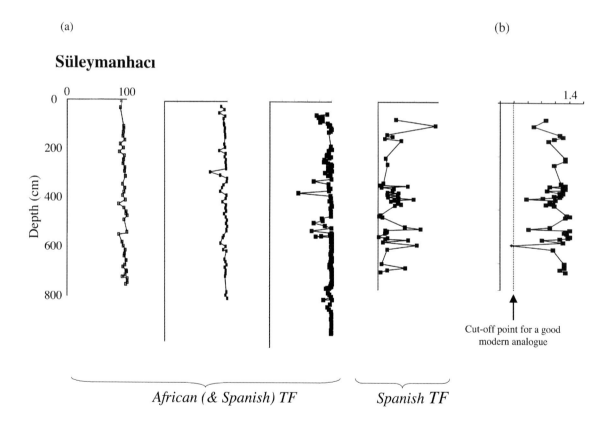

Fig. 17.1 (a) Comparison of the percentage of fossil species represented in modern data-sets, for conductivity reconstruction of Turkish fossil sequences based on the African TF (transfer function) (Gasse et al. 1995), with three optima from the Spanish TF (Reed 1998a), and a Spanish sequence based on the Spanish TF; (b) results of modern analogue matching of the Spanish Laguna de Medina sequence using MAT (Juggins, unpublished), where values in excess of 0.995 indicate a poor match between fossil and modern data.

Factors affecting representativity

Taphonomic effects

In the Laguna de Medina sequence, one of the most important factors affecting representativity is diatom dissolution. Diatom dissolution and/or fragmentation can be an important problem in saline lakes (*e.g.* Flower 1993, Ryves 1994, Reed 1998b), and in some fresh eutrophic lakes (*e.g.* lowland China; H. Bennion, unpublished data). It can result in the complete loss of the record or its dominance by robust taxa which are more resistant to dissolution and therefore over-represented in the fossil flora. In the Medina summary diatom diagram (Fig. 17.2), *Campylodiscus clypeus* and *Mastogloia braunii* are examples, being common in the modern flora but only as minor components (3.5% and 1.0% max., respectively), indicating that a large proportion of fragile taxa has been lost in the fossil record, and that a reconstruction for these samples will be highly biased as a consequence. The Medina record is exceptional in that the reconstruction is unreliable for all levels, while in other saline lake studies (*e.g.* Eski Acıgöl; Roberts *et al.* 2001) poor reliability is restricted to short-lived phases of high salinity where dissolution is pronounced, and the quantitative record gives a good indication of relative salinity change if it is accepted that the upper end of the salinity gradient is poorly modelled.

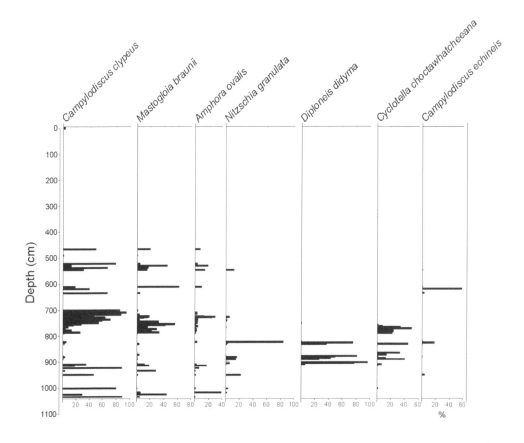

Fig. 17.2 Summary diatom diagram for the Laguna de Medina, southwest Spain, showing dominant taxa (present at >20% in >1 sample); diatoms are absent in other levels.

Lack of a modern analogue

A second factor affecting representativity is a true 'ecological' lack of modern analogues for fossil taxa, which have no representation at all in the modern data-set. Of the dominants in Medina (Fig. 17.2), these comprise *Nitzschia granulata*, *Diploneis didyma*, *Cyclotella choctawhatcheeana* and *Campylodiscus echineis*. Some of these species are common in African lake sediments (*e.g.* Fontes and Gasse 1991) and Holocene Dalmatian coastal lagoon sequences (Wunsam *et al.* 1999), but are absent both in the Spanish (Reed 1998a) and African (Gasse *et al.* 1995) modern flora. In the Medina sequence they occur with the ostracod *Cyprideis torosa*, which is now restricted to shallow coastal lagoons in Spain (Reed unpublished) and indicates an environment of permanent, shallow, chloride-dominated waters which no longer exists in inland lakes of these regions. In another example, nutrient reconstruction has been hampered by the lack of modern analogues for common Late Pleistocene *Cyclotella* species in a study of central Italian crater lakes (Ryves *et al.* 1996).

As a partial solution, transfer functions may be strengthened by the inclusion of optima from other regional data-sets, as in the example of the Turkish lakes above where African optima (from Gasse *et al.* 1995) were used in combination with a small number of Spanish optima (from Reed 1998a) for no-analogue taxa. This requires careful consideration, however, since there may be a lack of equivalence of optima between data-sets. Differences in ecological preferences are likely to be most significant in widely separated geographic regions, where variability in unmeasured environmental parameters could affect the niche space occupied by different species (Juggins *et al.* 1994). Equally, an individual region may not contain a sufficient range of lakes to span the entire ecological gradient being investigated, and the resultant transfer function (or individual optima) may be accurate for the type of lake on which they were based, but not for others. These issues are considered by comparison of some of the published regional data-sets listed in Table 17.1.

Comparison of regional modern data-sets

Salinity and conductivity

The comparability of different data-sets can be assessed by considering published optima for common taxa. In the case of salinity and conductivity, North American and Australian transfer functions are based on salinity, whereas African and Spanish transfer functions are based on conductivity; detailed discussion here will be restricted to conductivity optima. The optima of common taxa derived in conductivity transfer functions are compared in Fig. 17.3, and show that estimates are generally lower in the African data-set. This is probably a function of the inclusion of many freshwater sites (including mountain lakes) in the African data-set, and relatively few hypersaline sites. In contrast, shallow, freshwater lakes are rare in Spain, but the upper end of the salinity gradient is well represented. In spite of this, the data-sets show a similar ordering of the optima, which indicates high equivalence between them.

A detailed comparison of these data-sets with those of North America or Australia is beyond the scope of this paper. In a separate study, Reed (1998a) noted that the flora of Spain had closer affinities to the African data-set than to the North American data-sets, both in terms of the common taxa present, and of their estimated optima. This supports the contention that data-sets from widely separated geographic regions may not be equivalent. Fritz *et al.* (1999) suggest that North American and Australian sulphate-rich lakes provide a potential modern analogue for the aforementioned species, *Cyclotella choctawhatcheeana* but, as noted, associated species in the fossil record appear to indicate a chloride-dominated environment, and there is a danger that the modern environment inhabited by the species differs from that of the early Holocene. In contrast, the strong affinity between the Quaternary flora of the Turkish lakes (above), and the modern flora of the African and Spanish transfer functions, indicates that application to sequences in this case – *i.e.* in a region which is geographically close to that of the transfer function, and in which factors such as lake chemistry and regional climate are more similar, is valid.

Total phosphorus

For total phosphorus, the optima of common taxa are compared in Fig. 17.4. In contrast to conductivity data-sets, there are major differences in the magnitude of estimated optima in the UK transfer function (Bennion 1995) compared to the central European and North American transfer functions (Reavie *et al.* 1995, Wunsam and Schmidt 1995), and also few similarities in the relative ordering of common taxa. The former was based mainly on eutrophic-hypereutrophic ponds, whereas the latter two were in the oligotrophic-mesotrophic range. In contrast to the conductivity (and salinity) transfer functions, where full coverage of the ecological gradient was achieved with only minor weaknesses in parts of the range, the TP transfer functions do not individually span the entire nutrient gradient. Some taxa, such as *Aulacoseira granulata* or *Stephanodiscus hantzschii*, clearly have higher nutrient preferences, but the quantitative estimate varies according to the range covered in the modern data-set.

In addition, while the salinity/conductivity data-sets are derived from semi-arid regions with many similarities in climate, the nutrient data-sets are derived from a wider range of environments ranging from lowland shallow ponds (Bennion 1995) to high mountain, oligotrophic lakes (Wunsam and Schmidt 1995). As such, the nutrient data-sets are perhaps more likely to exhibit differences in unmeasured environmental parameters sufficient to have a significant effect on species distributions.

Total phosphorus reconstructions may therefore be unreliable outside the region in which they were derived. As an example of the limits of their applicability, on the

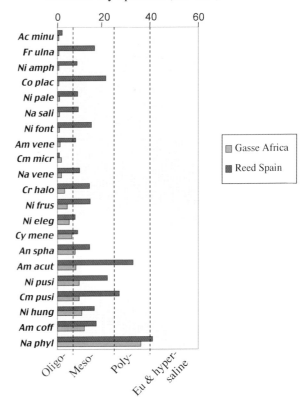

N occ > 20, N_2 >20 in Africa; N occ > 15, Spain

Fig. 17.3 Comparison of published conductivity optima for common taxa in the African (Gasse et al. *1995) and Spanish (Reed 1998a) transfer functions. Approximate salinity boundaries from oligosaline to hypersaline marked on the diagram. The African data-set is large, and common taxa are defined as those with both a high N occ (number of occurrences) and N_2 (N_2 diversity index), to restrict the number of species considered.*

mesotrophic Balkan Lake Dojran, Griffiths *et al.* (2002) derived a TP profile based on optima from the two latter transfer functions (*i.e.* which covered a similar nutrient range). However, its value was to demonstrate graphically the extreme complacency of the record. The absolute quantitative values were not relied upon, and it would not have been possible to use this technique to interpret a record with detailed fluctuations.

EDDI: The European Diatom Database Initiative (Addendum)

The output from an EU project, EDDI (the European Diatom Database Initiative; Battarbee *et al.* 2000) has been placed in the public domain since completion of the first draft of this paper. The project aimed at harmonising taxonomy to ensure regional consistency of species identification between different contributing diatom laboratories across Europe. The modern data-sets were then combined to produce three *combined* transfer functions for conductivity (based on the original Africa, Spain and Kazakhstan data-sets), total phosphorus (*e.g.* Ireland, SE England, Denmark and Alpine data-sets), and pH data-sets. On the assumption that the resulting improvement of representativity of individual species, from sites across a wider geographic region than that provided by any single data-set alone, would result in greater accuracy of estimated optima, the aim was to strengthen the interpretative power of transfer functions for the Europe-Africa region as a whole. The data may be accessed at http://craticula.ncl.ac.uk:8000/Eddi/jsp/index.jsp.

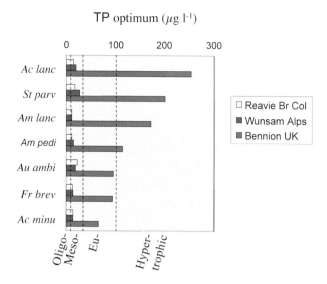

N occ. >20 in UK data-set, >15 in the rest

Fig. 17.4 Comparison of published TP optima for common taxa in the expanded British Columbian (Reavie et al. *1995), central Alpine (Wunsam and Schmidt 1995) and UK (Bennion 1995) data-sets. TP boundaries from oligotrophic to hypertrophic are marked on the diagram. Common taxa are defined with a high N occ (number of occurences) in the UK, and relatively high N occ in other data-sets.*

The project has been extremely valuable in allowing outside workers to access the modern data for the first time; information is provided on individual species and characteristics of sampling sites, and access to the modern data-sets allows reconstruction of past water chemistry using either the original transfer functions or the new, combined salinity, nutrient and pH transfer functions.

It was demonstrated above that differences in estimated optima between the largest two conductivity data-sets resulted mainly from differences in the coverage of the extremes of the salinity gradient (more fresh sites in Africa *vs.* more hypersaline sites in Spain), but that otherwise the optima showed similarities in their ranking according to conductivity preference. These differences stem from real geographic differences; fresh water sites are virtually non-existent in semi-arid regions of Spain (Reed 1998a), and there is little doubt that the combination of data-sets for the Europe-Africa region will have strengthened the interpretative power of the technique.

In the case of nutrient reconstruction, major discrepancies were noted above in the estimated TP preferences of many taxa. While the estimated optima for common planktonic species may have been improved, it is possible that fossil reconstruction for sites dominated by shallow water benthic taxa, which exhibit the highest differences in their optima and may be responding less strongly to nutrient availability, still cannot be reconstructed reliably in other geographic regions by using the combined transfer function. For scientific rigour, it may be advisable to compare representativity and estimated optima for dominant species within the smaller data-sets in order to assess how accurate a 'combined' optimum may be.

CONCLUSION

This paper has highlighted the potential of diatom-based transfer functions as a powerful tool for improving palaeoenvironmental reconstruction in an archaeological context. It has been stressed throughout that there is a need for caution in applying published transfer functions for a range of reasons, both taphonomic and ecological, and also due to potential technical limitations of individual data-sets. The recent publication of the results of a major EU project, EDDI, which allows access to individual data-sets and new, combined regional data-sets for the Europe-Africa region, has the potential to improve greatly the interpretative power of diatom transfer functions. Again, however, it was stressed that caution may sometimes be necessary even in applying the larger, combined transfer functions in other geographic regions, particularly in the case of nutrient reconstruction. The paper has also provided a framework for assessing the potential reliability of reconstructions by other users.

ACKNOWLEDGEMENTS

This paper was written while in receipt of a Leverhulme Special Research Fellowship, for which the Trust is gratefully acknowledged. Jon Garner is thanked for assistance with some of the figures. Huw Griffiths is also to be thanked for his comments on the manuscript; this is one of many papers dedicated to his memory.

REFERENCES

Anderson, N. J. & B. Rippey 1994. Monitoring lake recovery from point-source eutrophication: the use of diatom-inferred total phosphorus and sediment chemistry. *Freshwater Biology* 32, 625–39.

Battarbee, R. W. 1986. Diatom Analysis, in B. E. Berglund (ed.) *Handbook of Holocene palaeoecology and palaeohydrology*, 527–70. Chichester: John Wiley.

Battarbee, R. W. 1988. The use of diatom analysis in archeology: a review. *Journal of Archaeological Science* 15, 621–44.

Battarbee, R. W. 1999. The importance of palaeolimnology to lake restoration. *Hydrobiologia* 395/396, 149–59.

Battarbee, R. W., S. Juggins, F. Gasse, N. J. Anderson, H. Bennion & N. G. Cameron 2000. European Diatom Database (EDDI). An Information System for Palaeoenvironmental Reconstruction. *European Climate Science Conference*, Vienna City Hall, Vienna, Austria, 19–23 October 1998, 1–10.

Bennion, H. 1995. A diatom-phosphorus transfer function for shallow, eutrophic ponds in southeast England. *Hydrobiologia* 275/276, 391–410.

Bennion, H., S. Juggins & N. J. Anderson 1996. Predicting Epilimnetic Phosphorus Concentrations Using an Improved Diatom-Based Transfer Function and Its Application to Lake Eutrophication Management. *Environmental Science & Technology* 30, 2004–7.

Birks, H. J. B., J. M. Line, S. Juggins, A. C. Stevenson & C. J. F. ter Braak 1990. Diatoms and pH reconstruction. *Philosophical Transactions of the Royal Society of London B* 327, 263–78.

Birks, H. J. B., N. J. Anderson & S. C. Fritz 1995. Post-glacial changes in total phosphorus at Diss Mere, Norfolk inferred from fossil diatom assemblages, in S. T. Patrick & N. J. Anderson (eds) *Ecology and Palaeoecology of Lake Eutrophication*, 48–9. Copenhagen: Geological Survey of Denmark DGU Service Report 7.

Denys, L. & H. de Wolf 1999. Diatoms as indicators of coastal paleoenvironments and relative sea-level change, in E. F. Stoermer & J. P. Smol (eds) *The Diatoms: applications for the environmental and earth sciences*, 277–91. Cambridge: Cambridge University Press.

Flower, R. J. 1993. Diatom preservation: experiments and observations on dissolution and breakage in modern and fossil material. *Hydrobiologia* 269/270, 473–84.

Fontes, J. C. & F. Gasse 1991. Chronology of the major palaeohydrological events in NW Africa during the late Quaternary: PALHYDAF results. *Hydrobiologia* 214, 367–372.

Fritz, S. C. 1989. Lake development and limnological response to prehistoric and historic land-use in Diss, Norfolk, UK. *Journal of Ecology* 77, 182–202.

Fritz, S. C., S. Juggins & R. W. Battarbee 1993. Diatom Assemblages and Ionic Characterization of Lakes of the Northern Great Plains, North America: A Tool for reconstructing Past Salinity and Climate Fluctuations. *Canadian Journal of Fisheries and Aquatic Science* 50, 1844–56.

Fritz, S. C., B. F. Cumming, F. Gasse & K. R. Laird 1999. Diatoms as indicators of hydrologic and climatic change in saline lakes, in E. F. Stoermer & J. P. Smol (eds) *The Diatoms: applications for the environmental and earth sciences*, 41–72. Cambridge: Cambridge University Press.

Gasse, F., S. Juggins & L. Ben Khelifa 1995. Diatom-based transfer functions for inferring past hydrochemical characteristics of African lakes. *Palaeogeography, Palaeoclimatology, Palaeoecology* 117, 31–54.

Griffiths, H. I., J. M. Reed, M. J. Leng, S. Ryan & S. Petkovski 2002. The conservation status and recent palaeoecology of Balkan Lake Dojran. *Biological Conservation* 104, 35–49.

Hall, R. I. & J. P. Smol 1999. Diatoms as indicators of lake eutrophication, in E. F. Stoermer & J. P. Smol (eds) *The Diatoms: applications for the environmental and earth sciences*, 169–82. Cambridge: Cambridge University Press.

Hammer, U. T. 1986. *Saline Lake Ecosystems of the World*. The Hague: Dr. W. Junk.

Hardie, L. A. & H. P. Eugster 1970. The evolution of closed-basin brines. *Mineralogical Society of America Special Paper* 3, 273–90.

Harris, T. 1995. Eutrophication of Lakes and Reservoirs in the United Kingdom: Causes, effects and controls. *Geography* 80, 60–71.

Harrison, S. P., L. Saarse & G. Digerfeldt 1991. Holocene changes in lake levels as climate proxy data in Europe, in B. Frenzel (ed.) *Evaluation of climate proxy data in relation to the European Holocene*, 149–58. Stuttgart: Gustav Fischer Verlag.

Juggins, S. 1992. Diatoms in the Thames Estuary, England: Ecology, Palaeoecology, and Salinity Transfer Function. *Bibliotheca Diatomologica* 25, 1–216.

Juggins, S. & N. Cameron 1999. Diatoms and archeology, in E. F. Stoermer & J. P. Smol (eds) *The Diatoms: Applications for the environmental and earth Sciences*, 389–401. Cambridge: Cambridge University Press.

Juggins, S. & C. J. F. ter Braak 1992. *CALIBRATE – program for species-environment calibration by [weighted-averaging] partial least squares regression*, 1–20. London: Environmental Change Research Centre, University College London.

Juggins, S., R. W. Battarbee, S. C. Fritz & F. Gasse 1994. The CASPIA project: diatoms, salt lakes, and environmental change. *Journal of Paleolimnology* 12, 191–6.

Kashima, K. 1995. Sedimentary diatom assemblages in freshwater and saline lakes of the Anatolia Plateau, central part of Turkey: an application for reconstruction of palaeosalinity change during Late Quaternary, in D. Marino & M. Montresor (eds) *Proceedings of the Thirteenth International Diatom Symposium*, 93–100. Bristol: Biopress.

Lotter, A. F., H. J. B. Birks, W. Hofmann & A. Marchetto 1998. Modern diatom, cladocera, chironomid, and chrysophyte cyst assemblages as quantitative indicators for the reconstruction of past environmental conditions in the Alps. II. Nutrients. *Journal of Paleolimnology* 19, 443–63.

Mason, C. F. 1997. *Biology of Freshwater Pollution*. 3rd edition. Harlow: Longman.

Reavie, E. D., R. I. Hall & J. P. Smol 1995. An expanded weighted-averaging model for inferring past total phosphorus concentrations from diatom assemblages in eutrophic British Columbia (Canada) lakes. *Journal of Paleolimnology* 14, 49–67.

Reed, J. M. 1995. The potential of diatoms and other palaeolimnological indicators for Holocene palaeoclimate reconstruction from Spanish salt lakes, with special reference to the Laguna de Medina (Cádiz, southwest Spain). Unpublished PhD thesis, University College London.

Reed, J. M. 1998a. A diatom-conductivity transfer function for Spanish salt lakes. *Journal of Paleolimnology* 19, 399–416.

Reed, J. M. 1998b. Diatom preservation in the recent sediment record of Spanish saline lakes: implications for palaeoclimate study. *Journal of Paleolimnology* 19, 129–37.

Reed, J. M., N. Roberts & M. J. Leng 1999. An evaluation of the diatom response to Late Quaternary environmental change in two lakes in the Konya Basin, Turkey, by comparison with stable isotope data. *Quaternary Science Reviews* 18, 631–47.

Reed, J. M., A. C. Stevenson & S. Juggins 2001. A multi-proxy record of Holocene climate change in southwest Spain: the Laguna de Medina, Cádiz. *The Holocene* 11, 705–717.

Roberts, N. & H. E. Wright Jr. 1993. Vegetational, lake-level and climatic history of the Near East and southwestern Asia, in H. E. Wright Jr, J. E. Kutzbach, T. Webb III, W. F. Ruddiman, F. A. Street-Perrott & P. J. Bartlein (eds) *Global Climates since the Last Glacial Maximum*, 194–200. Minneapolis: Minnesota Press.

Roberts, N., C. Kuzucuoğlu & M. Karabiyikoğlu 1999. The Late Quaternary in the Eastern Mediterranean Region: An Introduction. *Quaternary Science Reviews* 18, 497–501.

Roberts, N., J. Reed, M. J. Leng, C. Kuzucuoğlu, M. Fontugne, J. Bertaux, H. Woldring, S. Bottema, S. Black, E. Hunt & M. Karabiyikoğlu 2001. The tempo of Holocene climatic change in the eastern Mediterranean region: new high-resolution crater-lake sediment data from central Turkey. *The Holocene* 11 (6), 721–736.

Ryves, D.B. 1994. Diatom dissolution in saline lake sediments. An experimental study in the Great Plains of North America. Unpublished PhD thesis, University College London.

Ryves, D. B., V. J. Jones, P. Guilizzoni, A. Lami, A. Marchetto, R. W. Battarbee, R. Bettinetti & E. C. Devoy 1996. Late Pleistocene and Holocene environmental changes at Lake Albano and Lake Nemi (central Italy) as indicated by algal remains. *Memorie dell'Istituto Italiano di Idrobiologia* 55, 119–48.

Saros, J. E. & S. C. Fritz 2000. Nutrients as a link between ionic concentration/composition and diatom distributions in saline lakes. *Journal of Paleolimnology* 23, 449–53.

Stoermer, E. F. & J. P. Smol 1999. Applications and uses of diatoms: prologue, in E. F. Stoermer & J. P. Smol (eds) *The Diatoms: applications for the environmental and earth sciences*, 3–8. Cambridge: Cambridge University Press.

Street-Perrott, F. A. & S. P. Harrison 1985. Lake levels and climate reconstruction, in A. D. Hecht (ed.) *Paleoclimate Analysis and Modelling*, 291–340. Chichester: John Wiley.

ter Braak, C. J. F. & H. van Dam 1989. Inferring pH from diatoms: a comparison of old and new calibration methods. *Hydrobiologia* 178, 209–23.

Verschuren, D., K. R. Laird & B. F. Cumming 2000. Rainfall and drought in equatorial east Africa during the past 1,100 years. *Nature* 403, 410–14.

Whitmore, T. J., M. Brenner, J. H. Curtis, B. H. Dahlin & B. W. Leyden 1996. Holocene climatic and human influences on lakes of the Yucatan Peninsula, Mexico: an interdisciplinary, palaeolimnological approach. *The Holocene* 6, 273–87.

Wunsam, S. & R. Schmidt 1995. A diatom-phosphorus transfer function for Alpine and pre-alpine lakes. *Memorie dell'Istituto Italiano di Idrobiologica* 53, 85–99.

Wunsam, S., R. Schmidt & J. Müller. 1999. Holocene lake development of two Dalmatian lagoons (Malo and Veliko Jezero, Isle of Mljet) in respect to changes in Adriatic sea level and climate. *Palaeogeography, Palaeoclimatology, Palaoecology* 146, 251–281.

Yang, J.-R. & M. Dickman. 1993. Diatoms as indicators of lake trophic status in central Ontario, Canada. *Diatom Research* 8, 179–93.

Chapter 18

Human Activity: An Allogenic Factor in Wetland Dynamics

M. J. Bunting and J. E. Schofield

INTRODUCTION

Wetlands have been a significant Holocene landscape element in much of the temperate zone, even in areas where recent drainage and reclamation has reduced wetland cover to insignificant levels (Mitsch and Gosselink 2000). They provide important resources for human populations, including water, food from water plants and animals living in the wetland or drinking from it, reeds for thatch or bedding, and multi-stemmed shrubs such as *Salix* (willow) and *Alnus* (alder) which provide 'natural coppice' for weaving and construction. Wetlands also seem to have had significance beyond these resources for some cultural groups, yielding 'votive offerings' such as the Roos Carr figures (Sheppard 1902), a wide range of metal artefacts, and possible victims of 'ritual murders' (*e.g.* Glob 1969, Chown 1984).

Humans are not biologically suited to live permanently in waterlogged environments and therefore, the wetland edge will always have been an important landscape element for them. Few, if any, people will have actually lived in the wetlands, but regular movement from the dry land (or dry islands within an expanse of wetland) into the wetland and back again is highly probable, and the wetland edge is a rich source of archaeological evidence (*e.g.* Gordon and McAndrews 1992). The accumulation of wetland sediments over time may obscure archaeological traces of earlier occupation, but these same sediments preserve within themselves an 'archive' of biotic and abiotic traces of the former environments, thus enabling the palaeoecologist to reconstruct landscape changes both within and beyond the wetland (*e.g.* Lowe and Walker 1997).

This paper focuses on topographically confined wetland systems and the landscapes around them, as distinct from extensive tracts of wetland such as those around estuaries or those associated with blanket peat. Many topographically confined wetlands are 'complex' systems (*sensu* Bunting and Warner 1998), where the sediment sequence shows that the basin was once an open lake which has since undergone hydroseral succession, infilling with sediment and occupation by a series of different wetland communities. Such systems are abundant in formerly glaciated landscapes where the topography varies markedly at the local scale; for example, hummocky moraine (where kettleholes and other collapse features provide frequent depressions where water can collect) or glacially scoured rock (where glacial erosion creates a series of depressions which fill with water).

A typical landscape of hummocky moraine is that of Holderness in eastern England; although today the meres (water-filled depressions) of Holderness are mostly infilled and drained, investigation of the sediment accumulations in landscape low points reveals many former lakes and wetlands in this area (Sheppard 1956, 1957, Dinnin 1995, Dinnin and Lillie 1995). In this paper, we present a summary of the results of palaeoecological analysis of sediments from Lambwath Mere, one such former lake in Holderness, and seek wider perspectives on the questions raised by the palaeoenvironmental record, by comparison with data from similar basins in southern Ontario, Canada.

HYDROSERAL SUCCESSION

'Complex' systems are defined as sites where the basin has clearly been occupied by different wetland types, beginning with open water, and later occupied by wetland vegetation (Bunting and Warner 1998). This process is termed terrestrialisation (Pons 1992, Heathwaite *et al.* 1993), or hydroseral succession. The succession concept has been a paradigm of great importance in twentieth century plant ecology (*e.g.* Mitsch and Gosselink 2000), and various theories have been advanced to explain the sequence of communities identified which, in broad terms at least, appear to be replicable in many different locations.

Wetland succession occurs as an inevitable consequence of internal ecosystem processes such as sediment accumulation in a permanently waterlogged environment, where decay rates are low (*e.g.* Burrows 1990, Mitsch and Gosselink 2000). This raises the sediment surface relative to the watertable and thus creates a new habitat for colonisation by different species. The vegetation sequences that develop as a result of these internal processes are termed 'autogenic'. However, succession can also be strongly affected by external factors such as changes in relative sea-level (RSL) or climate, or by human activity (which may be deliberate, *e.g.* creating drainage channels, or accidental); resultant vegetation sequences are termed 'allogenic' (*e.g.* Walker 1970, Burrows 1990).

Successional sequences are currently viewed as the outcome of plant population interactions in fluctuating environmental conditions, emphasising the role of repeated, relatively frequent disturbance of vegetation. It assumes neither long-term site stability nor the existence of a fixed endpoint to succession (Burrows 1990, Glenn-Lewin *et al.* 1992). The ecological concept of the hydrosere was first defined for British vegetation types by Tansley (1939), and has dominated ideas on the origin and development of mire vegetation (*e.g.* Walker 1970; Waller 1994). The hydrosere describes a typical primary succession in freshwater basins or along the margins of slow-flowing rivers. Vegetation changes within the hydrosere occur primarily in response to autogenic sediment accumulation, leading to a sequence of communities from open water to vegetated shallow water to mire.

There are few direct studies of succession owing to the longevity of critical plant species, especially trees (Yu *et al.* 1996), and therefore chronosequences, or space-for-time substitutions, have traditionally formed the basis for models of plant succession. This approach takes an extant sequence of plant communities across an environmental gradient (*e.g.* from open water to dry land), and assumes that there are no ecologically significant differences among sites in the spatial array except age or time since last disturbance (Jackson *et al.* 1988). Therefore the sequence along the gradient is assumed to represent the series of stages that will occur at one point on that gradient over time; the oldest (or longest-undisturbed) location is assumed to have moved in order through each of the younger stages in the sequence before attaining its current state. Palaeoecological techniques (*e.g.* pollen, plant macrofossil and lithostratigraphic analyses) provide a powerful means of independently testing the validity of chronosequences (Jackson *et al.* 1988, Yu *et al.* 1996).

Tansley (1939) used chronosequences to develop model hydroseres, and envisaged a 'climax community' of raised mire in the cool oceanic climates of north and west Britain, and *Quercus robur* (English oak) woodland in the drier south and east. Walker (1970) combined stratigraphic records from wetland sites throughout the British Isles to show that the transition from fen to oakwood is unsubstantiated, and that the

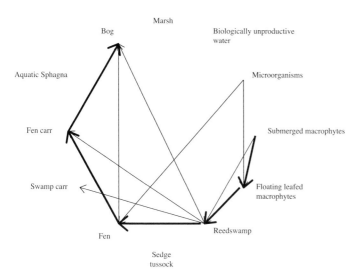

Fig. 18.1 Simplified polygon showing principal transitions in hydroseral successions in basin sites free from allogenic influences (after Walker 1970). Solid lines denote dominant pathways.

natural 'endpoint' of autogenic hydroseres throughout Britain is bog. He suggested a 'preferred' course of succession for hydroseres (see Fig. 18.1). This sequence is only recorded in 46% of his records, and alternative sequences, often skipping stages or showing successional reversals (*i.e.* one community replaced by another characteristic of wetter conditions), are common. More recent work (*e.g.* Wheeler 1980, Smith and Morgan 1989, Waller *et al.* 1999) has modified this model, suggesting that in base-rich wetland ecosystems *Alnus glutinosa* (alder) may form stable fen carr communities which can only be replaced through successional reversal (*i.e.* flooding of the carr surface).

THE REGIONAL ISSUE: THE MERES OF HOLDERNESS

There are over 70 former meres identified in the Holderness region of eastern England (Flenley 1987, Dinnin 1995). Archaeological and documentary evidence demonstrate the importance of the meres as a resource to human populations throughout the Holocene (Sheppard 1956, 1957, Dinnin 1995). The results from one of these sites, Lambwath Mere, are considered here.

Site description

Holocene deposits marking the former position of Lambwath Mere are located along the present line of Lambwath Stream near the village of Withernwick, approximately 15 km northeast of the city of Kingston upon Hull and 5 km inland from the East Yorkshire coast (latitude 0° 11' west, longitude 53° 50' north) (Fig. 18.2). The mere formed in a major depression in the irregular till surface of Holderness, exposed following the retreat of Devensian ice from the region *c.*13,000 yr BP, and at its maximum extent covered a surface area of approximately 3 km^2 (Dinnin and Lillie 1995).

It appears likely that from the fifteenth century AD onwards the mere had been reduced to the status of an ephemeral wetland. Despite the presence of a network of land drains, the area around Withernwick is still prone to extensive winter flooding and the watertable remains high (Dinnin and Lillie 1995). Today, most of the landscape is either under arable production or used as pasture (Schofield 2001). The site of the former mere contains the only surviving fragments of agriculturally unimproved, undisturbed semi-natural grassland in the whole of Holderness, and 0.25 km^2 of meadows centred around NGR TA204397 were designated a Site of Special Scientific Interest (SSSI) by English Nature in 1987. The area supports over

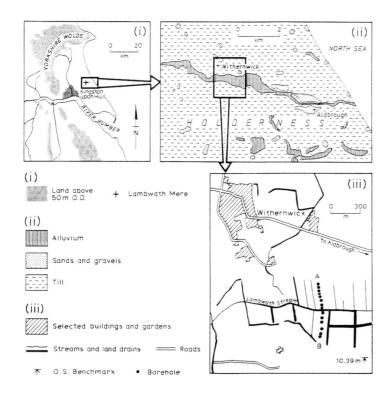

Fig. 18.2 Lambwath Mere:
(i) location within the region.
(ii) drift deposits in the area around Withernwick.
(iii) location of borehole transect across Lambwath Mere (after Schofield 2001).

50 native species of grasses and herbs, including *Thalictrum* spp., *Primula veris*, *Briza media* and *Conopodium majus*. Lithostratigraphic investigation of these deposits (Fig. 18.2) revealed a sequence of Holocene gyttjas and peats capped by a unit of clay, which extend to a maximum depth of *c.* 9 m. A sediment core recovered from the basin centre contains a palaeoecological record for the period *c.* 9500–1250 yr BP (Schofield 2001). In this paper we focus on the record since 5150 yr BP.

The methods used for pollen analysis, sequence chronology and other analyses are presented in detail in Schofield (2001). Summary pollen and spore diagrams are presented here – Fig. 18.3a shows the major wetland taxa identified and Fig. 18.3b shows common dryland types. Descriptions of local pollen assemblage biozones (lpazs) are given in Table 18.1.

Hydroseral succession at Lambwath Mere

Between *c.* 9500 and 2850 yr BP vegetation changes in the basin appear to have occurred along a progressive series in response to autogenic mechanisms, particularly basin infilling by detrital lake muds and peats. Between *c.* 9500 and 5150 yr BP the basin contained deep, open freshwater with floating algae (*Pediastrum boryanum*). Small populations of submerged, floating-leaved, and semi-emergent hydrophytes (*e.g.* *Nymphaea alba* (white water lily), *Potamogeton* spp. (pondweeds) and *Sparganium/Typha* spp. (bur-reeds and reedmace) probably grew in shallow water around the margins of the mere. From around 7940 ± 45 yr BP (SRR-6541) *Alnus glutinosa* fen carr woodlands appear to have formed over wet soils around the borders of the mere.

The record presented here begins *c.* 5150 yr BP (lpaz LM3), when the water in the mere had shallowed sufficiently to allow the widespread expansion of populations of submerged and floating-leaved hydrophytes.

Local pollen assemblage zone (lpaz)	Depth (cm)	Approx. age range (^{14}C yr BP)	Pollen and spore characteristics	Inferred dry land vegetation	Inferred wetland vegetation
LM3	328–230	5150–3700	Base defined by decline in *Ulmus* pollen from >5% base sum to <1%, recovering after 4500 BP to 2–3%. Dominant types *Alnus glutinosa*, *Quercus*, *Corylus avellana*, also *Tilia* and *Fraxinus*. Small increase in herb pollen, including *Plantago lanceolata*, plus pollen from floating leaved aquatic plants (e.g. *Trapa natans*, *Nuphar*, *Nymphaea alba*).	Mixed deciduous woodland; may be some limited human disturbance of woodland cover.	Shallow open water wetland communities similar to NVC A8, *Nuphar lutea* communities (Rodwell *et al*. 1995). Water depth reduces through zone. Alder fen carr fringe.
LM4	230–177	3700–2850	Dominated by *Alnus glutinosa*; *Quercus* decreases slightly from c.20% to 15% base sum, *Corylus avellana*-type and *Tilia* decrease throughout; *Fraxinus excelsior* remains near constant, *Salix* gradually increases, with a peak of 20% base sum at c.3,250 BP. Herb pollen generally increases, particularly Cyperaceae and Poaceae. Pollen of floating leaved aquatic plants less important relative to LM3; increases in semi-emergent hydrophytes, particularly *Sparganium*-type.	As above.	Dominated by emergent communities, possibly NVC S14, *Sparganium erectum* reedswamp (Rodwell *et al*. 1995), with some deeper pools containing floating-leaved aquatics, replaced by willow carr from around 3200 BP; alder carr still present at basin edge.
LM5	177–141	2850–2250	Marked reduction in trees & shrubs. *Alnus glutinosa* remains dominant but decreases from c.40% to 25%. Poaceae and Cyperaceae increase throughout, from c.5% base sum at the base to c.30%/c.15% respectively. Apiaceae exceeds 5% base sum; other herbs include *Plantago lanceolata* (2–3%) and *Filipendula* (up to 2%), and *Sparganium*-type. Pteropsida (monolete) indet. & *Pteridium aquilinum* also consistently present	Intensive deforestation; grazing, some cereal cultivation.	Reversal to reed swamp.
LM6	141–67	2250–1350	Herbs (especially Poaceae, Cyperaceae, *Plantago lanceolata*, Asteraceae (Lactuceae) and Asteraceae (Asteroideae) and aquatic (*Menyanthes trifoliata*, *Potamogeton*-type, *Sparganium*-type) dominatew, with Pteropsida (monolete) indet. and *Pteridium aquilinum*. Low tree & shrub values; minimum at c.1,750 BP, minor recovery thereafter (espec. in *Alnus glutinosa*). Some charcoal spikes noted.	Minimal woodland cover. Grazing, some arable crops.	Ongoing reversal/replacement of carr with reed-swamp.
LM7	67–49	1350–post 1250	Dominated by Poaceae, Cyperaceae and *Salix*, with *Filipendula* and Pteropsida (monolete) indet. spores. Other trees & shrubs negligible.	Open grassland.	Invasion of willow fen carr; possible return to reedswamp after this zone.

Table 18.1 Local pollen assemblage biozones (lpaz) at Lambwath Mere since c. 5000 ^{14}C yr BP.

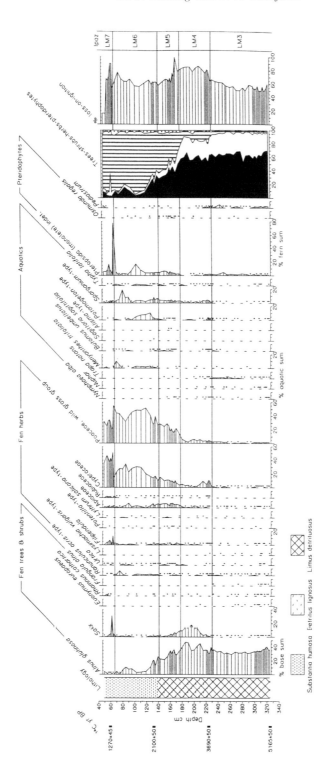

Fig. 18.3a Summary pollen diagrams for Lambwath Mere: Main wetland pollen and spore taxa. Base sum for percentage calculations: Total Land Pollen and Spores minus Pteropsida (monolete) indet. and obligate aquatics. Fern spore percentages are calculated on base sum plus Pteropsida (monolete) indet., and aquatic pollen percentages are calculated on base sum plus obligate aquatics.

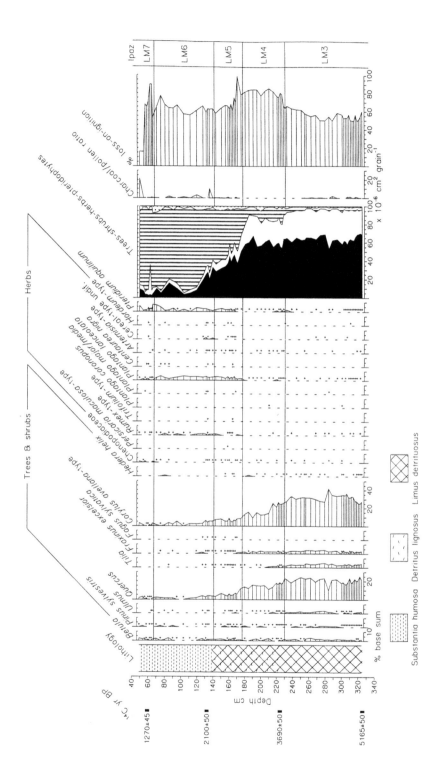

Fig. 18.3b Summary pollen diagrams for Lambwath Mere: Main dryland pollen and spore taxa. Base sum for percentage calculations: Total Land Pollen and Spores minus Pteropsida (monolete) indet. and obligate aquatics.

The water appears to have been warm and eutrophic, and a highly productive ecosystem characterised by *Nymphaea alba*, *Nuphar lutea* (yellow water lily), *Potamogeton* spp. and *Trapa natans* (water chestnut) appears to have developed. At the centre of the basin, the replacement of submerged and floating-leaved communities by a reedswamp, probably comprising *Phragmites australis* (reed), *Sparganium* spp., *Typha* spp. and *Butomus umbellatus* (flowering rush), occurs at 3690 ± 50 yr BP (SRR-6539) (lpaz LM4). Increasing *Salix* (willow) pollen percentages suggest that by *c.* 3250 yr BP this reedswamp had been replaced by a *Salix* swamp carr, although fen peats of similar age from the north basin margin are dominated by pollen of *Alnus glutinosa*, suggesting *Alnus* fen carr was locally present.

From *c.* 2850 yr BP *Alnus* fen carr and *Salix* swamp carr communities appear to have been gradually replaced by fen vegetation (reedswamp, tall-herb fen) more typical of the earlier stages of a hydrosere (lpaz LM5). Increases in pollen of *Potamogeton* spp. and *Menyanthes trifoliata* (bog bean) after *c.* 2100 yr BP (lpaz LM6) suggest that during the late Iron Age the site became increasingly wet and poor in nutrients. This community persisted for *c.* 750 years. In lpaz LM7, a brief increase in *Salix* pollen percentages (*c.* 1350 yr BP) suggests local re-establishment of tall shrub communities. Peat formation in the basin ceased shortly after 1270 ± 45 yr BP (SRR-6537), and is overlain by a clay-rich unit which may have supported a Cyperaceae (sedge) dominated marsh until the fourteenth century when the mere was drained by the monks of Meaux (Sheppard 1957).

Dryland vegetation history and human activity

The early to mid-Holocene vegetation history of well-drained sites around the mere at Lambwath reflects patterns of immigration, establishment and competitive interaction between species arriving in Holderness following their expansion from glacial refugia, and a typical mixed deciduous woodland of *Quercus*, *Ulmus* (elm), *Tilia* (lime), *Corylus avellana* (hazel) and *Fraxinus excelsior* (ash) developed around the mere. The mid-Holocene *Ulmus* decline is dated to 5165 ± 50 yr BP (SRR-6540) at this site, and coincides with the first palynological evidence from Lambwath Mere for prehistoric human impact on the vegetation cover. Pollen of *Plantago lanceolata* (ribwort plantain) and traces of pollen from other ruderal herbaceous taxa, which start to be recorded just before the *Ulmus* decline and are continually recorded in the following lpaz, may indicate that pastoral farming in small woodland clearings occurred throughout the late Neolithic and Bronze Age.

The main phase of woodland clearance at Lambwath began *c.* 2850 yr BP and continued throughout the Iron Age and into the Roman period, with some woodland regeneration during the Early Medieval period. Woodland cover around the site reached a minimum at *c.* 1750 yr BP, at which time the landscape appears to have supported predominantly open grassland communities. Increased numbers of cereal-type pollen grains and pollen of 'anthropogenic indicators' (Behre 1981) are found after *c.* 2850 yr BP, possibly recording arable farming systems. From the late Iron Age onwards, erosion of topsoils intensified, resulting in a large influx of minerogenic particles into the mere and ultimately a sedimentary change within the basin from peat formation to clay deposition.

ISSUES RAISED BY THE LAMBWATH SEQUENCE

Identifying the pollen source area in complex wetland systems

In order to interpret pollen assemblages and reconstruct past environments, the analyst needs to understand the origins of the pollen grains making up the assemblage. Pollen taphonomy is complex, but a key idea is that of the pollen source area (PSA). Two common definitions of this concept are found: (a) the area from which the majority of the pollen of a given type is derived, and (b) the area of surrounding vegetation for which compositional changes are most sensitively represented by the pollen assemblage.

PSA varies with plant species (*e.g.* Prentice 1985, Jackson 1990, 1994, Sugita 1994, Calcote 1995), but more importantly from our perspective it also varies according to the size of canopy opening, increasing for larger openings. It can also be affected by any vegetation growing on the receiving sediment surface, which both produces pollen itself (thus diluting any incoming signal) and acts as a physical filter to particles settling out of the air. For example the PSA for a pollen morphotype such as *Quercus* (oak), in a medium-sized lake (such as Lambwath Mere in the early Holocene) is in the order of 1000 m (Jackson 1990, Sugita 1994). In a closed canopy woodland (such as an alder carr) the PSA of *Quercus* is on the order of 30–100 m (Andersen 1973, Bradshaw 1981, Calcote 1995).

Therefore hydroseral change can have a profound effect on the pollen record of the surrounding dry land (Jacobson and Bradshaw 1981, Edwards 1991). Fig. 18.4 shows the variation in abundance of non-wetland pollen types at the present day in Oil Well Bog, a wetland in southern Ontario (Bunting *et al.* 1998a). The relationship between surface community and percentage pollen transported from the surrounding dry land is clearly complex. This variation may be particularly challenging when attempting to detect small-scale changes in the surrounding uplands, such as initial patchy clearance for agriculture.

A hypothetical model of trends in PSA for Lambwath Mere since 5000 yr BP is shown in Fig. 18.5, based on the pollen recruitment properties of the wetland communities occupying the basin over that period (discussed briefly above). This model can be used to inform the interpretation of the pollen record. For example, LPAZ LM4 records gradually falling pollen values for dryland trees, particularly *Quercus*, *Tilia* and *Corylus avellana*-type (see Fig. 18.3b), which might indicate loss of upland forest cover to purposive

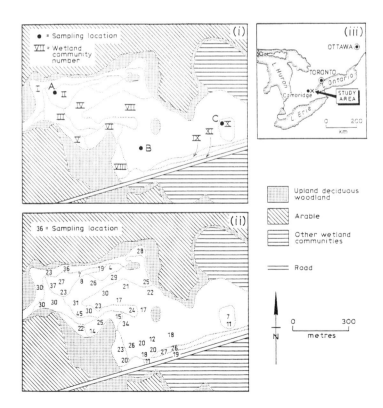

Fig. 18.4 Map showing (i) location of Oil Well Bog in southern Ontario; (ii) map of the present-day wetland vegetation (communities I and III are dominated by tall shrubs, II by low-shrubs, V and XI by herbaceous plants, IV, VI, VIII and IX by deciduous trees with a shrub understorey, VII by deciduous and coniferous trees and X by a dense conifer stand) and location of collection points for pollen cores summarised in Table 18.2 (after Bunting et al. 1998b); (iii) distribution of percentages of non-wetland pollen across the site at present day. These data represent a minimum count, since some pollen types (e.g. Poaceae) include plant species of both wetland and dryland; only taxa where no possible source species was recorded within the wetland were included in this figure.

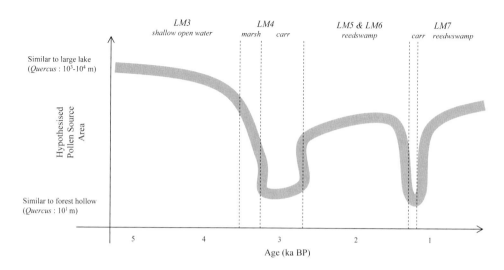

Fig. 18.5 Hypothetical model of changing pollen source area at Lambwath Mere (see text for details).

clearance by humans. The hypothetical model (Fig. 18.5) suggests that the effective PSA was lower during this lpaz, as marsh and fringing carr invade the basin; this would be sufficient to explain the decreasing proportions of pollen from dryland trees. The latter interpretation is supported by the lack of an associated increase in pollen from 'anthropogenic indicators' such as *Plantago lanceolata* or *Rumex*-type (docks), which would be expected if larger clearings were forming. Pollen data from sediments of roughly equivalent age from a core taken at the basin edge, nearer the dry land and, therefore, remaining within the PSA of upland woodland despite the changing vegetation cover, do not show a comparable decrease in arboreal pollen values. Hydroseral succession at the basin centre during LPAZ LM4 apparently reduced the input of dryland pollen to the sedimentary record without there having been any real changes to the dryland vegetation mosaic.

A similar pattern can be seen in records from wetlands in southern Ontario. For example, Fig. 18.6 shows two summary pollen sequences for the Holocene from Spiraea Wetland (Bunting and Warner 1999), a complex wetland (*sensu* Bunting and Warner 1998). The records are virtually identical in the lower part of the diagram, when the basin was occupied by a lake. An apparent opening up of the woodland canopy is recorded between 1.0 m and 0.7 m in the total land pollen and spores diagram, *c.* 5500–1600 yr BP. During this period, human presence in the area was extremely low, and no lake record from the region shows disturbance of the woodland canopy. The pollen and macrofossil evidence shows the site to be occupied by a complex of marsh, dwarf shrub and tall shrub swamp communities during this period, and the local pollen rain distorts the dryland record.

In this case, as there are very strong grounds for assuming the wider landscape to be entirely tree-covered, and there is no evidence for trees growing within the basin, it is possible to re-calculate the data from Spiraea Wetland and present it, as in Fig. 18.6(ii), as percentages of tree taxa only. This diagram shows no woodland clearance event, and the sequence of dryland tree communities is comparable with that from other sites in the region. However, this simple tactic for reducing the hydroseral distortion problem is not suitable for many sites. It cannot yield useful results where trees form part of the wetland vegetation, or, as in the British late Holocene, when woodland may not be stable and dominant in the wider landscape. Even in the simple case of Spiraea Wetland, the 'basis sum Arboreal Pollen' diagram may not be giving a

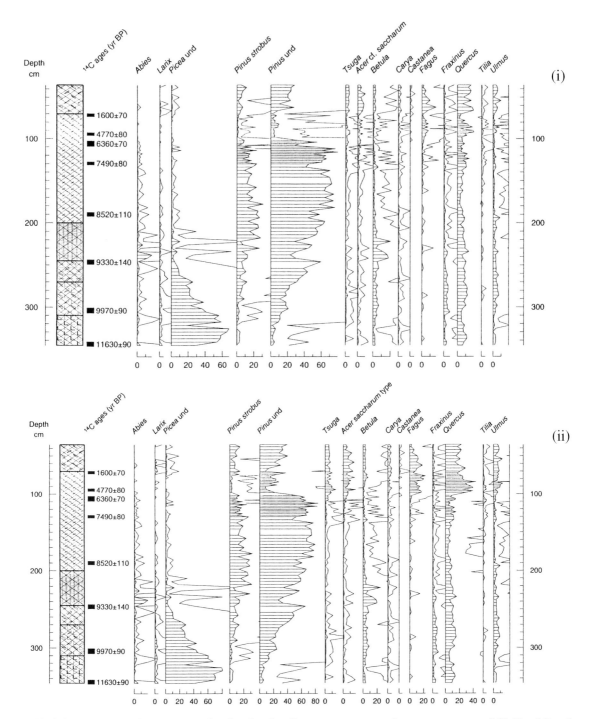

Fig. 18.6 Comparison of Spiraea Wetland upland pollen types presented as percentages of (i) Total Land Pollen and Spores and (ii) Arboreal Pollen (see also Bunting & Warner 1999).

clear picture of the upland pollen record since, as mentioned above, filtration by the surface vegetation may have a differential effect on the different tree types, leading to under-representation of some and over-representation of others.

Many of the pollen types produced in open agricultural communities, such as grasses, are also produced by wetland plants, and therefore mitigation of the hydroseral effect requires a more detailed understanding of the taphonomic characteristics of each stage in the wetland sequence, and of the spatial patterning of different seral communities within the basin.

SPATIAL HETEROGENEITY OF WETLAND COMMUNITIES

Basin wetlands are rarely uniformly vegetated (see for example the map of Oil Well Bog in Fig. 18.4), and the hydroseral stage can therefore vary both temporally and spatially within a basin. At Lambwath, data from two cores, one from the edge of the basin and one from the centre, shows that reedswamp encroached from the edge of the basin into the central, shallow, open water communities. Even once carr was established across the whole basin (in the latter part of lpaz LM4), the canopy consisted of *Salix* in the basin centre but *Alnus glutinosa* at the edge. Thus the biasing effects of the surface vegetation will not have been constant through either space or time.

This is also seen by comparing results from three short cores taken from different contemporary wetland communities within Oil Well Bog in southern Ontario (Fig. 18.4, Table 18.2, Bunting *et al.* 1998b). Reconstruction of wetland communities within the basin is based on a study of modern pollen-vegetation relationships within the wetland (Bunting *et al.* 1998a). Before *c.* 500 years BP, the overall wetland surface seems to have been wetter than today, with a relatively uniform dwarf shrub vegetation with some tall shrub (equivalent to 'carr') in more central locations. From *c.* 500 BP, coniferous swamp communities became established in part of the wetlands, implying that the system had become drier, possibly in response to the Little Ice Age climate event. With Euro-Canadian settlement in around AD 1830, the surface became wetter again (*cf.* Warner *et al.* 1989) and dwarf shrub swamp was established at all three coring points. This reversal was followed by succession to different stages at the three points, with the most rapid and marked changes occurring in the centre of the site, suggesting that the marked heterogeneity of the modern wetland reflects a continuing response to European settlement in the area.

In order to understand the nature of hydroseral succession within a wetland, and therefore to be able to use our understanding of the differing PSA associated with each community, we need to study the spatial and temporal variation across the system. A multiple-coring approach is essential; lithostratigraphy can identify broad trends, and expensive microfossil analysis and age estimation can then be used in carefully chosen contrasting locations.

Allogenic human activity and hydroseral change

The pattern and rate of hydroseral change at Oil Well Bog (Table 18.2) appears to have been strongly affected by European settlement in the surrounding uplands. This is not unexpected, as many wetlands in southern Ontario show marked changes in hydroseral stage coincident with European settlement (Bunting and Warner 1998), and indeed every example of floating mat (*schwingmoor*) wetlands studied in detail so far shows that mat expansion followed European settlement (*e.g.* Warner *et al.* 1989, Campbell *et al.* 1997).

European settlement in eastern North America involved rapid, large-scale clearance of the dryland forests for agriculture. The nature of vegetation cover within a catchment is an important variable governing the water budget for that system (Gregory and Walling 1973), since vegetation affects both total volume and rate of input, storage and output of water, via processes such as interception, throughfall, stemflow and transpiration. Plants also bind sediments, protecting soils from erosion. Trees fulfil a critical hydrological

	Upland vegetation	Core A – wetland 'peninsula'	Core B	Core C – central community
Modern vegetation	Arable/belt of deciduous forest	Dwarf shrub-*Sphagnum*	Mixed deciduous swamp	Coniferous swamp
2200–c.500BP	Mixed deciduous forest	Dwarf shrub swamp	From 1650 BP: dwarf shrub swamp with occasional *Salix*	From 1050 BP: tall shrub swamp
c.500–150 BP	Decline in *Fagus*; possible Little Ice Age response	Possible coniferous swamp nearby	Possible coniferous swamp nearby	Possible coniferous swamp nearby
Post-150 BP (1830 AD)	*Ambrosia* rise, deforestation and widespread agricultural clearance: European settlement	Dwarf shrub swamp with marshes nearby	Dwarf shrub swamp	Dwarf shrub swamp / Tall shrub swamp
?Post-50 BP (1900 AD)	*Ambrosia* levels fall: probable partial abandonment of fields, establishment of modern deciduous woodland 'fringe'	tall shrub boundary nears coring point	c.AD 1890: tall shrub species colonise / c.AD 1920: swamp community with pine, possible deciduous element	Coniferous swamp

Table 18.2 Summary of dryland and hydroseral change reconstructed from pollen records from Oil Well Bog, Southern Ontario (published in more detail in Bunting et al. *1998b).*

role by evapotranspiring at much higher rates than non-arboreal vegetation, by improving drainage via their rooting system, and by maintaining the circulation of nutrients (Roberts 1989). The removal of tree cover can enhance groundwater supplies and increase surface runoff by reducing transpiration losses, interception and re-evaporation of precipitation from leaf surfaces (*e.g.* Moore 1993). The subsequent increased total input, or increased variability of input, may 're-set' hydroseres to an earlier successional stage.

Increased erosion of catchment surfaces following the removal of trees, and the subsequent transport of dissolved nutrients and sediment into wetlands via surface runoff, could also have significant effects on succession through changes in water and sediment chemistry, water colour and transparency, and sediment accretion rates (Jackson *et al.* 1988). These hydrological connections between wetlands and the surrounding dryland catchment mean that human-induced changes in landuse practice in uplands can play a significant role in wetland vegetation dynamics.

The Holocene is characterised by increasing levels of human disturbance of the natural vegetation mosaic. Direct and purposive modification of wetland vegetation communities has occurred through the initiation of drainage schemes and human management of wetland ecosystems in order to provide societies with resources such as food, fuels and building materials (Rackham 1997). Mid- to late Holocene Old World deforestation and the development of 'cultural landscapes' (Faegri 1988) dominated by arable and pastoral farming systems change the supply of water, sediment and nutrients to wetlands, and have led to detectable hydroseral changes, which are seen in records throughout Britain.

Direction of hydroseral change

In southern Ontario floating mat wetlands, the disturbance is associated with 'progressive' successional changes (changes towards drier vegetation communities which usually occur later in the autogenic successional sequence). In Oil Well Bog, the initial response to human activity was a reversal in the sequence. The extant community was replaced by one characteristic of wetter conditions and thus of earlier stages in the autogenic sequence (a 'retrogressive' change), although subsequently the wetland again developed drier communities.

At Lambwath Mere, only limited disturbance of the dryland tree canopy is inferred in lpazs LM3 and LM4, during which progressive succession from shallow open water to carr communities is taking place. Relative and absolute declines in the values for pollen of deciduous woodland trees during lpaz LM5 (particularly *Quercus, Fraxinus excelsior* and *Corylus avellana*-type), and increases in pollen and spores from 'anthropogenic indicators' (Behre 1981), provide strong evidence for intensive deforestation of the landscape around the mere beginning at *c.* 2850 yr BP and continuing throughout the Iron Age. Woodland clearance may have resulted in the destabilisation and erosion of catchment soils, which is also indicated by increases in the mineral content of sediment, and by magnetic evidence from the sediments collected at the edge of the basin for increased input of iron minerals (*e.g.* haematite, geothite and magnetite) (Schofield 2001). A few *Hordeum*-type (barley-type) grains, occurring together with pollen of Chenopodiaceae (fat hen family), *Persicaria maculosa*-type, *Plantago major/media* (rat-tail or hoary plantain) and *Artemisia*-type (wormwood/mugwort), provide tentative evidence for low intensity arable activity. Most of the cleared areas created probably supported open damp grassland grazed by domesticated animals, shown by increases in pollen of *Plantago lanceolata* and the appearance of pollen of *Trifolium*-type (clovers) and *Centaurea nigra* (lesser knapweed). *Pteridium aquilinum* (bracken) appears to have flourished in this environment, possibly because the species is avoided by grazing animals.

Coincident with this extensive woodland clearance, a successional reversal from fen carr to fen or reedswamp appears to have occurred at *c.* 2850 yr BP. Throughout lpaz LM5 pollen of fen carr trees (particularly *Salix* and *Alnus glutinosa*) steadily decreases and is replaced by pollen of fen herbs and semi-emergent aquatic plants, such as Cyperaceae and Poaceae. High pollen frequencies of Apiaceae may represent common fen herbs such as *Peucedanum palustre* (milk parsley) or *Oenanthe* spp. (water dropwort), and other taxa such as *Ranunculus* (buttercups), *Filipendula* (meadow-sweet), and *Osmunda regalis* (royal fern) were probably also growing in the fen. Small increases in percentages of *Sparganium*-type and *Butomus umbellatus* pollen may imply localised re-expansion of reedswamp.

The onset of this reversal broadly coincides with late Holocene climatic deterioration (increasing wetness) throughout northwest Europe dated *c.* 2650 yr BP (van Geel *et al.* 1996), and with the earliest recurrence surface recorded by Smith (1985) from raised mires in the Humberhead Levels, an extensive tract of lowland peat some 50 km to the west of Lambwath Mere. Therefore climatic change could have been an additional factor behind this successional reversal.

Human utilisation of wetlands and intervention in the hydrosere

An alternative explanation for the hydroseral reversal in LM5 is utilisation of the mere surface by Iron Age people, since frequent mowing of sedges and reeds would inhibit the regeneration of woody species, and some domesticated herbivores, such as sheep, can also be effective causal agents of declining tree cover. Areas of the carr woodland may also have been deliberately cleared to provide wood for fuel and building materials, or to create openings to promote the growth of reed beds.

Fen carr failed to re-invade the mere surface from *c.* 2100 to *c.* 1350 yr BP, longer than the earlier reedswamp stage in the autogenic sequence (start of LM4), even though the site will have been relatively dry. In contrast, Table 18.2 shows the relatively short duration of the successional reversal at Oil Well Bog, which was also driven by upland clearance. The lack of fen colonisation has two possible causes; either the combination of climate and changed catchment vegetation kept the mere surface too wet to permit the growth of fen trees and shrubs, or tree regeneration was inhibited by some combination of management practices as outlined above.

Contemporary management activity, notably in the Norfolk Broads, uses approaches such as scrub-slashing and burning to restore and maintain reed beds in situations where carr woodland would become established without human intervention, thereby showing the effectiveness of determined human action in reversing hydroseres or preventing progressive vegetation change. The challenging nature of the wet, uneven substrate means that these interventions are largely achieved by hand, not by the use of heavy machinery. Historical data for Lambwath Mere shows both regular activity aimed at managing the mere to maintain specific resources, and the economic value of those resources (notably fish and reeds) (Sheppard 1956, 1957).

In Britain the management of wetlands can be traced back to at least Neolithic times. Coppicing and pollarding of trees, including wetland species such as *Alnus glutinosa,* was practised in order to produce poles for the construction of wooden trackways, which crossed extensive tracts of wetland, connecting dryland islands and peninsulas (*e.g.* the Sweet Track across the Somerset Levels; Coles and Coles 1989). Earlier wetland management is also sometimes inferred; for example, one explanation of the early Mesolithic use of fire at Star Carr in the Vale of Pickering is the maintenance of reedbed communities (Mellars and Dark 1998).

The arrival of the Romans in Britain (AD 43) was followed by the first known attempts at drainage of fens, and the land may have been used as settlement overflows for population from crowded uplands. Fens and saltmarshes were certainly used for the summer and autumn grazing of livestock (Rackham 1997). Documentary sources record the importance of wetland resources and revenue from the Early Medieval period onwards. Many fens were intensively managed for reeds (*Phragmites australis*) and sedge (*Cladium mariscus*) to provide thatch and fuel, and many Medieval estates in both Fenland and the Humber Wetlands were important eel fisheries (Sheppard 1957, Rackham 1997). Peat was also cut for fuel, and large-scale peat-digging was responsible for the creation of the Norfolk Broads, which now demonstrate many examples of secondary hydroseres (Wheeler 1984).

Fens and bogs were increasingly being perceived as inhospitable, and as barriers to settlement and agricultural progress, and were consequently affected by agricultural improvements and enclosure, which began in the late Medieval period, often leading to drainage and conversion of wetland to agricultural land. At Lambwath Mere, reedswamp apparently persisted until *c.* 1350 yr BP, probably as a result of human management via grazing and/or mowing. A brief increase in pollen of *Salix* after this time could reflect the temporary abandonment of such management practices, or the use of part of the basin as an osier bed. Peat formation in the basin ceased shortly after 1270 ± 45 yr BP (SRR-6537), following an increase in the influx of clay particles, presumably eroding from the farmed soils of the surrounding catchment. This mineral substrate may have supported a Cyperaceae dominated marsh until the fourteenth century when the mere was finally drained by the monks of Meaux (Sheppard 1957).

Other allogenic factors affecting hydroseral succession

Changes in climate and human activity are probably the most common external factors influencing vegetation change in British Postglacial wetlands, but other allogenic mechanisms may also contribute. These can include 'wildfire' (frequently caused by lightning strikes; Patterson *et al.* 1987) and extreme weather-related phenomena (*e.g.* storms or flooding). Wild or domesticated animals may also deflect successional series through grazing damage (Tansley 1939), and the dam-building activities of beavers (*Castor* spp.) may create small ponds and initiate paludification of adjacent habitats (Mitsch and Gosselink 2000). The detection of beaver impact in the palaeoecological record is difficult, but beaver-cut wood from the European species *Castor fiber* (present in Britain up to *c.* 800 yr BP) has been recorded from several sites in England and Wales, including Star Carr (Yorkshire), the Baker Platform (Somerset Levels) and Runnymede (Thames valley) (Coles 1992).

SYNTHESIS

The basin at Lambwath has supported a range of different wetland communities over the last 5000 years. In this paper, we have explored the consequences of these changes for palaeoenvironmental reconstruction of the surrounding dry land and the detection of prehistoric human activity, and the effects of human activity on the wetland, whether intentional or accidental.

'Complex' wetland systems like Lambwath Mere often occur in areas where there are few, if any, lakes and raised bogs, and thus provide a useful resource for palaeoecologists seeking to reconstruct past environments and human activity. The well defined edge and relatively small size of these systems means that the sedimentary archive will contain a strong signal from the surrounding dry land. However, unless the effects of hydroseral succession on that signal can be understood and compensated for, reliable reconstruction of human activity and dryland vegetation dynamics, and comparison with records from other types of sedimentary system, will not be possible. Modelling approaches and modern surface-sample studies enable direct comparison of pollen assemblages with the vegetation that produced them, and thus have an important contribution to make. However, the results cannot be applied to palaeoenvironmental reconstruction unless the past environments within the basin being studied are understood. The spatial and temporal variability of vegetation communities in topographically confined wetlands mean that a multiple-core approach is essential. Lithostratigraphy and pollen analysis provide a general outline, but may need to be supplemented with plant macrofossil analysis and other techniques at critical horizons, since some critical plant types are all but invisible to palynology (*e.g.* Juncaceae) or have low, variable palynomorph production (*e.g. Salix, Typha* and *Sphagnum*) and therefore the plants might have been present in abundance yet scarcely registered in the pollen record.

Complex systems and the hydroseral changes they have undergone should not simply be seen as a problem to be filtered out from the pollen record in order to 'read' the dryland activities of people, but as an important part of our understanding of the prehistoric landscape. Even where the wetland is not much used by the local population, hydroseral change provides supplementary information on the intensity and nature of the landscape response to changing behaviour in the dryland parts of the catchment. However, available evidence suggests that, as in historical periods, the wetland resource was often important to prehistoric communities both economically and 'ritually'.

Changes in the hydrosere will have had a marked effect on both environment and economy, and therefore provide additional insight into the context of prehistoric human behaviours. In some cases these changes are met with a direct response, as humans intervene in the hydrosere to restore or maintain a desirable seral stage using a variety of management techniques. Therefore reconstructing hydroseral changes can provide guidance in the search for archaeological sites, or provide greater insight into the landscapes of human occupation.

ACKNOWLEDGEMENTS

The authors would like to thank Keith Scurr for drawing Fig. 18.2. Additional thanks go to the Carstairs Countryside Trust (CCT), for access to their land at Lambwath Grasses. The full economic costs of radiocarbon dates from Lambwath Mere were covered by a grant to JES from the NERC.

REFERENCES

Andersen, S. T. 1973. The differential pollen productivity of trees and its significance for the interpretation of a pollen diagram from a forested region, in H. J. B. Birks & R. G. West (eds) *Quaternary Plant Ecology*, 108–114. Oxford: Blackwell.

Behre, K. E. 1981. The interpretation of anthropogenic indicators in pollen diagrams. *Pollen et Spores* 23, 225–245.

Bradshaw, R. H. W. 1981. Modern pollen-representation factors for woods in south-east England. *Journal of Ecology* 69, 45–70.

Bunting, M. J. & B. G. Warner 1998. Hydroseral development in southern Ontario: patterns and controls. *Journal of Biogeography* 25, 3–18.

Bunting, M. J. & B. G. Warner 1999. Late Quaternary vegetation dynamics and hydroseral development in a shrub swamp in southern Ontario, Canada. *Canadian Journal of Earth Sciences* 36, 1603–1616.

Bunting, M. J., B. G. Warner & C. R. Morgan 1998a. Interpreting pollen diagrams from wetlands: pollen representation in surface samples from Oil Well Bog, southern Ontario. *Canadian Journal of Botany* 76, 1780–1797.

Bunting, M. J., C. R. Morgan, M. van Bakel & B. G. Warner 1998b. Pre-European settlement conditions and human disturbance of a coniferous swamp in southern Ontario. *Canadian Journal of Botany* 76, 1770–1779.

Burrows, C. J. 1990. *Processes of Vegetation Change*. London: Unwin Hyman.

Calcote, R. 1995. Pollen source area and pollen productivity: evidence from forest hollows. *Journal of Ecology* 83, 591–602.

Campbell, D. R., H. C. Duthie & B. G. Warner 1997. Post-glacial development of a kettle-hole peatland in southern Ontario. *Ecoscience* 4, 404–418.

Chown, M. 1984. Cheshire's oldest inhabitant? *Nature* 311, 290.

Coles, B. 1992. Further thoughts on the impact of beaver on temperate landscapes, in S. Needham & M. G. Macklin (eds) *Alluvial Archaeology in Britain*, 93–99. Oxford: Oxbow Books.

Coles, B. & J. Coles 1989. *People of the Wetlands: Bogs, Bodies, and Lake-Dwellers*. London: Thames & Hudson.

Dinnin, M. H. 1995. Introduction to the Palaeoenvironemtal Survey, in R. Van de Noort & S. Ellis (eds) *Wetland Heritage of Holderness*, 27–48. Hull: Humber Wetlands Project.

Dinnin, M. H. & M. Lillie 1995. The palaeoenvironmental survey of the meres of Holderness, in R. Van de Noort & S. Ellis (eds) *Wetland Heritage of Holderness*, 49–86. Hull: Humber Wetlands Project.

Edwards, K. J. 1991. Using space in cultural palynology: the value of the off-site pollen record, in D. R. Harris & K. D. Thomas (eds) *Modelling Ecological Change*, 61–73. London: Institute of Archaeology.

Faegri, K. 1988. Preface, in H. H. Birks, H. J. B. Birks, P. E. Kaland, P.E. & D. Moe (eds) *The Cultural Landscape: Past, Present and Future*, 1–4. Cambridge: University Press.

Flenley, J. R. 1987. The meres of Holderness, in S. Ellis (ed.) *East Yorkshire Field Guide*, 73–81. Cambridge: Quaternary Research Association.

Glenn-Lewin, D. C., R. K. Peet & T. T. Veblen 1992. *Plant Succession: theory and prediction*. London: Chapman & Hall.

Glob, P. V. (trans. R. Bruce-Mitford) 1969. *The Bog People: Iron-Age Man Preserved*. London: Faber & Faber.

Gordon, D. L. & J. H. McAndrews 1992. Field-testing a model of paleohydrology for prehistoric site prediction at Lake Teragami, northeastern Ontario. *Annual Archaeological Report, Ontario New Series* 3, 80–85.

Gregory, K. J. & D. E. Walling 1973. *Drainage Basin Form and Process: a geomorphological approach*. London: Edward Arnold.

Heathwaite, A. L., K. Göttlich, E. G. Burmeister, G. Kaule & T. Grospietsch 1993. Mires: Definitions and Form, in A. L. Heathwaite & K. Göttlich (eds) *Mires: Process, Exploitation, and Conservation*, 1–75. Chichester: John Wiley & Sons.

Jackson, S. T. 1990. Pollen Source Area and Representation in Small Lakes of the Northeastern United States. *Review of Palaeobotany and Palynology* 63, 53–76.

Jackson, S. T. 1994. Pollen and spores in Quaternary lake sediments as sensors of vegetation composition: theoretical models and empirical evidence, in A. Traverse (ed.) *Sedimentation of Organic Particles*, 253–286. Cambridge: Cambridge University Press.

Jackson, S. T., R. P. Futyma & D. A. Wilcox 1988. A palaeoecological test of a classical hydrosere in the Lake Michigan dunes. *Ecology* 69, 928–936.

Jacobson, G. L. & R. H. W. Bradshaw 1981. Selection of Sites for Palaeovegetational Studies. *Quaternary Research* 16, 80–96.

Lowe, J. J. & M. J. C. Walker 1997. *Reconstructing Quaternary Environments*. 2nd edn. Harlow: Longman.

Mellars, P. A. & P. Dark 1998. *Star Carr in Context*. Cambridge: MacDonald Institute for Archaeological Research.

Mitsch, W. J. & J. G. Gosselink 2000. *Wetlands*. 3rd edn. New York: John Wiley & Sons.

Moore, P. D. 1993. The origin of blanket mire, revisited, in F.M. Chambers (ed.) *Climate Change and Human Impact on the Landscape*, 217–224. London: Chapman & Hall.

Patterson, W. A.III., K. J. Edwards & D. J. Maguire 1987. Microscopic charcoal as a fossil indicator of fire. *Quaternary Science Reviews* 6, 3–23.

Pons, L. J. 1992. Holocene peat formation in the lower parts of the Netherlands, in J. T. A. Verhoeven (ed.) *Fens and Bogs in the Netherlands: vegetation, history, nutrient dynamics and conservation*, 7–79. Dordrecht: Kluwer Academic Publishers.

Prentice, I. C. 1985. Pollen Representation, Source Area, and Basin Size: Toward a Unified Theory of Pollen Analysis. *Quaternary Research* 23, 76–86.

Rackham, O. 1997. *The Illustrated History of the Countryside*. London: Phoenix Illustrated.

Roberts, N. 1989. *The Holocene*. Oxford: Blackwell.

Schofield, J. E. 2001. Vegetation Succession in the Humber Wetlands. Unpublished PhD Thesis. University of Hull.

Sheppard, J. A. 1956. The draining of the marshlands of East Yorkshire. Unpublished PhD thesis, University of Hull.

Sheppard, J. A. 1957. The Medieval Meres of Holderness. *Transactions of the Institute of British Geographers* 23, 75–86.

Sheppard, T. 1902. Notes on the ancient model of a boat, and warrior crew, found at Roos, in Holderness. *Transactions of the East Riding Antiquarian Society* 9, 62–74.

Smith, A. G. & L. A. Morgan 1989. A succession to ombrotrophic bog in the Gwent Levels, and its demise: a Welsh parallel to the peats of the Somerset Levels. *New Phytologist* 112, 145–167.

Smith, B. 1985. A palaeoecological study of raised mires in the Humberhead Levels. Unpublished PhD thesis, University of Wales.

Sugita, S. 1994. Pollen representation of vegetation in Quaternary sediments: theory and method in patchy vegetation. *Journal of Ecology* 82, 881–897.

Tansley, A. G. 1939. *The British Islands and Their Vegetation*. Cambridge: Cambridge University Press.

van Geel, B, J. Buurman & H. T. Waterbolk. 1996. Archaeological and palaeoecological indications of an abrupt climatic change in the Netherlands, and evidence for climatological teleconnections around 2650 BP. *Journal of Quaternary Science* 11 (6), 451–460.

Walker, D. 1970. Direction and rate in some British Post-glacial hydroseres, in D. Walker & R. G. West (eds) *Studies in Vegetational History of the British Isles*, 117–139. Cambridge: Cambridge University Press.

Waller, M. P. 1994. *The Fenland Project, Number 9: Flandrian environmental change in Fenland*. Cambridge: Cambridgeshire Archaeological Committee, East Anglian Archaeology Report No. 70.

Waller, M. P., A. J. Long, D. J. Long, & J. B. Innes 1999. Patterns and processes in the development of coastal mire vegetation: Multi-site investigations from Walland Marsh, Southeast England. *Quaternary Science Reviews* 18, 1419–1444.

Warner, B. G., H. J. Kubiw & K. I. Hant 1989. An anthropogenic cause for quaking mire formation in southwestern Ontario. *Nature* 340, 380–384.

Wheeler, B. D. 1980. Plant communities of rich-fen systems in England and Wales. III. Fen meadow, fen grassland and fen woodland communities, and contact communities. *Journal of Ecology* 68, 761–788.

Wheeler, B. D. 1984. British Fens: A Review, in P.D. Moore (ed.) *European Mires*, 237–281. London: Academic Press.

Yu, Z., J. H. McAndrews & D. Siddiqi 1996. Influences of Holocene climate and water levels on vegetation dynamics of a lakeside wetland, *Canadian Journal of Botany* 74, 1602–1615.

Chapter 19

Recent Research from the Peatlands of the North York Moors: Human Impact and Climatic Change

J. J. Blackford, J. B. Innes, A. E. Kelly and S. L. Jones

INTRODUCTION: THE NORTH YORK MOORS AREA

The north east (England) region contains a wide range of wetlands, either present or past, in an equally wide range of geomorphic settings (Innes 1999). These provide a palaeoecological record that covers the whole period from the Late-Glacial through to the present. The North York Moors area consists of tabular, sloping hills and steeper escarpments, incised by deep valleys. The Cleveland hills run east-west to the north of this block (Fig. 19.1). The blanket peats of the central watershed have been the main focus of palynological study, although additional sites in meltwater channels such as Ewe Crag Slack (Jones 1978) and in valley bottoms such as Fen Bogs (Atherden 1976) have also been exploited (Innes 1999).

Pollen diagrams from the area show a replicated picture of Holocene vegetation change, with some general patterns that are common to much of central Britain but with some regional anomalies. The general pattern includes a succession of thermophilous trees, arriving in a sequence that can be regarded as typical of the Holocene period in the region (Innes 1999). *Alnus* (alder) however, arrived later to the moorland areas than to lowland areas (dated at 6,650 ± 290 BP) and in places *Pinus* remained an important component of the early mid-Holocene for longer than in the north of England as a whole (Innes 1999). There are also some aspects of mid-Holocene vegetation change that occurred on a local scale, usually interpreted as being of human origin, and these are discussed below.

PRE-AGRICULTURAL HUMAN IMPACT

The more specific, small-scale patterns comprise vegetation changes in the late-Mesolithic period, defined here as the palynological phase between the rise in *Alnus* at 7,500 radiocarbon years BP in the north-east region and the fall in *Ulmus* (elm), dated to shortly after *c.* 5,000 radiocarbon years BP in the North York Moors area. During this period short-term vegetation disturbances have been identified and attributed to the actions of Late Mesolithic people (Simmons 1996, Simmons and Innes 1996a–c). The evidence for this can be summarised as follows:

– Deposits of exceptionally high levels of micro-charcoal, with often visible macro-charcoal fragments, indicative of burning in the area and at that precise location respectively.

–   Changes in tree and shrub pollen that suggest woodland disturbances, accompanied by shifts towards species favoured by increased light at ground level.

–   Increases in shrubby and herbaceous species thought to be favoured by fire and those benefiting from disturbed ground.

–   Increases in taxa that would have been of benefit to hunter-gatherers, particularly Gramineae/Poacaeae (grasses) and *Corylus avellana* (hazel).

Interpretations of these changes based on human impact have been supported by the high incidence of finds of Mesolithic artefacts, mostly flint scatters, across the central watershed area (Spratt 1993). The pattern of vegetation change is exemplified by the pollen and charcoal data from North Gill on Glaisdale Moor (NG3, Simmons and Innes 1988) where peaks in *Corylus* were inferred as anthropogenic disturbance phases- particularly, the "fire removal of woodland" (Simmons and Innes 1988, 264) (Fig. 19.2).

Further work at North Gill has shown the benefit of two relatively recent practices in pollen and charcoal research; a multiple-core approach and fine-resolution pollen analysis (FRPA) (Innes and Simmons 1999, 2000). Charcoal data from a sequence of cores has for the first time revealed the spatial extent of the fire events. Pollen data show a degree of diversity in the nature of the inferred disturbance events. This diversity

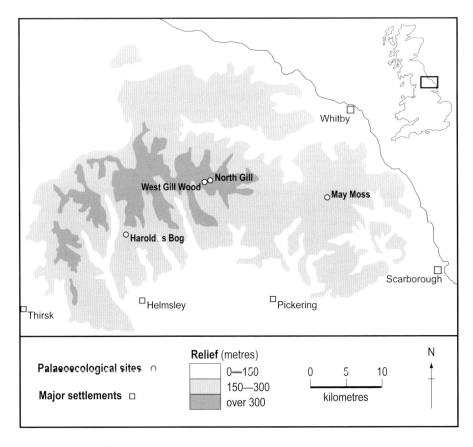

Fig. 19.1 Map showing location of sites discussed in the text.

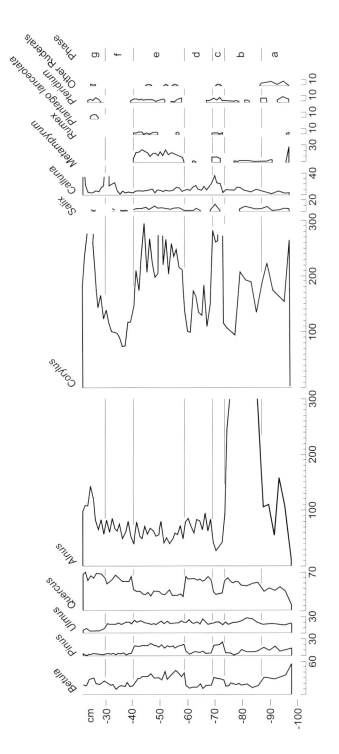

Fig. 19.2 Selected pollen data from North Gill B (after Simmons and Innes 1988: 264).

includes both variability in the duration of the events and in the nature of the vegetation change – with different inferred anthropogenic disturbance events resulting in different end-products. In some cases *Corylus* was favoured and in others, Gramineae/Poaceae (grasses). Some involved fire, while others have been interpreted as non-fire canopy-manipulation events. This diversity is illustrated by two further examples below.

From North Gill profile NG4 (Simmons and Innes 1996a), there is evidence of 'canopy manipulation'; the creation of a temporary change in the relative balance of tree and shrub types, with a little more light reaching ground level promoting increased representation of the ground flora, especially *Pteridium*. This event, detected by 1mm sample intervals, occurred shortly before 5,315 ± 45 radiocarbon years BP (Simmons and Innes 1996a). The changes involved are subtle. *Quercus* pollen percentages decline from around 55% to 45% of a tree-pollen sum with *Alnus* removed (Fig. 19.3). *Alnus* itself declines and the less vigorous pollen producers *Tilia* and *Fraxinus* show an increase in representation. *Rumex*, present before the event, increase slightly in frequency, but only to a maximum of three grains.

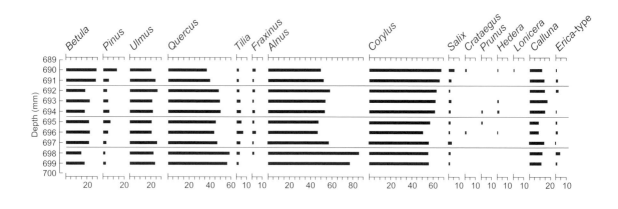

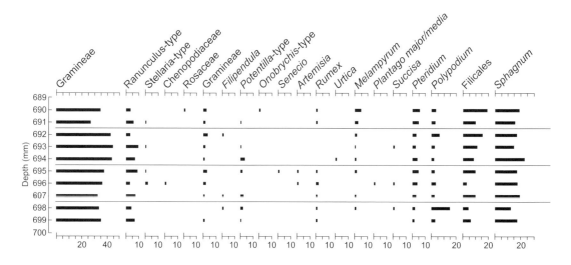

Fig. 19.3 Selected pollen and charcoal data from North Gill 4 (after Simmons and Innes 1996a-c).

At North Gill 5 (Fig. 19.4), also analysed at a time resolution in the order of 2–3 years per sample, a slightly different form of forest change is recorded. *Quercus* and *Alnus* pollen decline, as at NG4, but in the initial disturbance phase (Zone 2 in Innes and Simmons 1999) there is no response from either *Pteridium* or *Rumex*, considered key indicators in the previous example. *Salix* and *Corylus* increase, perhaps benefiting from more rapid regeneration than other taxa, an increase in fringe habitat and increased light penetration for sub-canopy species. *Pinus* frequencies increase, possibly due to the influx of long-travelled grains, increasing in relative abundance due to the removal or suppression of local trees, but neither *Tilia* or *Fraxinus* respond in a way likely to suggest a relative increase. Gramineae, *Calluna* and *Melampyrum* are the indicators of open ground that expand most (Fig. 19.4).

In each of these examples it is evident that something happened to alter the relative abundance of different tree species, the presence and relative abundance of ground taxa and the representation of overall tree cover. How can these inferred canopy fluctuations be safely interpreted as anthropogenic in origin, however? Through the recognition of such *differences* between disturbance events, high-resolution palynology (temporal and spatial) has to an extent undermined the standard interpretations. If a human-induced manipulation can leave a charcoal record or not charcoal, can show an increase in *Rumex*, *Plantago* sp. and *Pteridium* or not, and can – but does not have to – show an increase in pioneering woody species, what remaining criteria can be considered diagnostic? If the criteria are purely 'structural', that is, the replacement of canopy-forming trees with open-ground indicators, then any cause of the clearance, including wind-throw, wild animal herds or natural fires, will produce a similar signal (Tipping 1996, Brown 1997). Indicators of disturbed ground (for example *Rumex* and *Plantago lanceolata*) may be more useful in detecting human impact than simply indicators of light reaching ground level. Even these weed species, however, are endemic in a human-free forest and would benefit quickly from the uprooting of trees by wind-throw, puddling by wild animal herds or rootling by wild pigs, all of which can cause ground disturbance (Rowley-Conwy 1982, Grigson 1982).

An important question remains, therefore; how can an anthropogenic disturbance phase be identified with any certainty? A link between vegetation change and archaeological finds would help, for example a concentration of evidence for burning and clearance episodes clustered around settlement sites (Edwards 1990). However, although archaeological finds relating to this period are widespread, they are usually uninformative. Settlement and midden sites are extremely rare, with flint scatters forming the sole Mesolithic artefactual assemblage in many areas (Spratt 1993). This is true of the late Mesolithic throughout north-west Europe as well as in the North York Moors area (Bonsall 1989, Edwards 1996). Finds of tools, or the debris remaining at possible tool-making sites, do coincide spatially with disturbance features in pollen diagrams; but even then the question remains as to which came first (Brown 1997). Did the Mesolithic foragers use naturally-occurring disturbed areas, or create specific types of disturbance themselves? How can such events, especially when the changes are within a continuing woodland and shrub vegetation type, be distinguished from the natural variability present in all ecosystems?

While there are clearly no instant answers, there are two possible strategies that may be used to address these questions. First, broadening the range of sub-fossil types used in reconstruction and second, using modern, near-analogue sites in an attempt to determine microfossil 'signatures', which may be representative of different land-use practices.

FUNGAL SPORES

One group of sub-fossils with the potential to help explain the mid-Holocene vegetation changes discussed above are the fungi. Fungi are found in many current environments, including soil, live and dead plant material, animals and animal wastes and in many human environments. Their spores, fruit bodies and

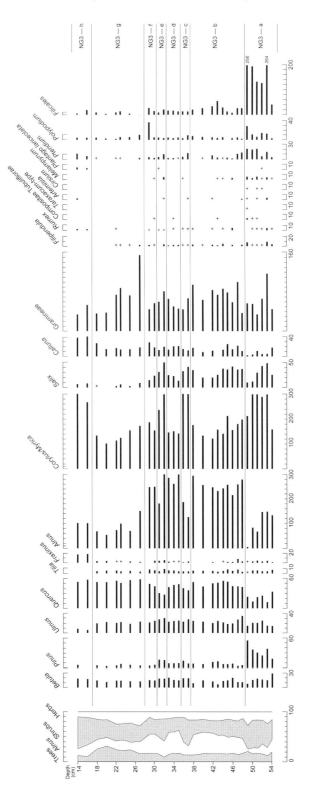

Fig. 19.4 Selected pollen and charcoal data from North Gill 5 (after Innes and Simmons 1999).

hyphae are present in palaeoecological samples, especially in pollen preparations where they sometimes outnumber pollen grains. Some are environment-specific and others host-species specific (van Geel 1986). Simmons and Innes (1996c) quantified the spores of *Neurospora* and *Gelasinospora*, previously thought indicative of burning (van Geel 1978) and showed a correlation between charcoal concentrations and fungal spore percentages. In the absence of macroscopic, or visible charcoal layers, these fungi may be effective indicators of burning at the coring or sampling site. Species of dung fungi have been found in a variety of contexts, including lake and bog sediments (van Geel *et al.* 1989, Blackford *et al.* 1996), where evidence for increased numbers of mammals has been inferred (Speranza *et al.* 2000). Fungi may also be indicative of rotting wood, stored food products or land-use practices such as cereal growing or hay production.

SURFACE SAMPLING AND ANALOGUE STUDIES

Many interpretations of fossil or sub-fossil fungi come from a process of association within the palaeoecological data sets. In order to fully understand and exploit the record of preserved fungal spores, it is necessary to investigate current environments and identify individual taxa or assemblages that may be indicative of particular environments, land use processes or surface conditions.

As part of a broader-scale programme of surface sampling, including sites in Cumbria, Devon, Orkney and the Scottish Highlands, near-analogue sites from the North York Moors were selected with the aim of identifying fungal and pollen assemblages, or indicator species, diagnostic of particular processes. The North York Moors sites include West Gill Wood, Rosedale Head, Hawdale Griff, Water Griff, Hole of Horcum, Ugglebarnby Moor and Rosedale Moor. These sites include grazed and ungrazed moorland, heavily and lightly-grazed woodland, and burned and unburned moorland. From each site, between 6 and 12 surface and shallow sub-surface samples have been taken, in order to count the fungal remains, pollen and charcoal. The eventual aim is then to link these surface samples to the fossil record, either through a quantified analysis of assemblages or by the use of indicator taxa.

One of these sites, West Gill Wood, is close to the pollen sites studied at North Gill described above (Fig. 19.1). It is a small area of quite open, mixed woodland, with a mixed age-range of trees, growing in a zone between the *Calluna*-dominated moorland and improved valley pastures. The woodland is currently lightly grazed by sheep, fallow and roe deer, and by rabbits, although at the time of sampling (August, 1999) there was no evidence of recent grazing, with no dung, trails or the cropped ground flora that characterise grazed woodland. The moorland area surrounding the woodland is periodically burned to maintain the dominance of *Calluna*, although the woodland itself has no recent fire history. The trees around the sampling site were dominated by *Alnus* (alder), with *Quercus* (oak), *Fraxinus* (ash), *Betula* (birch), *Sorbus* (rowan), *Ilex* (holly), *Crataegus* (hawthorn) and *Corylus* (hazel) also present. The ground flora within 10m was dominated by Gramineae (grasses) and ferns including *Pteridium*. Other herbs close to the sample sites included *Cirsium* (thistles) Rubiaceae (bedstraws), *Rumex* (sorrel), *Potentilla erecta* (tormentil), *Plantago lanceolata* (ribwort plantain) and *P. media* (lesser plantain), a range not dissimilar to that recorded in the mid-Holocene pollen diagrams from North Gill and elsewhere on the moors (Innes 1999, figures 2–4).

Fig. 19.5 shows the results of 12 samples each from moss polsters and from a sub-surface humus sample taken from between 1 and 2 cm beneath the ground surface. The assemblage is dominated by *Alnus* pollen, which represents 45–95% of the total land pollen (TLP). Of the remaining trees and shrubs, only two types exceed 10% of TLP; *Betula* (max. 10.5%) and *Corylus* (max. 12%). Ground flora pollen frequencies are dominated by *Calluna* and Gramineae. *Calluna* dominates large areas of land around the woodland, although it is scarce within it. The pollen assemblage overall, then, is comparable to that from North Gill in the pre-Neolithic period, but with some important differences. These include the relatively small percentages of *Quercus* and *Corylus* in the near-analogue site, where *Alnus* is more dominant (even

West Gill Wood

Fig. 19.5 Selected pollen and fungal remains in surface samples from West Gill Wood. Percentages are expressed as percentage total land pollen, with those taxa outside the pollen sum calculated on the same basis. Samples are alternately from surface moss and from humus layer sediments at 1–2 cm depth from the same point.

considering the pollen sums used), and the complete absence of *Melampyrum* (cow-wheat), thought to be favoured by burning in Mesolithic disturbance phases (Simmons and Innes, 1996a–c, Innes and Simmons 2000). *Filipendula* (meadowsweet), a potentially dominant tall herb, is also absent from the modern record. Charcoal frequencies are variable in the surface and sub-surface samples, but are consistent with an interpretation as showing fire in the area, although not at the point of sampling (Blackford 2000a).

Fungal spores from West Gill Wood include many different types, including several that have been assigned indicator values in previous work. Types 55A and B, for example, have been considered indicative of burnt ground and/or grazing (dung), 55A is present in all of the samples from West Gill Wood. Two types of dung fungus, *Sporormiella* and *Cercophera* (Type 112, after van Geel 1978) are present, but both are quite rare. *Gelasinospora* and *Neurospora* are absent, which does not contradict their previously assigned indicator value of showing burnt ground. Other common types present include Type 10, Type 18A and Type 19. Type 10 occurs on the roots of *Calluna,* but 18A and 19 have been assigned no indicator value as yet, despite being present in a wide range of peats and in other sediments (van Geel 1978, Blackford 1990, Long 1994). These spores could be part of a general wetland and humic-soil fungal flora, with a widespread and apparently ubiquitous distribution.

Taken together, the surface samples can help in the interpretation of sub-fossil data. For example, a fossil slide with 60–70% trees, a mixture of shrubs including *Corylus* and a wide range of herbaceous types including species associated with agriculture, might be interpreted as a woodland landscape, with small clearings used for pastoral agriculture. In fact, the landscape that has produced this pollen spectrum is largely open, but with the samples coming from a small area of quite open woodland. The fungal spores would suggest no burning on site, whereas the microcharcoal record (5–40% of TLP) suggests burning within the local area. Fungal records also suggest light grazing, and future work aims to improve this indicator value by looking at more surface samples, leading to a grazing index based on pollen and spore frequencies. Studies such as this may be able to resolve some of the issues of causality and consistency in prehistoric pollen diagrams. For example, knowledge of when grazing took place, whether *in-situ* burning occurred or when dead wood was left standing, adds new detail to the overall reconstruction. Clearly a 'reference database' of surface samples on a regional basis would aid the accurate interpretation of sub-fossil palynological data.

This approach depends entirely on the successful identification of analogue sites. It seems unlikely that real modern analogues exist, given the immense pressures on, and changes to, the vegetation through at least 6,000 years of fire, farming, wood collecting and the introduction of non-native species. In addition, climatic conditions are not the same as they were in the Mesolithic or Neolithic periods. However, 'near-analogue' studies can perhaps identify signatures of pollen, charcoal and fungi that might be indicative of comparable environments. In addition, surface-studies based on significant *processes* relevant to the interpretation of ancient landscapes, rather than specific environments, may be less affected by the inevitable changes in background conditions.

CLIMATE CHANGE

Until recently, the only attempt to reconstruct the past climate of the area considered here came from 'Harold's Bog', a small area of blanket peat on East Bilsdale Moor (Fig. 19.1). The record from Harold's Bog includes a replicated peat humification curve, limited mire stratigraphy, pollen, charcoal and fungal spore analysis (Blackford and Chambers 1999). Blackford and Chambers (1999) showed six periods when changes to wetter conditions can be inferred (Fig. 19.6), dated to approximately 2,630 ± 60 BP, 1,985 ± 60 BP, 1,390 ± 65 BP, 1,020 ± 60 BP, 900 ± 60 BP (all uncalibrated radiocarbon years BP) and AD 1480–1590 (an interpolated age-estimate). Time-series analysis of the record from Harold's Bog over the last 2,000 years has shown a periodic element in the data. There is evidence for cycles of approximately 200, 100 and 60 years length, of varying degrees of statistical significance, with that at around 100 years being the most secure (Chambers and Blackford 2001).

This single record has now been greatly supplemented. A strategy of employing a wider range of palaeoecological indicators has been adopted by Chiverrell and Atherden (1999) and Chiverrell (2001).

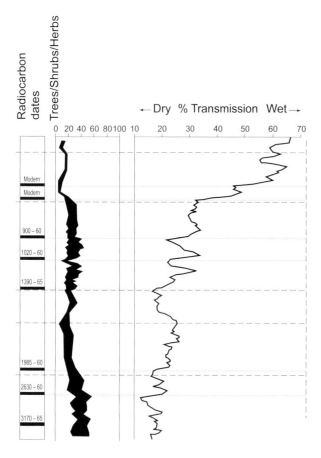

Fig. 19.6 Peat humification record and summary pollen data from Harold's Bog, showing dates of main humification changes (after Blackford and Chambers 1999). Data are smoothed using a 3-sample running mean.

Over the past 30 years, peat-based palaeo-climatic reconstructions have mostly used plant macrofossils, especially species of *Sphagnum* (Barber *et al.* 2000) and 'degree of humification', an indicator of decomposition, as relative measures of former surface wetness (Blackford 2000b). However, more recent work has included the analysis of testate amoebae, whose remains are indicative of water table depth (Charman *et al.* 1999). While the records from peat humification and macrofossils only show relative increases or decreases in water table depth, leading to inferences such as 'a change to wetter conditions', environmental reconstruction based on amoebae can now be quantified- and translated into estimates of water table depths. This is achieved via a set of transfer functions developed by Woodland *et al.* (1998), based on a comprehensive survey of the modern distributions of the different species and their current water-table ranges. The application of this technique to the North York Moors (Chiverrell 2001, Fig. 19.7) has substantially improved the record of past climatic change from the area. Minimum, mean and maximum water table depths are calculated for each sample depth. This improvement has come about by (1) introducing a new indicator group to the problem and (2) applying the results of near-analogue studies to the sub-fossil data. It is also possible that fungal spores may also be used to reconstruct past surface wetness, with some types being more common in dry phases of peat growth and others in wetter phases (van Geel 1978). At Harold's Bog, for example, fungal spores Type 10 and Type 12 increase in those samples that have higher degrees of decomposition, while *Geoglossum* and *Anthostomella fuegiana* spores are more common in wetter samples (Blackford and Chambers 1999).

CONCLUSION: FUTURE RESEARCH

The North York Moors area has been subjected to concentrated palaeoecological study, especially along the central watershed. At least six PhD projects have concentrated on the vegetation history of the Moors, and over fifty publications discuss the events. It would seem, therefore, that little or no further research would be required, and indeed a substantial synthesis of the vegetation history of the area can be achieved (Innes

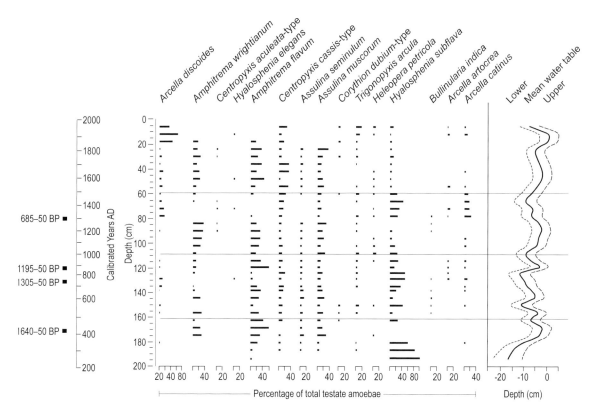

Fig. 19.7 Selected testate amoebae from May Moss core D2, with water table reconstructions (in depth below surface) and 68% upper and lower confidence intervals. This curve and others in Chiverrell (2001: 18) represent the first quantified palaeohydrological reconstruction from the region. Reproduced in a modified form with permission from the author and John Wiley and Son Ltd.

1999). However, as discussed in part above, several unanswered questions remain regarding the detection of human impact and its separation from natural factors of woodland change.

Other significant, unanswered questions include the following:

–   How widespread were the late-Mesolithic disturbances?

While there is clear evidence of Mesolithic-aged disturbances, it is not known whether they were restricted to the hill-tops, where most peat deposits occur, or whether valleys sides and bottoms were similarly affected. So,

–   How did the climatic changes alter the vegetation?

Palaeoecological evidence has been used to infer both human impact and climatic change from the North York Moors region. The two aspects are likely to be related, however, and this relationship remains largely unexplored. For example, could the drier phases in the late Holocene, as inferred by Chiverrell (2001), have caused increased fire frequency, or allowed more grazing on the uplands? Alternatively, the switches to enhanced summer wetness suggested by peat humification records may have reduced the agricultural productivity of the area and subdued natural and anthropogenic fires.

The North York Moors area can be seen as a suitable location for further research for another reason. As an area where palaeoecological work has concentrated, it can be used as a testing ground for new techniques. Many of the same sites can be re-analysed using new approaches, so that the sedimentary sequence, time-scale and vegetation history are already known; if these had to be determined from the outset it could add years to new projects. Overall, we suggest that great benefits can come from the analysis of near-analogue sites or process-based analogues, and through the use of new indicator types wherever possible.

ACKNOWLEDGEMENTS

Work at West Gill Wood was funded by the Leverhulme Trust. Ed Oliver produced the figures. Permission to reproduce the figures was given by authors and publishers (Academic Press, the Quaternary Research Association, John Wiley and Blackwell Science Ltd).

REFERENCES

Atherden, M. A. 1976. Late Quaternary vegetation history of the North York Moors, III. Fen Bogs. *Journal of Biogeography* 3, 115–124.

Barber, K .E., D. Maddy, N. Rose, A. C. Stevenson, R. E. Stoneman & R. Thompson 2000. Replicated proxy-climate signals over the last 2000 years from two distant peat bogs: new evidence for regional palaeoclimate teleconnections. *Quaternary Science Reviews* 19, 481–487.

Blackford, J. J. 1990. Blanket mires and climatic change: a palaeoecological study based on peat humification and microfossil analyses. Unpublished PhD Thesis, University of Keele.

Blackford, J. J. 2000a. Charcoal fragments in surface samples following a fire and the implications for interpretation of subfossil charcoal data. *Palaeogeography, Palaeoclimatology, Palaeoecology* 164, 33–42.

Blackford, J. J. 2000b. Palaeoclimatic records from peat bogs. *Trends in Ecology and Evolution* 15, 193–198.

Blackford, J. J. & F. E. Chambers 1999. Harold's Bog, East Bilsdale Moor, in D. R. Bridgland, B. P. Horton. & J. B. Innes (eds) *The Quaternary of North-East England*, 91–98. London: Quaternary Research Association Field Guide.

Blackford, J. J., K. J. Edwards, P. C. Buckland & K. Dobney. 1996. Keith's Peat Bank, Hoy: Mesolithic Human Impact? in A. Hall (ed) *The Quaternary of Orkney*, 62–68. London: Quaternary Research Association Field Guide.

Bonsall, C. 1989. (ed.) *The Mesolithic in Europe*. Edinburgh: John Donald.

Brown, A. G. 1997. Clearances and clearings: deforestation in Mesolithic/Neolithic Britain. *Oxford Journal of Archaeology* 16, 133–146.

Chambers, F. M. & J. J. Blackford 2001. Mid- and late-Holocene climatic changes: a test of periodicity and solar forcing in proxy-climate data from blanket bogs. *Journal of Quaternary Science* 16, 329–338.

Charman, D. J., D. Hendon, & S. Packman 1999. Multiproxy surface wetness records from replicate cores on an ombrotrophic mire: implications for Holocene palaeoclimatic records. *Journal of Quaternary Science* 14, 451–464.

Chiverrell, R. C. 2001. A proxy record of late Holocene climate change from May Moss, north east England. *Journal of Quaternary Science* 16, 9–29.

Chiverrell, R. C. & M. A. Atherden 1999. Climate change and human impact – evidence from peat stratigraphy at sites in the eastern North York Moors, in D. R. Bridgland, B. P. Horton & J. B. Innes (eds) *The Quaternary of North-East England*, 113–130. London: Quaternary Research Association Field Guide.

Edwards, K. J. 1990. Fire and Scottish Mesolithic: evidence from microscopic charcoal, in P. Vermeesch & P. van Peer (eds) *Contributions to the Mesolithic in Europe*, 71–79. Leuven: Leuven University Press.

Edwards, K. J. 1996. A Mesolithic of the Western and Northern Isles of Scotland? Evidence from pollen and charcoal, in T. Pollard & A. Morrison (eds) *The Early Prehistory of Scotland*, 23–38. Edinburgh: Edinburgh University Press.

Grigson, C. 1982. Porridge and pannage: pig husbandry in Neolithic England, in M. Bell & S. Limbrey (eds) *Archaeological Aspects of Woodland Ecology*, 297–314. Oxford: BAR International Series 146.

Innes, J. B. 1999. Regional vegetation history, in D. R. Bridgland, B. P. Horton. & J. B. Innes (eds) *The Quaternary of North-East England*, 21–34. London: Quaternary Research Association Field Guide.

Innes, J. B. & I. G. Simmons. 1999. North Gill, in D. R. Bridgland, B. P. Horton. & J. B. Innes (eds) *The Quaternary of North-East England*, 99–112. London: Quaternary Research Association Field Guide.

Innes, J. B. & I. G. Simmons. 2000. Mid-Holocene charcoal stratigraphy, fire history and palaeoecology at North Gill, North York Moors, UK. *Palaeogeography, Palaeoclimatology, Palaeoecology* 164, 151–166.

Jones, R. L. 1978. Late Quaternary history of the North York Moors VI. The Cleveland Moors. *Journal of Biogeography* 5, 81–92.

Long, D. J. 1994. Prehistoric field systems and the vegetation development of the gritstone uplands of the Peak District. Unpublished PhD Thesis, University of Keele.

Rowley-Conwy, P. A. 1982. Forest grazing and clearance in temperate Europe with special reference to Denmark: an archaeological view, in M. Bell & S. Limbrey (eds) *Archaeological Aspects of Woodland Ecology*, 199–215. Oxford: BAR International Series 146.

Simmons, I. G. 1996. *The Environmental Impact of Later Mesolithic Cultures*. Edinburgh: Edinburgh University Press.

Simmons, I. G. & J. B. Innes. 1988. Studies in the late Quaternary vegetation history of the North York Moors VIII. Correlation of Flandrian II litho- and pollen stratigraphy at North Gill, Glaisdale Moor. *Journal of Biogeography* 15, 273–297.

Simmons I. G. & J. B. Innes 1996a. An episode of prehistoric canopy manipulation at North Gill, North Yorkshire, England. *Journal of Archaeological Science* 23, 337–341.

Simmons I. G. & J. B. Innes 1996b. Disturbance phases in the mid-Holocene vegetation at North Gill, North York Moors: form and process. *Journal of Archaeological Science* 23, 183–191.

Simmons I. G. & J. B. Innes 1996c. Prehistoric charcoal in peat profiles at North Gill, North Yorkshire Moors, England. *Journal of Archaeological Science* 23, 193–197.

Speranza, A., J. Hanke, B. van Geel & J. Fanta 2000. Late-Holocene human impact and peat development in the Černá Hora bog, Krkonoše Mountains, Czech Republic. *The Holocene* 10, 575–585.

Spratt, D. A. 1993. The Upper Palaeolithic and Mesolithic Periods, in D. A. Spratt (ed) *Prehistoric and Roman archaeology of North-East Yorkshire*, 51–78. York: Council for British Archaeology.

Tipping, R. 1996. Microscopic charcoal records, inferred human activity and climatic change in the Mesolithic of northernmost Scotland, in T. Pollard & A. Morrison (eds) *The Early Prehistory of Scotland*, 39–61. Edinburgh: Edinburgh Univeristy Press.

Van Geel, B. 1978. A palaeoecological study of Holocene peat bog sections in Germany and Netherlands, based on the analyses of pollen, spores and macro- and microscopic remains of fungi, algae, cormophytes and animals. *Review of Palaeobotany and Palynology* 25,: 1–120.

Van Geel, B. 1986. Application of fungal and algal remains and other microfossils in palynological analyses, in B. E. Berglund (ed) *Handbook of Palaeoecology and Palaeohydrology*, 497–505. Chichester: Wiley.

Van Geel, B., R. S. Coope & T. van der Hammen 1989. Palaeoecology and stratigraphy of the Lateglacial type section at Usselo (The Netherlands). *Review of Palaeobotany and Palynology* 60, 25–129.

Woodland, W. A., D. J. Charman & P. C. Sims 1998. Quantitative estimates of water tables and soil moisture in Holocene peatlands from testate amoebae. *The Holocene* 8, 261–273.

PART 4
HUMAN-LANDSCAPE
INTERACTIONS

Prehistoric Hunter-Gatherers in Wetland Environments: Mobility/Sedentism and Aspects of Socio-Political Organisation

George P. Nicholas

INTRODUCTION

Wetlands have been a long overlooked dimension of the cultural geography of hunter-gatherers. As is becoming increasingly evident, the archaeology and human ecology of wetland environments provides many opportunities to increase our understanding of past hunting and gathering peoples (for overview, see Nicholas 1998). Much attention, for example, has been devoted to the role of wetlands within human landuse patterns, revealing new insights into the productivity, scheduling and procurement of wetland resources (*e.g.* Simms 1999). However, some important dimensions of past human behaviour have not been fully explored. This paper expands on an earlier chapter (Nicholas *this volume*) to review two significant aspects of hunter-gatherer lifeways, namely degree of mobility/sedentism and expressions of socio-political organisation.

MOVING AROUND – STAYING PUT

An important dimension in hunter-gatherer studies is their use of space and commitment to place. The high degree of mobility attributed to many of these groups results in a widely travelled and utilised landscape (Binford 1982). Not surprisingly, there is a broad literature relating to mobility (see Kelly 1995, 111–60 for a review). The spectrum of mobility-sedentism is not only expressed in different ways (*e.g.* collectors versus foragers [Binford 1980]), but also influences many elements of cultural systems, including population size and density, birth rate, material culture and political organisation.

If we accept the argument that hunter-gatherers are mobile primarily because of their need (a) to obtain food and other resources, and (b) to move before depleting the local environment, then it follows that under conditions where either the resources come to the people (*e.g.* coastal settings, anadromous fish runs or migratory caribou) or where the resource base is both productive and resilient, mobility should decrease. Of course the reality is more complex than this, as Kelly (1995) and Binford (2001) discuss, and a variety of issues need to be factored in (*e.g.* return rates, risk assessment). Nonetheless, the simple statement presented here is a useful starting point for examining the role of wetlands in the use of space by hunter-gatherers.

The attraction of humans to wetlands has been documented in many locations around the world and for many different periods of time, as previously noted (Nicholas 1998, *this volume*), but the idea is not a new one. For example, the notion that the Great Basin wetlands of western North America were important to past hunter-gatherers goes back almost a century (Thomas 1985, 18–19). What is sometimes referred to as the limno-sedentary hypothesis emerged from the work of Heizer and Napton (1970) and others, a key element of which was a semi-sedentary lifestyle supported by a productive lake and marsh environment. In contrast, the limno-mobility hypothesis (Thomas 1985, 20, and others) asserts that while wetland resources were both important and intensively used, they were utilised in a more restricted way. However, Kelly (1995, 159) maintains that 'it is not useful to think of mobility in terms of either a single dimension of group movement or as a dichotomy (mobile versus sedentary)'.

Much of Kelly's own research at Stillwater Marsh, Nevada (see Figs 20.1 and 20.2 for location of this and other sites mentioned in the text), has addressed the question of whether this wetland was a primary base for relatively sedentary hunter-gatherers, part of a seasonal round or a location used only when resources elsewhere were scarce. In his most recent work on the subject (Kelly 1995, 2001), he suggests that mobility decreased as the productivity of wetlands in the area increased (relative to other ecological

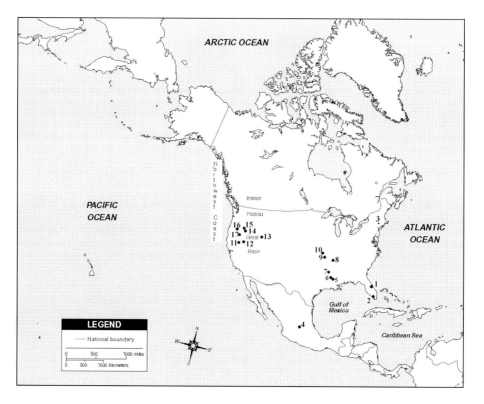

Fig. 20.1 Prominent North American sites and locations mentioned in the text.
Key: 1. Windover, 2. Calusa, 3. Robbins Swamp, 4. Teotihuacan, 5. Watson Brake, 6. Poverty Point, 7. Sloan, 8. Carrier Mills, 9. Cahokia, 10. Koster, 11. Stillwater Marsh, 12. Spirit Cave, 13. Great Salt Lake, 14. Diamond Swamp, 15. Malheur Lake, 16. Lake Abert/Chewaucan Marsh, 17. Nightfire Island.

settings) during the late Holocene, at which time they were used more intensively. He thus proposes that the Stillwater Marsh people fall somewhere in between, suggesting that the Heizer-Thomas models, while useful as heuristic or explanatory tools, are just that, and not necessarily a reflection of reality.

Based on examples drawn from the Great Basin and other locations in North America (Jeffries 1987, Nicholas 1988), Australia (Williams 1988), Europe (Coles and Lawson 1987, Coles and Coles 1989) and elsewhere, it is evident that wetlands did serve as attractors. However, as the Great Basin debate indicates, the nature of the attraction remains difficult to interpret, and we clearly need to look at the broader picture. Areas with rich wetlands often appear as special cases, with changes occurring over time in response to their individual developmental histories, as well as to the degree of contrast between different regions (Nicholas 2001).

This is the approach taken at Robbins Swamp, a large wetland system in the northeastern United States, where the author has explored the idea of core/peripheries in the early Postglacial, and subsequent landuse patterns (Nicholas 1988, 1992). Among the hundreds of archaeological sites identified at Robbins Swamp are a significant number containing definite or possible Paleoindian and Early Archaic artefacts that would have been associated with what is termed a wetland mosaic, consisting of swamp and marsh, riverine, upland and other elements. The number and distribution of these sites reveal a relatively distinct concentration of sites within the region at that time, perhaps representing one of a series of contemporary core areas of settlement (often associated with other large wetland systems, elsewhere in the region).

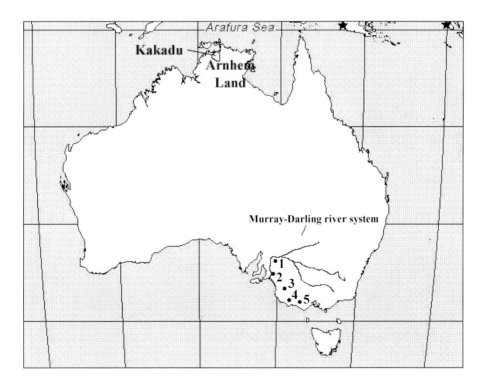

Fig. 20.2 Prominent Australian sites and locations mentioned in the text.
Key: 1. Lake Victoria, 2. Roonka Flat, 3. Toolondo, 4. Caramut, 5. Mt. Eccles, Lake Condah, MacArthur Cree area (all locations are approximate).

These wetland mosaics appear to have been ecologically distinct and culturally recognised places in the landscape until about 8000 years ago, when the increasing attractiveness of river valleys, lake systems, the coastal plain, upland areas and other non-wetland biomes encouraged or influenced significant changes in settlement patterns (see also Kuehn 1998, Stafford *et al.* 2000). Places like Robbins Swamp continued to be an important resource base, although its role may have changed from a core area to one used more seasonally or on a resource-specific basis.

Another element of hunter-gatherer landuse is the nature of the settlements themselves. The size of sites and duration of occupation are clearly tied to degree of mobility-sedentism. Although it has long been recognised that substantial variation exists in the number of residential moves and distances travelled by hunter-gatherers during the course of a year (Kelly 1995, table 4.1), variation also occurs in the nature of their settlements. The terms most frequently utilised – *camps, base camps, satellites, special activity sites* – reflect the transitory nature of the hunter-gatherer lifestyle. The term *village* is usually restricted to sedentary or seasonally sedentary groups, as found, for example, in the Northwest Coast and Interior Plateau regions of North America. Most archaeological sites associated with wetlands are best described using the usual terminology, but there are locations where the nature of the settlements and structures contained therein does not match the usual expectations for hunter-gatherers. The presence of numerous, relatively substantial structures suggests both longer-term occupations by larger than usual numbers of people, and may also reflect a greater array of activities and possibly even social statuses than is usually associated with hunter-gatherers.

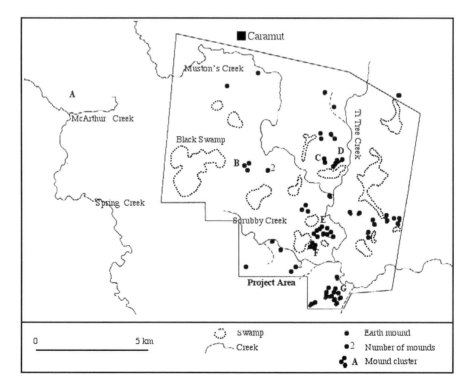

Fig. 20.3 Constructed earthen mounds in the Caramut area, southwestern Victoria, Australia (after Williams 1988: 77).

In southwestern Victoria, Australia, impressive concentrations of villages, earth mounds, artificial canals and assembly places (Williams 1988, Lourandos 1997, 216) are associated with swamps. For example, the largest mound cluster at Caramut (Fig. 20.3) contains 28 mounds; ethnohistorical accounts describe an associated settlement containing 20–30 'well-built huts, situated near a congregational area where up to 2,500 participants were recorded' (Lourandos 1997, 216). Other wetland-related examples of semi-sedentary populations in southeastern Australia are found at Mt. Ecles, Lake Condah and MacArthur (Flood 1995, 214–19, Builth 1999). Altogether, these are very impressive site numbers and building accomplishments, comparable in some respects to Archaic and Woodland Tradition settlements in North America.

In eastern North America, the middle Holocene is characterised as a period of increasing local intensification (*e.g.* Phillips and Brown 1983, Neusius 1986). Wetland resources certainly contributed to the longevity of occupation at many sites in the region, such as Koster (Struever and Holton 1979) and Carrier Mills (Jeffries 1987), some of which became quite large. One of the more unusual manifestations of the Late Archaic period was the Poverty Point culture (*c.* 4200–2700 BP). Located in the Mississippi River drainage system in Louisiana, the Poverty Point site consists of a large U-shaped series of six concentric earthen ridges (Fig. 20.4), with a central plaza and mound, altogether covering an area of about 200 hectares (Fagan 2000, 399). Dwellings were likely to have been constructed along the ridges.

The local environment, which included floodplain wetlands and oxbow lakes, as well as extensive forested areas, was apparently very productive. Little is actually known about the Poverty Point peoples, but the site evidently served as a trading centre. The extent of the community, and the labour involved in constructing the earthworks, begs the question what type of socio-political organisation is represented, as in the absence of burials there is no direct evidence of social classes. Wetlands were major contributors to the high productivity and resource diversity found in this region, and also that of the American Bottomlands (Bareis and Porter 1984), representing a resource base that contributed to the subsequent development of the sedentary and stratified Mississippian tradition.

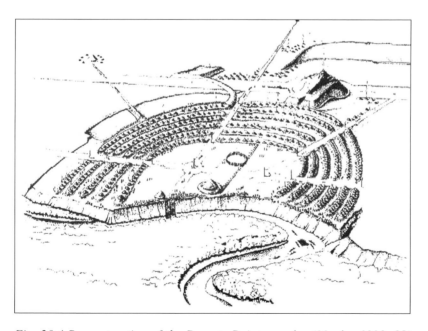

Fig. 20.4 Reconstruction of the Poverty Point complex (Manley 1993: 88).

In contrast to the Eastern Woodlands, the Great Basin is a much drier and more sparsely vegetated environment; the difference in carrying capacity of each region is substantial. Nonetheless, there is evidence that the Stillwater Marsh people were semi-sedentary, possibly for months at a time, supported by upland game hunting and pinyon nut harvest, and by lake and wetland fish and other resources (Fowler 1982). Some of the larger Stillwater sites include both shallow and deep house-pit depressions (Kelly 2001, 294), along with a variety of other features. The presence of large house-pit settlements near water and the wetlands resource is a consistent pattern found throughout the Great Basin during the late Holocene (Oetting 1989, 271), including the Malheur Lake wetlands (Oetting 1999), Diamond Swamp (Musil 1995) and Nightfire Island (Sampson 1985), all in Oregon. More than 580 depressions and 70 rock circles have been identified in the Lake Abert-Chewaucan Marsh area, with the number of depressions per site ranging from three to 68 (Oetting 1989, 127). The majority of these are likely to be house pits, but some may represent caches or other special purpose structures. The name 'Chuwaucan', as in Lake Abert-Chuwacan Marsh, is derived from the Klamath term *tcuwakan* ('where wapato [*Sagitaria latifolia*] is gathered', Spier, cited in Oetting 1989, 20).

The examples presented here reiterate an important point made by Robert Kelly and others. Our tendency to apply too casually the terms *sedentary* or *mobile* to particular peoples can obfuscate aspects of past hunter-gatherer behaviour, as well as obscure the diversity of lifeways that were once present. That the examples presented here do not fit comfortably within the usual descriptive/explanatory models of hunter-gatherers suggests that they may be too confining. Such a position is further strengthened when we examine the corresponding socio-political elements of some wetland hunter-gatherers.

EXPLORING SOCIO-POLITICAL ORGANISATION IN WETLAND ENVIRONMENTS

Why did hunter-gatherers stop being hunter-gatherers and start being something else? This has long been one of the key questions in both Old and New World archaeology, and the causes for the transition remain unclear. Population pressures, climate change, environmental fluctuations and other factors resulted in new opportunities and new challenges that different groups met in different ways. Some people continued as hunter-gatherers, others eventually became part- or full-time farmers.

The advent of farming is associated with the appearance of sedentary lifestyles, increases in population size and density, the rise of ranked or stratified societies, the elaboration of material culture and the development of cultural complexity. Amidst the considerable diversity that makes up hunter-gatherers, there is a sub-set that has become known as 'complex hunter-gatherers' – groups exhibiting traits associated with horticultural societies (*e.g.* Price and Brown 1985). I have refrained from using this term here because aspects of such 'complexity' may be a more normal characteristic of prehistoric hunter-gatherers than we now think.

What opportunities do wetlands provide to explore questions about socio-political organisation? There are a number of topics that can be investigated in those wetland areas that have the type of extensive and impressive archaeological records noted above. In addition, because they do not constitute a major biome in their own right, the aboriginal exploitation of swamp, marsh and other settings may go unnoticed or be underrepresented in cross-cultural surveys of hunter-gatherer societies (*e.g.* Murdock 1967, Binford 2001) and their prehistoric counterparts (see Nicholas 1998, 721). For the purpose of illustration, not explanation, three areas of interest are considered below: population size and territoriality; public works; and interpersonal and intergroup relations.

Population size and territoriality

Within resource-rich areas occupied by hunter-gatherers, territory size is often smaller than it is elsewhere. This is evident in the number and size of tribal boundaries in different environmental zones of Australia

(Tindale 1974). In addition, in more densely populated areas, territories may be more clearly demarcated and even defended against trespass (Kelly 1995, 23, 203). As Lourandos (1997, 296) notes for southwestern Victoria, Australia, 'There was intense competition for land and natural resources...especially at the borders between social groups. Disputes over fixed territorial rights and accusations of breached agreements could lead to bloodshed. The wide range of spears, clubs, and shields from the region suggests that combat, or at least antagonistic display, was frequent'.

Territoriality can be evident in types of social relationships, by evidence of violence or by territory markers (*e.g.* prominent land features, cemeteries and perhaps even oral history). Cemeteries and recognised burial areas may indicate claims to particular territory because they are culturally recognised places in the landscape which are returned to. The Murray-Darling River area, southeastern Australia, was by all accounts very densely populated, evidenced not only in historic observations and modern demographic reconstructions but also by archaeological data (Williams 1988). Mulvaney and Kamminga (1999, 306) note that there is evidence of burial areas along the central Murray River from 18,000 years ago (Lourandos 1997, 233). The earliest of the Lake Victoria burials (O'Neill 1994) may extend back to 10,000 BP, but what is even more important in this case is the reported number of burials. If there are indeed 10,000 individuals represented (Littleton *et al.* 1994), this would be extraordinary but would not be at odds with some of the archaeological remains noted above.

Hunter-gatherer cemetery sites associated with wetlands are also found in North America. The Paleoindian Sloan cemetery site in Arkansas is situated in an area that would have had extensive backswamp and other aquatic zones present when the site was in use during the late Pleistocene-early Holocene (Delcourt *et al.* 1997). In contrast, the early Holocene Windover cemetery (Doran and Dickel 1988) is actually within a marsh-bordered pond, indicating that wetlands had special spiritual (or practical) attributes. Wetlands may also have been seen as places to be avoided. Even when burials are not located near wetlands, there still may be a connection. For example, the 9400 year-old Spirit Cave burial in Nevada (Tuohy and Dansie 1997) was wrapped in a shroud of extraordinarily closely woven tule. The workmanship is so fine as to be comparable to, if not actually done on, a loom (C. Fowler pers. comm. 1997).

Public works

A variety of mounds, earthworks and canals is now known to have been constructed by hunter-gatherers in or near wetlands. The term 'public works' is applied generously; some of the low mounds dispersed throughout certain areas of Arnhem Land, for example, were likely the result of families increasing or maintaining an elevated area in an intermittently flooded area, perhaps over several generations. The larger and more numerous mounds in the Caramut region, however, are suggestive of a different use, and certainly a more organised effort. The presence of nearby villages containing numerous and relatively substantial structures, plus accounts of seasonal congregations of several thousand people, could reflect scheduled clan activities, or even the possibility of Big Men-like roles.

The Poverty Point group is approximately 50 times larger than the largest of the mound groups in the Caramut area, and would have required not only a substantial labour force but also considerably more direction, especially to construct such a carefully planned configuration. Likewise, the numerous channels and rock alignments (Fig. 20.5) constructed in and around wetlands in southeastern Australia, which were used for trapping or managing eels, probably represent a range of family- to clan-based labour efforts.

Throughout much of eastern North America, thousands of mounds are associated with the Hopewell and Mississippian cultures, constructed by sedentary, socially ranked farming peoples. However, a small number of large mound sites in the southeastern United States are now known to date to between *c.* 6500 and 3300 years ago (Hamilton 1999, 346), well before the advent of farming. Poverty Point falls within the later range. The best known of the earlier mound sites is Watson Brake, Louisiana (Fig. 20.6), situated on a terrace that would have overlooked the palaeo-Arkansas River and an extensive wetland (Saunders *et al.* 1997); initial mound construction began *c.* 5400 BP. Even if constructed over many years, the accomplishment is no less impressive, both for the labour involved and the organisation and motivation. Saunders (1999, 1) wryly notes that 'Archaic period mounds are significant if they were constructed by societies with social inequality, and they are equally significant if they were constructed by egalitarian societies. On the one hand, we extend traditional models (social inequality) back almost two thousand years earlier than thought; or on the other hand, we break the traditional bond between monumental architecture and social inequality, an element often considered necessary for the construction of public architecture'.

Interpersonal and intergroup relations

Identifying status, gender relations and other social aspects in ancient populations is notoriously difficult, not only because it is contingent upon certain types and contexts of material culture, but also because it is subject to our interpretations as outsiders. Within these limits, however, we can at least have a glimpse into the social lives of past societies through mortuary practices and bioarchaeological investigations. Information is available on this from a number of wetland-associated burial sites.

Fig. 20.5 Rock alignments and channel in wetland near Mt. Ecles, southwestern Victoria, Australia (photo: G. Nicholas).

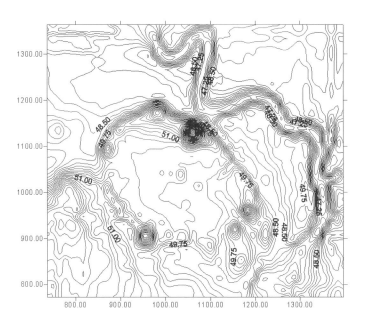

Fig. 20.6 The Watson Brake mounds, Louisiana (figure courtesy of Joe Saunders).

There is some evidence of social differentiation on the basis of grave goods. For example, at the Roonka Flat cemetery on the Murray River, southeastern Australia, 30% of the 216 interments include grave goods (Mulvaney and Kamminga 1999, 36). Grave 108, which dates to *c.* 5500 BP, is distinguished by the remains of a wallaby teeth headband (Fig. 20.7) and other ornamentation (Pretty 1986). In the Great Basin, grave goods are present in some burials, but there is no clear pattern as to what their presence or absence, or the inclusion of red ochre staining, means (Simms 1999). In the Stillwater Marsh area, the recovery and study of 416 individuals provides an important skeletal series for a wetland environment (Hemphill and Larsen 1999). A variety of grave goods is present (Raymond 1992), usually in small numbers per interment, which may reflect cherished items or materials required for the afterlife.

The Middle Archaic population in the Carrier Mills cemetery consists of 400 to 500 individuals (Jeffries 1987, 70); grave goods were found in only 27% of the 150 burials excavated. Items included projectile points, shell pendants and beads, bone pins and even a copper wedge (Jeffries 1987, 63). The type and distribution of these goods may be linked to sex and age, but may also reflect the gradual increase in socially achieved or acquired social status during the middle Holocene.

Gender relations may be reflected in mortuary practices by the types of material found with individuals. In some cases, it may also be possible to infer a sexual division of labour based on differential pathologies. For example, in the Malheur Lake, Stillwater Marsh and Great Salt Lake wetlands, the incidence of osteoarthritis is widespread (Hemphill 1999). Differences in the location and degree of osteoarthritis between men and women, suggests that males were engaged regularly in long-distance travel over rough terrain, while women were less mobile. There are also differences between the populations of these different wetland systems; unilateral osteoarthritis is high in the Malheur Lake group and may be linked to the use of spear-throwers (Larsen 2000, 58–59). Bioarchaeological data, such as occlusal grooves across the lower teeth (Larsen 2000, 24), may also hint at craft specialisation, or at least repetitive wear patterns associated with making lines for nets.

A very different line of evidence about life in wetland environments is provided by possible evidence of violence in skeletal pathology and rock art. At a number of locations in the Great Basin, individuals are found with parry-type fractures to the arm, while at Malheur Lake there is also a relatively high incidence of broken noses (B. Hemphill pers. comm. 1998) and other evidence of violent confrontations (Bettinger 1999, 330). One older male at Stillwater was killed or severely

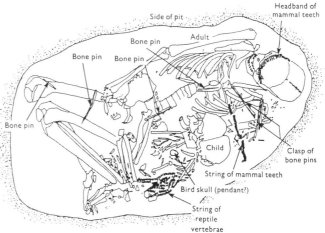

Fig. 20.7 Roonka Flat, Australia (photo: G. Nicholas), and Roonka Flat Grave 108 (after Manley 1993: 86).

injured after being struck by two projectile points (Raymond 1992, 16). In the Middle Archaic component of the Carrier Mills cemetery, not only were fractures more prevalent in males than females, but 'parry fractures' were commonly found in males (Jeffries 1987, 77). One individual at the Windover cemetery had a fractured skull (Dickel *et al.* 1989).

Taçon and Chippendale (1994) have proposed that there is substantial and relatively unambiguous evidence of violence in rock art from Arnhem Land, perhaps as early as 10,000 years ago. If they are correct, the reasons may be linked to the defence of territory, to territorial expansion due to environmental factors, or to explicitly social needs or desires. They also suggest that a change in the style and grouping of rock art figures, occurring around 6000 years ago, may be synchronised to an increase in social complexity, and possibly to increases in population density. At this point, there are no corresponding bioarchaeological data indicative of actual fighting.

CONSIDERING ISSUES OF SCALE AND REPRESENTATIVENESS

If we accept that the nature of wetland ecosystems is qualitatively different from other settings in the landscape, at least at certain times and places, then several things need to be considered. Firstly, wetlands may have been an attractive habitat for both historic and prehistoric hunter-gatherers, as their resource productivity, diversity and reliability may equal or in some cases exceed that of other environmental settings. This has implications for population size and density, degree of mobility/sedentism, division of labour, landuse practices, degree of territoriality and a host of other factors, all of which may influence or encourage certain types and trajectories of socio-political organisation.

Second, certain types of behaviours and activities are found in wetland areas that may not be found elsewhere. These include unusual types of cemeteries, for example, such as that found at the Windover site, or those that contain thousands of interments, such as found at Lake Victoria; the construction of earthen mounds at Watson Brake and the Caramut; the development of large-scale resource management systems (Nicholas *this volume*); and evidence of intergroup violence in Arnhem Land and the Great Basin. Regardless of whether such practices were relatively uncommon or simply poorly represented archaeologically, their recognition in wetland (and non-wetland) settings increases significantly the spectrum of hunter-gatherer behaviours.

Different scales of human-environmental interaction need to be considered in investigating hunter-gatherers' use of wetlands, especially within the context of the larger landscape they occupied. Most archaeological investigations occur at the micro-scale level (*i.e.* < 1 km^2 and <10 years (Dincauze 2000, 379)), where the focus is on what happened at a particular location at certain points in time. An example would be the excavation of a site like Star Carr in England (Clark 1971), where group/s of hunter-gatherers focused their activities alongside a reedswamp over 9000 years ago. The foremost concern here is cultural history and chronology, usually with the goal of reconstructing the social, economic and other dimensions of people's lives in the past (Mellars and Dark 1998). Shifting to the meso-scale (*i.e.*, < 1–10^4 km^2 and 10^2–10^4 years [Dincauze 2000, 377]), the larger area and longer-term focus provides access to other dimensions of past human ecosystems, cultural change and demographic trends. Not only does research at this scale reveal the broader patterns of landscape utilisation, but it also illuminates the processes of cultural and environmental change.

The element of time is particularly important in evaluating the type of public works discussed earlier in this paper. In terms of socio-political organisation, it makes a great deal of difference if we know that such constructions as the Watson Brake mounds or the Toolondo eeling canals were constructed over ten years or a thousand years. Although these are equally impressive regardless of the time involved, a shorter and more intensive episode of building obviously requires different types and degrees of organisation, planning and motivation, human resources and logistical support than do longer, more intermittent episodes. In the case of the extensive wetland canals and other constructions found in southwestern Australia (or, in non-wetland contexts, large fish weirs (Dincauze and Decima 1998)), it is also important to determine whether individual sections were constructed, utilised and maintained independent of each other, or as part of one large, integrated system.

CONCLUSIONS

Since little attention has been paid by archaeologists to wetlands (with some notable exceptions), they remain a largely unknown element in our understanding of prehistoric peoples. However, the impetus to investigate the archaeology and human ecology of these settings should be substantial given the often atypical sites and other records of past behaviour associated with them. Some of these are truly impressive – the swamp management of southwestern Australia; the 'sedentary, politically powerful, centralised, stratified' Calusa of south Florida (Marquardt 1988, 161); the Archaic mound builders of Louisiana; the Lake Victoria burial grounds – and rival the accomplishments of horticultural societies. Greater variation in, or accuracy of our knowledge of, past hunter-gatherer lifeways is revealed in the degree of mobility/sedentism and types of landuse patterns reflected in the archaeological record of the Great Basin wetlands and other localities.

If wetlands are not seriously considered in the scope of archaeological surveys or cultural resource management policies, then the archaeological record will not be representative of the full range of human activities that occurred in the past. This is especially important with the archaeology of hunter-gatherers, whose lifeways represent the baseline for modern humanity, and whose sites dominate the archaeological record worldwide.

ACKNOWLEDGEMENTS

I thank Malcolm Lillie for his invitation to participate in this project, and also Robert Kelly, Brian Hemphill, Heather Builth, Donald Pate, Joe Saunders, Catherine Fowler and Colin Pardoe for providing information and helpful insights into my on-going research on wetlands. In Australia, I am particularly grateful to Heather Builth with whom I spent several days touring wetland sites in the Mt. Condah/Mt. Eccles area of western Victoria, and to Donald Pate for organising a visit to Roonka Flat.

REFERENCES

Bareis, C. J. and J. W. Porter (eds) 1984. *American Botton Archaeology: a summary of the FAI-270 Project contribution to the culture history of the Mississippi river valley.* Urbana: University of Illinois Press.
Bettinger, R. L. 1999. Faces in prehistory: Great Basin wetlands skeletal populations, in B. E. Hemphill & C. S. Larsen (eds) *Prehistoric Lifeways in the Great Basin Wetlands: bioarchaeological reconstruction and interpretation*, 321–32. Salt Lake City: University of Utah Press.
Binford, L. R. 1980. Willow smoke and dogs' tales: hunter-gatherer settlement systems and archaeological site formation. *American Antiquity* 45, 4–20.
Binford, L. R. 1982. The archaeology of place. *Journal of Anthropological Archaeology* 1, 5–31.
Binford, L. R. 2001. *Constructing Frames of Reference: an analytical method for archaeological theory building using ethnographic and environmental datas sets.* Berkeley: University of California Press.
Builth, H. 1999. The connection between the Gunditjmara people and their environment: the case for complex hunter-gatherers in Australia. Unpublished ms. in author's possession.
Clark, J. G. D. 1971. *Excavations at Star Carr: an Early Mesolithic site at Seamer near Scarborough, Yorkshire.* Cambridge: Cambridge University Press.

Coles, B. & J.M. Coles 1989. *People of the Wetlands: bogs, bodies and lake-dwellers*. New York: Thames and Hudson.

Coles, J. M. and A. J. Lawson (eds) 1987. *European Wetlands in Prehistory*. Oxford: Clarendon Press.

Delcourt, H. R., P. A. Delcourt & P. D. Royall 1997. Late Quaternary vegetational history of the Western Lowlands, in D. F. Morse (ed) *Sloan: a Paleoindian Dalton Cemetery in Arkansas*, 103–22. Washington, D.C.: Smithsonian Institution Press.

Dickel, D. N., C. G. Aker, B. R. Barton & G. H. Doran 1989. An orbital floor and ulna fracture from the Early Archaic of Florida. *Journal of Paleopathology* 2, 165–70.

Dincauze, D. F. 2000. *Environmental Archaeology: principles and practice*. Cambridge: Cambridge University Press.

Dincauze, D. F. and E. Decima 1998. The Boston Back Bay fish weirs, in K. Bernick (ed) *Hidden Dimensions: the cultural significance of wetlands Archaeology*, 157–72. Vancouver: UBC Press.

Doran, G. H. and D. N. Dickel. 1988. Multidisciplinary investigations at the Windover Site, in B. A. Purdy (ed) *Wet Site Archaeology*, 263–90. Caldwell, NJ: Telford Press.

Fagan, B. M. 2000. *Ancient North America*. New York: Thames and Hudson.

Flood, J. 1995. *Archaeology of the Dreamtime*. New Haven: Yale University Press.

Fowler, C. S. 1982. Food-named groups among Northern Paiute in North America's Great Basin: an ecological interpretation, in N. M. Williams & E. S. Hunn (eds) *Resource Managers: North American and Australian hunter-gatherers*, 113–29. Boulder: Westview Press, .American Association for the Advancement of Science, Selected Symposium 67.

Hamilton, F. E. 1999. Southeastern Archaic mounds: examples of elaboration in a temporally fluctuating environment? *Journal of Anthropological Archaeology* 18, 344–55.

Heizer, R. F. & L. K. Napton 1970. *Archaeology and the prehistoric Great Basin lacustrine subsistence regime as seen from Lovelock Cave, Nevada*. Berkeley: Contributions of the University of California Archaeological Research Facility 10.

Hemphill, B. E. 1999. Wear and tear: osteoarthritis as an indicator of mobility among Great Basin hunter-gatherers, in B. E. Hemphill & C.S. Larsen (eds) *Prehistoric Lifeways in the Great Basin Wetlands: bioarchaeological reconstruction and interpretation*, 241–89. Salt Lake City: University of Utah Press.

Hemphill, B. E. and C. S. Larsen (eds) 1999. *Prehistoric Lifeways in the Great Basin Wetlands: bioarchaeological reconstruction and interpretation*. Salt Lake City: University of Utah Press.

Jeffries, R. W. 1987. *The Archaeology of Carrier Mills: 10,000 years in the Saline Valley of Illinois*. Carbondale: Southern Illinois University Press.

Kelly, R. 1995. *The Foraging Spectrum: diversity in hunter-gatherer lifeways*. Washington, D.C.: Smithsonian Institution Press.

Kelly, R. 2001. *Prehistory of the Carson Desert and Stillwater Mountains: environment, mobility, and subsistence in a Great Basin wetland*. Salt Lake City: University of Utah Anthropological Paper 123.

Kuehn, S. R. 1998. New evidence for Late Paleoindian-Early Archaic subsistence behavior in the Western Great Lakes. *American Antiquity* 63, 457–76.

Larsen, C. S. 2000. *Skeletons in Our Closet: revealing our past through bioarchaeology*. Princeton, NJ: Princeton University Press.

Littleton, J., H. Johnston & C. Pardoe 1994. Lake Victoria Lakebed. Report to National Parks & Wildlife Service, NSW.

Lourandos, H. 1997. *Continent of Hunter-Gatherers*. Cambridge: Cambridge University Press.

Manley, J. 1993. *The Atlas of Past Worlds: a comparative chronology of human history 2000 BC–AD 1500*. London: Cassell Publishers.

Marquardt, W. H. 1988. Politics and production among the Calusa of south Florida, in T. Ingold, D. Riches & J. Woodburn (eds) *Hunters and Gatherers 1: history, evolution and social change*, 161–88. New York: Berg.

Mellars, P. & P. Dark 1998. *Star Carr in Context: new archaeological and palaeoecological investigations at the Early Mesolithic site of Star Carr, North Yorkshire*. Cambridge: McDonald Institute Monographs.

Mulvaney, J. & J. Kamminga 1999. *The Prehistory of Australia*. St. Leonards, NSW: Allen & Unwin.

Murdock, G. P. 1967. *Ethnographic Atlas*. Pittsburgh: University of Pittsburgh Press.

Musil, R. R. 1995. *Adaptive Transitions and Environmental Change in the Northern Great Basin: a view from Diamond Swamp.* Eugene: University of Oregon Anthropological Papers 51.

Neusius, S. W. (ed.) 1986. *Foraging, Collecting, and Harvesting: Archaic Period subsistence and settlement in the Eastern Woodlands.* Carbondale: Center for Archaeological Investigations, Occasional Paper 6. Southern Illinois University.

Nicholas, G. P. 1988 Ecological leveling: the archaeology and environmental dynamics of early Postglacial land-use, in G. P. Nicholas (ed) *Holocene Human Ecology in Northeastern North America*, 257–96. New York: Plenum Press.

Nicholas, G. P. 1992. Directions in wetlands research. *Man in the Northeast* 43, 1–9.

Nicholas, G. P. 1998. Wetlands and Hunter-Gatherers: a global perspective. *Current Anthropology* 39, 720–33.

Nicholas, G. P. 2001. Time, Scale, and Contrastive Ecologies in Human Land-Use Studies. Paper presented at the 24th Society for Ethnobiology Conference, Durango, Colorado.

Oetting, A. C. 1989. *Villages and Wetlands Adaptations in the Northern Great Basin: chronology and land use in the Lake Abert-Chewaucan Marsh—Lake County, Oregon.* Eugene: University of Oregon Anthropological Papers 41.

Oetting, A. C. 1999. An examination of wetland adaptive strategies in Harney Basin: comparing ethnographic paradigms and the archaeological record, in B. E. Hemphill & C.S. Larsen (eds) *Prehistoric Lifeways in the Great Basin Wetlands: bioarchaeological reconstruction and interpretation*, 203–18. Salt Lake City: University of Utah Press.

O'Neill, G. 1994. Cemetery reveals complex aboriginal society. *Science* 264, 1403.

Phillips, J. L. and J. A. Brown (eds) 1983. *Archaic Hunters and Gatherers in the American Midwest.* New York: Academic Press.

Price, T. D. and J. Brown (eds) 1985. *Prehistoric Hunter-Gatherers: the emergence of cultural complexity.* San Diego: Academic Press.

Pretty, G. 1986. Australian history at Roonka. *Journal of the Historical Society of South Australia* 14, 107–22.

Raymond, A. W. 1992. *Who Were the Ancient People of Stillwater National Wildlife Refuge, Nevada?* Fallon: Fish and Wildlife Service, Stillwater Wildlife Refuge.

Sampson, C. G. 1985. *Nightfire Island: later Holocene lakemarsh adaptations on the western edge of the Great Basin.* Eugene: University of Oregon Anthropological Papers 33.

Saunders, J. W. 1999. Are We Fixing to Make the Same Mistake Again? Paper presented at the 60th Southeastern Archaeological Conference, Pensacola, Florida.

Saunders, J. W., R. D. Mandel, R. T. Saucier, E. T. Allen, C. T. Hallmark, J. K. Johnson, E. H. Jackson, C. M. Allen, G. L. Stringer, D. S. Frink, J. K. Feathers, S. Williams, K. J. Gremillion, M. F. Vidrine & R. Jones. 1997. A mound complex in Louisiana at 5400–5000 years Before Present. *Science* 277, 1796–9.

Simms, S. R. 1999. Farmers, foragers, and adaptive diversity: the Great Salt Lake Wetlands Project, in B. E. Hemphill & C.S. Larsen (eds) *Prehistoric Lifeways in the Great Basin Wetlands: bioarchaeological reconstruction and interpretation*, 21–54. Salt Lake City: University of Utah Press.

Stafford, C. R., R. L. Richards & C. M. Anslinger 2000. The Bluegrass fauna and changes in middle Holocene hunter-gatherer foraging in the southern Midwest. *American Antiquity* 65, 317–36.

Struever, S. & F. Holton 1979. *Koster: Americans in search of their prehistoric past.* New York: Anchor Press.

Taçon, P. & C. Chippindale 1994. Changing depictions of fighting in the rock art of Arnhem Land, N.T. *Cambridge Archaeological Journal* 4, 211–48.

Thomas, D. H. 1985. 'Why Bother Digging Hidden Cave Again???', in Thomas, D.H. (ed.) *The Archaeology of Hidden Cave, Nevada*, 17–38. New York: Anthropological Papers 61, American Museum of Natural History.

Tindale, N. B. 1974. *Aboriginal Tribes of Australia: their terrain, environmental controls, distribution, limits, and proper names.* Berkeley: University of California Press.

Tuohy, D. R. & A. Dansie 1997. New information regarding early Holocene manifestations in the Western Great Basin. *Nevada Historical Society Quarterly* 40, 24–53.

Williams, E. 1988. *Complex Hunter-Gatherers: a late Holocene example from temperate Australia.* BAR International Series 423, Oxford.

Chapter 21

Estuarine Development and Human Occupation at Bobundara Swamp, Tilba Tilba, New South Wales, Australia

G. S. Hope, J. J. Coddington and D. O'Dea

INTRODUCTION

In Australia little direct information about past human activity has been obtained from wetlands, but they are a valuable archive of change through time against which the archaeological record from middens and rock shelters can be compared. The constant movement of small bands of hunter-gatherers has meant that there is little sign of intensive occupation, and the use of wetlands lay in their attraction as sources of water, plant foods and water birds, for which little archaeological trace can be expected. Moreover, since Australia is generally warm and dry, peat accumulations are rare and often oxidised, leading to poor preservation of features and artefacts. One exception to this, however, was the discovery of boomerangs and digging sticks preserved by peat in South Australia (Luebbers 1975).

The infilled estuaries and barrier-dammed swamps of the relatively humid southern coast of New South Wales and eastern Victoria include many significant peatlands. The continental shelf of southeastern Australia is narrow and stream valleys are incised well below modern sea-level (Chapman *et al.* 1982). The coast of southern New South Wales has a poor sediment supply and consists of quartz sand beaches and low cliffed headlands. There is a microtidal regime that maintains stable water tables so that the lakes and swamps have persisted since the early Holocene. There are few large streams on this coast and open river estuaries are rare. Instead there is a spectrum of sand barrier-impounded saline lakes through to freshwater swamps behind beach ridges or dunes. In many areas the swamps provide the only access to freshwater at the coast, the streams feeding them being some kilometres inland.

This paper reports on a combined archaeological survey and study of vegetation history from Bobundara Swamp, a major coastal peatland (Fig. 21.1). The swamp basin sediments have been dated and analysed to provide the chronology of local environmental change to set against the archaeology. A preliminary analysis of an offshore section (Narooma 3b) provides a more general history of environmental change in the area after 14,000 BP.

THE TILBA TILBA AREA

Bobundara Swamp occurs in the Tilba Tilba area which is midway between the Narooma and Bermagui townships at 36° 25'S, 150° 06'E. An unusual Mesozoic eruptive centre, Mount Dromedary (797 m), the highest mountain in the area, dominates the region, lying only 5 km from the sea. Tilba is an Aboriginal word meaning windy, and the mountain does induce a high wind regime when compared with the undulating coastal plain to the north and south. The andesitic flows from the mountain extend to the sea and there are essexite, banatite and monzonite igneous rocks and Ordovician quartzites, graywacke and shales forming low headlands.

In this area there are few deposits of pre-Holocene sand. After sea-level rise around 6000 BP, beach ridges infilled old embayments with siliceous sand, and relatively active progradation occurred from 6–4000 BP. This is believed to be due to onshore movement of a sand reservoir (Chapman *et al.* 1982, Thom and Roy 1985) which eventually became ex-

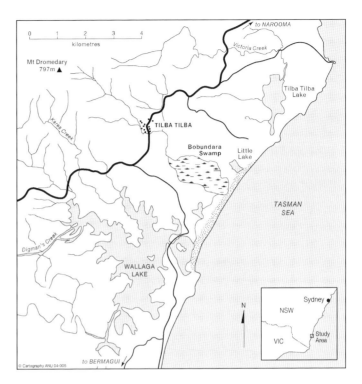

Fig. 21.1 The Tilba Tilba area.

hausted. After this time long beaches formed between the headlands, and low dunes then built behind them. Relative stability is indicated by a low carbonate content and a widespread humic soil, about 0.30 m deep, on these dunes. In many places coastal retreat and dune disturbance can be seen in truncated soil sections, sometimes buried beneath more recent sand. The dunes and beach barriers also enclose saline lakes, saltmarshes and freshwater swamps.

The climate is maritime and relatively humid, with an annual precipitation of 998 mm and a slight summer maximum, but the variability is high and evaporation is roughly equal to precipitation. Temperatures average 19.4 °C in summer and 10.1 °C in winter, when light frosts are common. The sea temperatures can be quite variable due to the incursion of cold southern water when westerly winds force the warm northern current to peel off into the Tasman Sea.

The pre-European vegetation is principally a mosaic of diverse forests dominated by different species of eucalypts (Keith and Bedward 1999), with increasingly closed canopies in shaded gullies where there occur a range of warm temperate rainforest elements, climbers and ferns such as lilly pilly (*Eugenia smithii*), *Pittosporum undulatum*, *Elaeocarpus reticulatus*, cabbage tree palm (*Livistona australis*), blackwood (*Acacia melanoxylon*), *Polyscias* and rough tree fern (*Cyathea australis*). Above 500 m on Mt Dromedary is a closed forest of plumwood (*Eucryphia moorei*), black sassafras (*Doryphora sassafas*) and coachwood (*Ceratopetalum apetalum*). On the coastal headlands the forest was dominated by potted gum (*Eucalyptus maculata*) and silver top ash (*E. seiberi*), with a seaward woodland of coast mahogany (*E. botryoides*), sheoke (*Casuarina glauca*) and swamp paperbark (*Melaleuca ericifolia*). On the dunes, woodlands of *Banksia integrifolia* and *E. botryoides* occur behind shrublands of *Acacia longifolia*, *Leptospermum laevigatum* and *Leucopogon australis*, and grasslands of *Spinifex sericea* and *Austrofestuca littoralis* fringing the beach.

Wetlands include saltmarsh dominated by samphire (*Sarcocornia quinquefolia*) and sea rush (*Juncus kraussii* ssp. *Australiensis*), with the sedges *Baumea juncea* and *Isolepis nodosa*. Freshwater swamps are dominated by the tall aquatic grass *Phragmites australis*, bullrushes (*Typha domingensis*), the sedges *Carex, Cyperus, Eleocharis* and *Isolepis* species and ribbon weed *(Triglochin striatum)*. Aquatic habitats support *Triglochin proceru,, Myriophyllum* spp., *Ranunculus inundatus* and water lilies such as *Ottelia ovalifolia* and *Nymphoides crenata*.

The coastal plains have been partly cleared, particularly on the volcanic soils, and the forests logged very widely, since 1835. Mines were operated at Tilba for several decades after 1870. European arrival can be detected by the appearance of the pollen of introduced weeds such as *Hypochoeris* spp. and *Pinus* after about 1890, together with changes in grassland extent.

Aboriginal occupation of Australia is now firmly established as having occurred more than 40,000 years ago (*e.g.* Mulvaney and Kaminga 1999), although Pleistocene sites have remained rare on the eastern coast. Lampert (1971) records a basal age of 21,000 BP at Burrill Lake rockshelter, about 120 km north of Tilba, and this remains the oldest dated site in the coastal region south of Sydney. Shell middens are Holocene, with basal dates of *ca* 5000 BP in coastal sands and increasingly common shell sequences spanning the last 3000 years (Hughes and Lampert 1982). Within the archaeological sequences there is a shift at around 5–3000 BP with the appearance of backed blades added to core and scraper types. Backed blades of chert or silcrete are absent from sites after 19–1600 BP and quartz is increasingly used. Fish hooks made from shell seem to be introduced about 5–800 years ago, and their appearance may be associated with a change in the species of shell exploited (Bowdler 1976) and more diverse resource gathering. Some have argued that this is evidence for intensification of coastal use in the last two millennia (Mulvaney and Kaminga 1999, 289), possibly a result of increased extractive efficiency. Hughes and Sullivan (1981) have suggested that an increasing use of fire led to increased rates of geomorphic infill over the past 2000 years.

Prior to European settlement, Aboriginal use of the coast was relatively intense, as it was an important resource zone. The most notable evidence for occupation are the shell scatters behind beaches and on headlands, but wherever surveys have been undertaken in forests (*e.g.* Sullivan 1983, Boot 1994), small sites of hearths with a few artefacts have been found, so it appears that all habitats were exploited. The visibility of the shell and paucity of bone or plant material has led to a concentration on its role in coastal subsistence, and represents a clear bias in the record.

Ethnographic evidence is restricted to a few explorers' accounts and settlers' letters, although the area was described in 1844 by G. A. Robinson (Mackaness 1941) and later studied by Howitt (1904) and Mathews (1904). The Tilba area is thought to have been occupied by clans of the northern group of the Yuin tribe (Howitt 1904), and Tindale (1974) records the area as a Djiringanj language group, although the location of the language boundaries has been disputed. Attenbrow (1976) interpreted the limited data on contact time ethnohistory to propose that bands were quite mobile, tending to occupy the coast during winter and to spread into the hinterland during summer. Aboriginal people live in the Tilba area today, some still at Wallaga Lake Reserve, which was established in 1891 (McKenna 2002). They are descended from Yuin peoples, mainly from the Bega Valley to the south. Some traditional knowledge is intact and is being preserved by the Umbarra Cultural Centre, Wallaga Lake.

Fishing from canoes with spears or lines and shell fish gathering were important marine activities, the resources being supplemented by seals, stranded whales and dolphins from time to time. A wide range of fauna, including kangaroo, possum, echidna, frogs, snakes and birds, were exploited, and visits made to offshore island sea bird nesting sites (rookeries). Insect grubs, worms and honey also made an unquantified contribution to the diet. A wide range of plant foods were available (Cribb and Cribb 1974, Gott 1982,

Hardwick 2001) from the rainforest (shoots of the palm, *Livistona australis,* growing tips of some ferns, bracken rhizomes, fruits from trees and shrubs such as *Eugenia smithii, Ficus* spp. and *Citriobatus spinescens*), coastal woodlands and dune scrub (fruits of Epacridaceae, *Exocarpos cupressiformis, Persoonia* and *Solanum, Banksia* flowers, orchid tubers, daisy yam tubers and leaves of New Zealand spinach, *Tetragonia tetragonioides* and coastal saltbush *Atriplex cinerea*). The seeds of the common cycad, *Macrozamia communis,* which requires careful pre-treatment before cooking, were a seasonally important starch source. Several *Acacia* species provide seeds that were roasted and ground.

Possibly more important were freshwater swamp plants with starchy rhizomes and tubers. These include *Phragmites australis,* which has delicious young shoots and a thick underground rhizome, as does *Typha domingensis* (Gott 1999). The aquatic *Triglochin procerum* and *T. microtuberosum* has tubers, starchy rhizomes and edible fruits (Hardwick 2001). Sedge species include *Schoenoplectus validus* (rhizomes, seeds and young shoots) and water chestnut-like tubers from several *Bolboschoenus* species such as *B. caldwellii, B. medianus* and *B. fluviatilis.* Thus swamps provided an important staple that may have been managed by fire and husbandry (Hope and Coutts 1971, Clark 1983, Head 1989). In addition, swamps provided fish and crustacea, as well as being important hunting grounds for water birds and animals coming to drink. By contrast the saline swamps around estuaries are less rich although Hardwick (2001) notes that the leaves of samphire (*Sarcocornia quinquefolia*) and some saltbush species are eaten.

BOBUNDARA SWAMP STUDY SITE

Tilba Creek drains 1425 ha from the southeastern slopes of Mt Dromedary, eastwards in a broad valley south of Little Dromedary Mountain. Bobundara Swamp is a freshwater tall grassland/sedgeland of *Phragmites australis* and *Typha domingensis* about 120 ha in extent that follows the lower valley and backs against a seaward dune (Fig. 21.2). This sand barrier is linked to the Tilba Tilba headland and an over-steepened bank on the northern margin of the swamp, but southward it is separated from former headlands to the outlet of Wallaga Lake 3km to the south. Water draining from the swamp runs south-wards behind the beach barrier into Merriwinga Swamp, a saltmarsh that is probably an infilled arm of Wallaga Lake. This large estuarine lake is intermittently connected to the sea across the sand barrier, and it provides a habitat model for former lake phases at Bobundara. The beach barrier was able to form without being breached

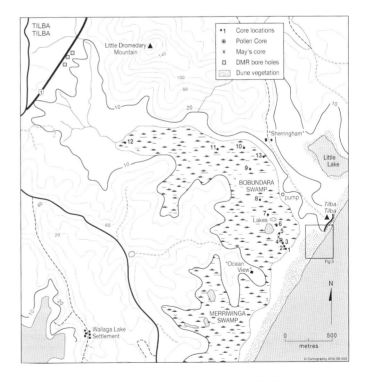

Fig. 21.2 Bobundara Swamp study site.

by water building up in the swamp due to a separate channel to the larger estuary to the south. North of the headland, Little Lake regularly breaches the barrier to the sea, despite having a very small catchment.

ARCHAEOLOGICAL SURVEY

An archaeological survey was carried out around the swamp, on the Tilba Tilba headland and on the beach barrier complex to the north and south (Coddington 1983). In addition, the Dibden artefact collection, made some time ago by a local farmer from the study area and the next beach to the north, was inspected. This included hammerstones, pebble percussion stones (showing an anvil pit), pebble choppers, both bifacially and unifacially flaked, edge-ground axes, possible 'oyster picks', horsehoof cores, scrapers, thumbnail scrapers, bondi points and other points and blades, knives, fabricators and elouras. Several playstones (very round pebbles) and quartz crystals (of possible ritual significance) were found from the barrier north of the headland. Stone type varied and included silcrete, chert and silicified tuff used for blades, and ferruginised sandstone pebbles used for anvils, hammerstones and edge-ground axes. Quartz pebbles were uncommon, possibly having been neglected by the collectors.

A large number of sites, principally shell scatters preserved in sand, were found along the beach barrier to the north and south of the swamp. Only a few sites were found which contained stone such as quartz, silcrete and chert, and these were usually close to headlands. A major blowout in the dunes near the entrance to Tilba Tilba lake in the north of the survey area, has exposed a wide range of stone artefacts, but elsewhere sand erosion has buried the older dune surface. Hence many sites near the sea are probably concealed beneath recent sand accumulations or by dense vegetation. The dune behind the northern beach is made from recent sand and only a few shell scatters were found. Mr Norm Hoyer, the landowner of Tilba Tilba headland, reported that in the past, ploughing there had thrown up abundant shell and hearthstones. However, a careful search and several auger holes found nothing, even though shell middens are known from headlands at Narooma and Bermagui. Eroded parts of the southern beach dune revealed a dark humified layer containing charcoal and shell, usually about 1 m below the surface layer of yellow sand. Most of the shellfish consisted of *Anadara trapezia*, although towards Wallaga Lake there is an increase in oysters (*Crassostrea* and *Ostrea* species) and the mud whelk *Pyrazus ebeninus*.

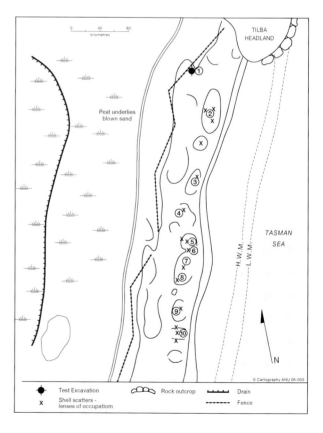

Fig. 21.3 Plan of archaeological deposits on the Bobundara barrier.

The land around Bobundara Swamp was also surveyed, but only scattered artefacts were found, mainly on the slope to the south of the swamp. The slopes and swamp margin are grass-covered, so artefacts were mainly located on eroded or bare track edges. These included several quartz flakes, one silcrete and one chert flake, a large edge-ground axe of banatite and a pitted anvil stone. The axe was large and flat, and resembles several that have been found over time around Sherringham farmhouse on the northern side of the swamp (N. Hoyer pers. comm.). To the west of the swamp only two flakes were found, perhaps because post-clearance slopewash has covered the original surface.

Blowouts seaward of the swamp revealed numerous sections with shell and charcoal eroding from a buried organic-rich soil horizon (Fig. 21.3). A reconstruction of several sections suggests that the deposit in this area is complex, with several discontinuous lenses separated by pale white sand. Disturbance of the dune may have been associated with occupation. The major shellfish recovered was *Anadara trapezia*, although *Cabestana spengleri*, *Ostrea angas,, Crassostrea commercialis, Plebidonax* and *Velectuctea flammea* were also present. Size measurements of the *Anadara* suggest that they were 3–4 years old when collected.

ARCHAEOLOGICAL EXCAVATION

Site 1, close to the headland, contained abundant artefacts eroding from a 0.8 m thick organic-rich horizon, which was subsequently selected for excavation. An adult human skeleton was temporarily exposed nearby in the same horizon. The excavation was sited in the deepest section available, determined by augering. This revealed that the organic-rich section thinned away from the exposed face, indicating that the deposit is lens-like, eroded, and possibly infilling a small hollow. A 0.5 × 1.0 m test pit was excavated in spits and the material sieved using 9, 5 and 2 mm sieves (Table 21.1). The stratigraphy is shown in Fig. 21.4.

The excavation recovered limited cultural material. Quartz becomes more common in the upper spits, coinciding with a decrease in average stone size. *Anadara trapezia* was the dominant shellfish found (Table 21.2).

Bone is rare but includes fragments of bird or mammal and two fish vertebrae, possibly from snapper or bream, which were found in Spit 2. Three fragments of the chiton *Ischnoradsia australis* were recovered from Spit 3. Several seeds of swamp rush, *Juncus* sp. (possibly *Juncus kraussii*) were found in Spits 2 and 3. *Juncus* seeds were ground to a paste and eaten by Aborigines (Cribb and Cribb 1974), and the plant is particularly common in saline marshes such as Meriwinga Swamp, to the south of Bobundara Swamp.

The material from the site is not very diverse but it does reveal that the diet included plants and animals other than shellfish. The *Anadara* were most likely brought from Wallaga Lake, 2–3 km distant, since at

Spit	Depth cm	Sediment
1	0–30	Clean very pale brown dune sand without artefacts
2	30–48	Mottled light yellowish brown sand with some cultural material
3	48–66	Dark grayish organic rich sand with shell, stone and charcoal fragments
4	66–85	Dark brown organic rich sand with occasional cultural material and charcoal Dates of 320 ± 110 yr. BP (ANU 3678) on charcoal and 730 ± 70 yr. BP (ANU 3708) on *Anadara* shell were obtained
5	85–>95	Light yellowish brown sand with occasional charcoal

Table 21.1 Stratigraphy of dune deposit near Tilba Tilba headland. Shell dates are liable to a marine reservoir correction of ca –350 years.

SPIT	Artefacts	Stone fragments	*Anadara* (g)	Other shell (g)	*Juncus* seeds	Bone
2	Scraper (Q)	38Q1S	17	1.4	3	-
3	Scraper (QZ) 3Flakes (Q)	42Q	72.3	5.8	1	4
4	Nil	10Q 2S	47.9	9.8	-	4
5	Scraper(S)	4Q 3S	6.0	3.0	-	1

Table 21.2 Materials from test excavation. Q quartz, QZ quartzite, S silcrete. Other shell includes Cabestana spengleri, Ninella torquata, Dicanthis orbita *and* Ostrea angasi.

the time of occupation it seems unlikely that there was a local supply, because Bobundara was already a fresh water lake. The deposit thus provides evidence that people visiting the Bobundara area brought shells and seeds, or that they were based there and sought food some distance to the south. In either scenario it is likely that it was the resources of the swamp that bought them there. The dune soil indicated by the organic-rich horizon suggests that dune stability was maintained through the occupation, and that the covering of sand is the result of recent dune disturbance by cattle, rabbits and fire.

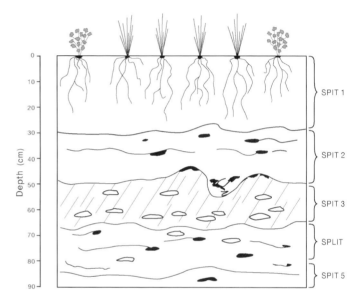

Fig. 21.4 Stratigraphy of the northern face of the archaeological section.

SEDIMENT CORING

Bobundara swamp was cored with a D-section Russian corer along an east-west transect 2.2 km in length, revealing that the swamp peats form a 2–2.5 m layer over lacustrine sediments. Road test bores a further 2 km up valley found a 3 m layer of fibrous peat. This was buried beneath 3 m of recent sands and clays, probably derived from slope erosion following clearance for pasture. Thus the swamp at one time extended 4 km up valley. In the easternmost 100 m of the transect, only shallow organic sections were obtained as beach sands of the former sea spit were encountered a few metres down. Near the dunes even these shallow peats are buried beneath recent sand. The lacustrine sediments thin up valley from the deepest section, located about 400 m from the northern and eastern margin. Levelling from the trigonometric station on Tilba Tilba headland showed the modern swamp surface to be close to mean sea-level. Repeated coring here provided a 14 m sequence of peats, lake muds and estuarine sandy muds covering the last 8000 years.

The record can be extended back to 14,000 BP by including data from a marine core. The Narooma NB3b core is a 343 cm gravity core taken by Peter Roy in 1988 at 36° 08.2'S, 150° 12.3'E in 89 m of water on the continental slope, 7.2 km offshore and about 24 km northeast of Bobundara swamp. The core

Site	Depth cm	Stratigraphy and dating
BOB	0–210	Brown fibrous telmatic peat with sedge stems, pH3.5. The base at 190–210cm dated as 460 ± 130 BP (ANU 3546)
BOB	200–520	Dark brown nekron mud with scattered charcoal fragments at 300cm and sand lenses around 400cm, increasingly sandy to base. pH 3.5. A date from 390–410 is 2540 ± 190 BP (ANU 3545)
BOB	520–610	Transition from sandy clayey nekron mud to grey clay containing organic muds, with shells of *Parcanassa buchardi* and *Notospisula parva*. The end of the transition at 520–540cm is dated at 3590 ± 150 BP (ANU 3544)
BOB	610–1320	Gray clay with abundant shell, becoming sandy below 1100cm pH 2.5–3.5, LOI 9–12% above 1130cm, <0.5% below. The level at 830–845cm is dated at 5410 ± 110 (ANU 9474) and the top of the sand at 1120–1130 is dated at 6940 ± 170 BP (ANU 9475)
BOB	1320–1385	Grey clayey medium sand with low organic content
NB3b	155–180	Olive grey muds with very fine sand and scattered bivalves. Shells at 155cm give an uncorrected age of 8140 ± 115 BP
NB3b	180–335	Dark grey very muddy very fine sand with 40% shell. Shell dates of 10,100 ± 160 BP at 215cm and 13,880 ± 160 BP at 330cm were obtained

Table 21.3 Stratigraphy and dating of Bobundara Swamp and the lower part of the Narooma-Bermagui (NB3b) core. The dates from Bobundara are acid washed organic fines (<500μm) and those from NB3b are shell dates, from which a marine correction of 350 years should be subtracted.

consists of olive-grey fine calcareous sandy muds and muddy sands, with scattered shell fragments. It covers the period from 15,000 to *ca* 1700 calendar years ago (P. Roy pers. comm.), thus marking the transgression and deepening water across the Pleistocene-Holocene boundary. The lower part of the core (160–340 cm) thus reflects shallow nearshore marine conditions before 8000 BP which complement the open shore conditions encountered in the lowest part of the Bobundara Swamp record (Table 21.3).

The dates noted above were adjusted for marine carbon or carbonate effects and calibrated using CALIB 4.2 (Stuiver and Braziunas 1993). They provide an almost linear accumulation model for the NB3b core, averaging 2.35 cm per 100 calibrated years. The Bobundara estuarine phase accumulates at 18.35 cm per 100 calibrated years above 1125 cm, and the sands below this point may have built up even faster, since the dates from 1125 and 1345 cm overlap. The reducing clay content suggests that the accumulation rate slowed gradually to a linear 9.7 cm per 100 calibrated years in the lacustrine phase above 600 cm. However, the peat above 200 cm accumulated extremely quickly, averaging 38.05 cm per 100 calibrated years, but possibly accumulated even faster since forest clearance began.

PALYNOLOGY

Samples at approximately 20 cm intervals were subjected to standard pollen preparation methods. Pollen was extremely sparse in the marine core and in some samples from both the basal sands and near-surface peats at Bobundara. Microscopic charcoal was estimated using counts of the particles >5 μm as a proportion of the pollen sum, as well as the point count method of Clark (1982). Pollen diagrams (Figs 21.5 and 21.6) show the changes through time from the two sections, with the lowest levels being represented by the marine core (N3b on both diagrams).

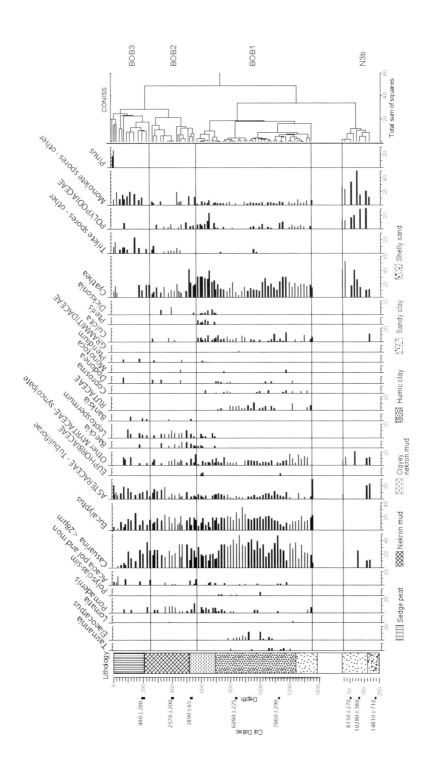

Fig. 21.5 Relaive pollen diagram of Bobundara Swamp and the NB3b lower section showing results of CONISS cluster analysis. The pollen sum is dryland taxa, excluding Melaleuca and aquatic herbs and grass pollen resembling that of Phragmites.

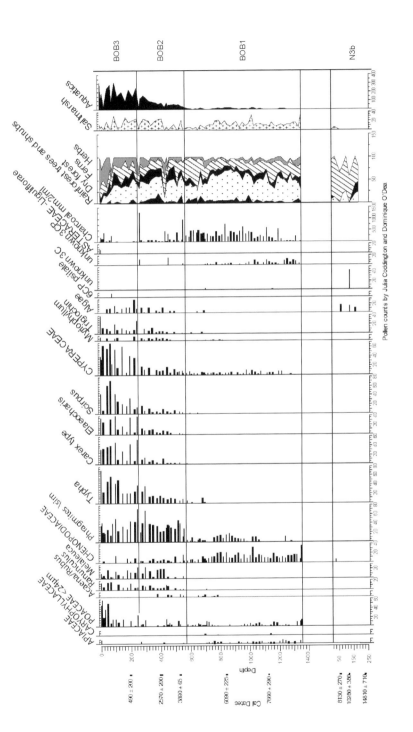

Fig. 21.6 Relative pollen diagram of Bobundara Swamp and the NB3b lower section showing vegetation type and changes over time. The pollen sum is dryland taxa, excluding Melaleuca and aquatic herbs and grass pollen resembling that of Phragmites.

CONISS analysis of common, non-aquatic pollen types distinguishes the estuarine muds and the marine core from the sediments above 570 cm (Fig. 21.5). At a lower level of significance the marine core is separated from the lower Bobundara section, and the upper peat phase is distinguished from the shallow lacustrine unit at 250 cm. Although aquatic pollen types were excluded from the analysis, they show significant responses at the same boundaries, indicating that changing aquatic environments might be affecting the influx of dryland pollen, particularly the input of pollen carried by streams.

NB3b (15,000–8130 cal yr BP)

The marine core is dominated by fern spores from rainforest gully and stream bank habitats, yet does not record any wet forest elements except possibly in the non-eucalypt Myrtaceae. The *Casuarina* and Asteraceae percentages are high, and may reflect riparian or estuarine margin taxa, since eucalyptus is entirely absent. This is supported by the presence of Chenopodiaceae in one sample, possibly derived from saltmarsh. The spectra are typical of those with water-carried pollen sources, and are in turn characteristic of marine samples. Charcoal is abundant in all samples. *Casuarina* declines and ferns increase after 10,300 cal BP, possibly reflecting changes in the vegetation or stream sources due to continuing transgression and climate change. Carbonised plant material is common in all samples.

BOB1 1380–570 cm (7600–4175 cal yr BP)

The sandy sediments at the base of the Bobundara core are derived from the final phase when the valley was open to the sea, a time when the over-steepened shoreline developed. A sand spit must have been forming at this time and finally cut off a large coastal lagoon, connected perhaps at high tide to Wallaga Lake. This long zone has lower levels of ferns such as *Blechnum* but significant wet forest elements such as *Elaeocarpus, Tasmannia* and possibly some Asteraceae and Rutaceae. Coddington (1983) reports occasional *Eucryphia* at 650 and 700 cm. With the decline in *Elaeocarpus* above 780 cm, *Pomaderris*, a wet forest shrub that responds to fire, increases. A ground fern of shaded forest, *Culcita dubia*, and the gully treefern, *Cyathea*, are present in large numbers, probably indicating a water-borne component. Dry forest taxa, particularly *Casuarina*, are prominent although *Eucalyptus* declines to a low at 1200 cm, being replaced by other Myrtaceae, a group that could include both dry and wet forest taxa. *Casuarina glauca* forms stands around Wallaga Lake at present, so at least some of this pollen type may indicate stands marginal to a developing saltmarsh. Chenopodiaceae, also indicating saltmarsh, are at a maximum in this zone, declining above 800 cm as other aquatic taxa such as *Typha* and sedges appear. *Phragmites*, an aquatic grass, becomes common above 940 cm, suggesting that a freshwater swamp may have been developing up valley as pollen inputs from the seaward saltmarsh diminished. Hystrichospheres and saline-tolerant snails indicate that the lake was saline or brackish throughout the zone. Charcoal is common throughout, with a relatively constant input above 1200 cm other than single peaks that may represent fire events.

BOB2 570–250 cm (4170–1010 cal yr BP)

Pomaderris remains at medium levels and *Casuarina* fluctuates in this zone. Asteraceae and shrub Myrtaceae increase while ferns decrease. This is consistent with disturbance of the forests by fire, although charcoal declines markedly in the middle of the zone. The aquatic and mire plants show a steady increase in taxa such as *Phragmites, Typha, Melaleuca,* sedges, *Myriophyllum, Ranunculus* and *Triglochin*. Shells are absent from the increasingly organic sediment and algae increase, but hystrichospheres are also present below 400 cm, indicating brackish episodes, although freshwater conditions with an expanding marginal swamp are gradually established by 400 cm. A *Phragmites* swamp around the inlet stream may have lowered the influx of water-borne pollen from streams. This may also explain the marked decline in charcoal early in the zone, with occasional peaks reflecting fire in the marginal swamps.

BOB3 250–0 cm (1010 cal yr BP–Present)

The rise in sedges, *Typha* and *Phragmites*, at the start of the zone and the presence of well preserved fibrous peat after 500 BP demonstrates that a marginal swamp expanded over the entire shallow lake during this phase, leaving only scattered ponds today. This would have changed the way in which pollen reaches the site, as surface water is only present after heavy rain. There is a steady decline in *Casuarina*, and other forest types with wet forest elements, except ferns, become rare. Dune vegetation inputs are indicated by *Banksia* and *Leptospermum*, while *Acacia* becomes more common at some levels. *Melaleuca* pollen, indicating the development of stands on the southern margin of the swamp, remains important until the most recent levels. Above 40 cm grass (non-*Phragmites*) increases greatly and *Eucalyptus* declines still further. Although *Pinus* is restricted to the top 20 cm, these changes may represent European clearing of the swamp margin. Pictures taken in 1890 show a much larger central lake, which today is reduced to two ponds. Charcoal, which is at very low levels since the swamp became vegetated, rises at this time, perhaps reflecting drainage and deliberate burning.

DISCUSSION

Environmental history

As noted above, each of the four zones represents a different environment of pollen deposition, from open sea, open to gradually enclosed estuarine lake, freshwater or brackish lake with marginal swamps, and finally closed tall graminoid peat swamp. This progression means that between-zone comparisons of the regional vegetation are difficult because each has different pollen and charcoal catchments. Given that many of the changes must be due simply to a progression from regional water-borne deposition to wind-carried and local swamp deposition, the evidence supports a picture of relative environmental stability in the area. *Eucalyptus* forest has probably remained the dominant type and there have been only minor changes in the extent of wet forest elements such as ferns. Although eucalypt pollen was not seen in the NB3b section, the upper Holocene does not differ markedly from the spectra shown here, and hence their presence can be assumed throughout. The only trend noted in the marine core is an increase in fern spore frequencies from Pleistocene to Holocene and a decline in *Casuarina*. This lack of obvious change around 10,000 years ago contrasts with evidence for increasingly warm and wet conditions further inland in New South Wales (*e.g.* Martin 1986, Hope 1994, Kershaw 1998).

The NB3b core was analysed for marine ostracods by Dr Iraj Yasseni, who found that a significant shift from warm water faunas commenced at about 12,700 cal years ago (P. Roy pers. comm.). By the Holocene-Pleistocene boundary at 10,350 cal BP, cold water faunas dominate, and continue to do so through the Holocene. These unexpected results suggest that in the glacial period, the average penetration of the warm eastern Australian boundary current was further south than today. The cause is unknown, but may reflect a reduced westerly flow of water into the Tasman Sea when a land bridge connected Australia to Tasmania. The lack of major change in the coastal vegetation (and the high level of diversity in the modern eucalypts) may thus be a measure of the stability caused by relatively mild maritime conditions near the coast during glacial times that changed to cooler water but warmer land conditions in the Holocene.

Within the Holocene the decline in moist forest may reflect slightly drier conditions after 3000 cal BP. However, *Pomaderris* has its best representation from 6000 to 2000 cal BP, somewhat later than in some Victorian and New South Wales sites (Kershaw 1998). *Pomaderris* may be as much a measure of fire frequency as climate, in which case its replacement of *Elaeocarpus* around 5900 cal BP may represent an increased frequency of large fires on the slopes of Mt Dromedary. Changes in the frequency of dry forests coincide with fern peaks; this too may reflect successional events following substantial fires. The microscopic charcoal record is strongly affected by taphonomic shifts, and can only be interpreted to show that fire has been a constant feature of both Pleistocene and Holocene vegetation. During the final swamp phase it is possible that the reeds were not commonly burnt, otherwise charcoal influxes would have been much higher. However, Aborigines are reported by eyewitness account to have burnt the swamp deliberately in the 1870s (N. Hoyer pers. comm.), therefore some fire peaks may be anthropogenic.

The best comparative record, from Holocene dune-dammed deposits at Kurnell, south of Sydney (Martin 1994), also shows relatively minor changes through the Holocene. Burning appears to increase at Kurnell after 5000 BP but the vegetation becomes more forested initially. After *ca* 2000 BP there is evidence of dune erosion and more seral vegetation, which Martin suggests may reflect more intense occupation. Other Holocene records from the east coast include Bumbo Lake (Tibby 1992), Lake Curlip (Ladd 1978), Loch Sport Swamp and Hidden Swamp (Hooley *et al.* 1980), Wilson's Promontory (Ladd 1979) and the southwestern coast of Victoria (Head 1988, 1989). These sites all record relatively stable dryland vegetation, although there is some evidence for a drying trend after 5000 years ago. Ladd (1979) found eucalyptus forest from before 13,000 BP on Wilson's Promontory but suggests that rainforest became more common from 11–7000 BP, while Hope (1974) found that elevated rainforest inputs continued on the west coast of the Promontory until *ca* 4500 BP. This supports the notion that coastal vegetation was fairly stable despite regional changes, even across the Pleistocene-Holocene boundary. Lake Bumbo, only 30 km north of Bobundara Swamp, appears to have slightly higher rainforest inputs prior to 3000 years ago, but this may have reflected changing source areas. This site remained a saline lake through its 5500 year history, although diatom studies show an increase in salinity after 4000 years, supporting the possibility of less freshwater flushing of the system (Tibby 1992). At Loch Sport Swamp and Hidden Swamp, closed dune-dammed hollows in southern Victoria, water levels were lower and the swamps became more saline after 4000–3000 BP.

The Lake Curlip sequence, about 150 km to the south, starts as a saline lake, but shallows rapidly and continues the successional trajectory via *Phragmites* swamp to a closed tall *Melaleuca ericifolia* thicket. This would have had a very low value as a resource area, as most other species are excluded or rare. The site received sediment from the floods of the large Snowy River, and hence had a short history of lacustrine and open swamp phases. At Wilson's Promontory, the inter-dune Cotters Lake has only recently become a sedgeland, with a probable reduction in freshwater resource (Hope and Coutts 1971). In western Victoria, Head (1988) found a relatively early appearance of freshwater lakes, and also evidence that the *Typha* reedswamps were burnt, probably as a result of deliberately lit fires, suggesting some manipulation of the hydrosere.

Resource availability

Fig. 21.7 outlines the four major phases that preceded the present swamp in the Bobundara area. From the point of view of resources for human settlement, the first phase of marine flooding in the late Pleistocene would have meant little change in access to marine resources as water advanced across a relatively flat topography. *Casuarina* seems to have formed prominent coastal woodlands, and temporary saltmarsh areas would have developed. Once the water approached modern sea-level, large numbers of calm bays would have appeared and shellfish such as oysters, *Anadara trapezia* and rock platform species would have become abundant. The death of vegetation exposed the soil to erosion and wave action over-steepened slopes around the bays.

As sea-level stabilised, sand spits and barriers started to form, cutting off embayments from wave activity and creating quiet estuaries or in some cases beach ridge plains. Many rock platforms tended to become buried in sand and quiet shallow water was invaded by saltmarsh or mangrove vegetation, bordered by *Casuarina glauca*. Continued growth of the sand barriers would have made them more durable by developing

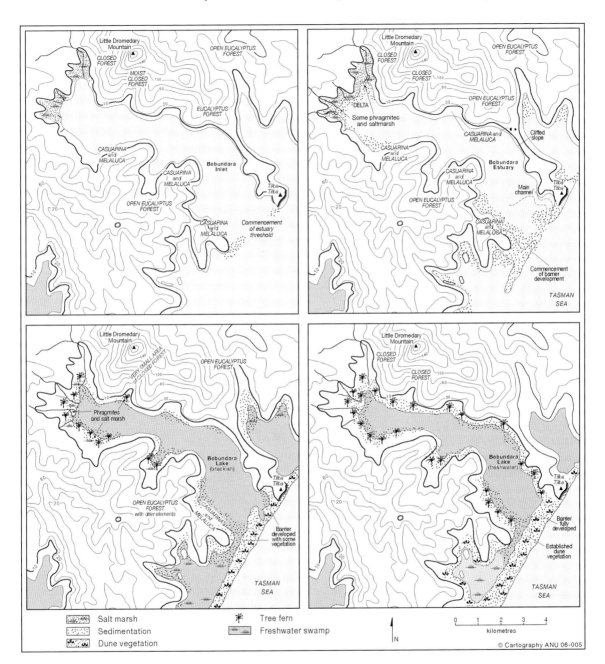

Fig. 21.7 Reconstruction of landscape changes around Bobundara Swamp.

flood tide deltaic deposits and dunes, and beach and dune vegetation resources would therefore have become available. The change from sandy to organic muddy sediment seems to have occurred by 7500 cal BP at Bobundara, suggesting that the barrier kept pace with continuing sea-level rise. After this time Bobundara Lake may have resembled the modern Wallaga Lake as a rich source of estuarine *Anadara trapezia*. This species has been recovered from 4 m depth on the northern margin of the swamp. Some saltmarsh and *Casuarina* may have occupied the shoreline. The permanent stream entering the lake may have been an important source of freshwater and access to the rainforest, as well as a source of brackish-tolerant *Phragmites* and *Bolboschoenus caldwellii* plant foods.

During the transition from estuary to freshwater lake, freshwater swamps expanded, presumably at the inland end of the lake. The development of a saltmarsh caused Bobundara Lake to be cut off from Wallaga Lake except during exceptionally high tides, and the lake was probably brackish or freshwater for most of the time after 3800 cal BP. After 2500 cal BP marginal swamps expanded. Disturbance of the barrier and dunes is evidenced by sand lenses in the lake at 2400 cal BP and bands of macro charcoal at 1500 cal BP. During this phase the lake would have been at its most productive for food plants and lake fauna, and had the potential to be a major source of starchy food in summer and autumn. The archaeological site was occupied at 400 cal BP, after the lake had partially infilled but when habitat diversity was possibly at its peak. People brought shells, probably from Wallaga Lake, to the site. They certainly burned parts of the swamp, and may have controlled encroachment by the next stage in the hydrosere – a scrub of *Melaleuca ericifolia*, ferns and cutting sedge. The dunes were used for human burial as well as a dry habitation site.

CONCLUSIONS

The southern coast of New South Wales has many types of estuaries, coastal lakes and scattered peatlands that have formed since sea-level rose to present levels, and Bobundara Swamp, one of the largest coastal peatlands, has formed as the latest successional stage from phases of an open inlet, closed estuary, saline lake and freshwater lake over the last 7500 years. A marine core, Narooma 3b, taken in 89 m of water, has extended the record by providing a general picture of coastal environments and fire history since the marine transgression after 14,000 BP. Aboriginal artefacts are common on the beach barrier enclosing the lake, and a burial has been dated at 350 BP, although pollen analysis has shown no distinctive changes in occupation history. The basin must have formed an important resource for Aboriginal groups and provided changing resources related to the stages of coastal development. People possibly manipulated vegetation successions through time, but the phase of swamp with freshwater ponds seems to have provided the highest resource diversity of staple plant starch sources and hunting opportunities.

Although the history and archaeology of the swamp lack resolution, the two records provide complementary evidence for change and human interaction with the environment. The time depth of human settlement of the area is certainly greater than that so far established, and the lake history clearly provides considerable evidence of the available resource potential through time.

ACKNOWLEDGEMENTS

We thank Mr Norm Hoyer of 'Sherringham' for bringing the site to our attention and, together with the Dibden family of 'Coolabah Park' and Mr Young of 'Ocean View', for information about past finds on their properties and the history of the swamp and dunes. Permission for the excavation was obtained from NSW National Parks and Wildlife Service and we benefited from the guidance of Gubboo Ted Thomas of Wallaga Lake community. Michael Green, Michael Hermes, Jutta Jaunzemis, Andrew MacIntyre, the late Peter May, Sharon Stokeld and members of several geomorphology classes provided help with fieldwork and analyses. Peter Roy kindly provided samples and allowed us to include unpublished data from the Narooma-Bermagui 3b core. Kari Barz identified fish remains, Wilfred Shawcross checked the artefactual identifications and Beth Gott provided information on potential plant food species.

REFERENCES

Attenbrow, V. 1976. Aboriginal subsistence economy on the Far South Coast of NSW, Australia. Unpublished BA dissertation, Department of Anthropology, University of Sydney.

Boot P. G. 1994. Recent research into the prehistory of the hinterland of the south coast of New South Wales, in M. Sullivan, S. Brockwell & A. Webb (eds) *Archaeology in the North*, 319–40. Darwin: North Australia Research Unit, Australian National University.

Bowdler, S. 1976. Hook, line and dilly bag: an interpretation of an Australian coastal shell midden. *Mankind* 10, 248–58.

Chapman, D. M., M. Geary, P. S. Roy & B. G. Thom 1982. *Coastal Evolution and Coastal Erosion in New South Wales*. Sydney: Coastal Council of NSW.

Clark, R .L. 1982. Point count estimation of charcoal in pollen preparations and thin sections of sediments. *Pollen et Spores* 24, 523–35.

Clark, R. L. 1983. Pollen and charcoal evidence for the effects of Aboriginal burning on the vegetation of Australia. *Archaeology in Oceania* 18, 32–7.

Coddington, J. J. 1983. Landscape evolution and its effect on Aboriginal occupation at Tilba Tilba, south coast New South Wales. Unpublished BA dissertation, Department of Geography, Australian National University.

Cribb, A. B. & J. W. Cribb 1974. *Wild Food in Australia*. Sydney: Collins.

Gott, B. 1982. Ecology of root use by the Aborigines of Southern Australia. *Archaeology in Oceania* 17, 59–67.

Gott, B. 1999. Cumbungi, *Typha* species: a staple Aboriginal food in southern Australia. *Australian Aboriginal Studies* 1, 33–50.

Hardwick, R. J. 2001. *Nature's Larder, a Field Guide to the Native Food Plants of the NSW South Coast*. Jerrabomberra: Homosapien Books.

Head, L. 1988. Holocene vegetation, fire and environmental history of the Discovery Bay region, southwestern Victoria. *Australian Journal of Ecology* 13, 21–49.

Head, L. 1989. Prehistoric Aboriginal impacts on Australian vegetation: an assessment of the evidence. *Australian Geographer* 20, 37–46.

Hooley, A. D., W. Southern & A. P. Kershaw 1980. Holocene vegetation and environments of Sperm Whale Head, Victoria, Australia. *Journal of Biogeography* 7, 349–62.

Hope, G. S. 1974. The vegetation history from 6000 B.P. to Present of Wilson's Promontory, Victoria, Australia. *New Phytologist* 73, 1035–53.

Hope, G. S. 1994. Quaternary vegetation, in R. Hill, R. (ed) *History of the Australian Vegetation*, 368–89. Cambridge: Cambridge University Press.

Hope, G. S. & P. J. F. Coutts 1971. Past and present aboriginal food resources at Wilson's Promontory, Victoria. *Mankind* 8, 104–14.

Howitt, A. W. 1904. *The Native Tribes of South-east Australia*. London: Macmillan.

Hughes, P. J. & R. J. Lampert 1982. Prehistoric population change in southern coastal New South Wales, in S. Bowdler (ed) *Coastal Archaeology in Eastern Australia*, 16–28. Canberra: Department of Prehistory, Australian National University.

Hughes, P. J. & M. E. Sullivan 1981. Aboriginal burning and late Holocene geomorphic events in eastern New South Wales. *Search* 12, 277–8.

Keith, D. & M. Bedward 1999. Vegetation of the South East Forests region, Eden, New South Wales. *Cunninghamia* 6, 1–218.

Kershaw, A. P. 1998. Estimates of regional climatic variation within southeastern mainland Australia since the last glacial maximum from pollen data. *Palaeoclimates* 3, 107–34.

Ladd, P. G. 1978. Vegetation history at Lake Curlip in lowland Eastern Victoria, from 5200 B.P. to Present. *Australian Journal of Botany* 26, 393–414.

Ladd, P. G. 1979. A Holocene record from the eastern side of Wilson's Promontory, Victoria. *New Phytologist* 82, 265–76.

Lampert R. J. 1971. *Burrill Lake and Currarong: coastal sites in southern N.S.W. Terra Australis 1*. Canberra: Department of Prehistory, Australian National University.

Luebbers, R. 1975. Ancient boomerangs discovered in South Australia. *Nature* 253, 39.

Mackaness, G. 1941. George Augustus Robinson's journey into southeastern Australia, 1844. *Journal of the Proceedings of the Royal Australian Society* 27, 318–49.

McKenna, M. 2002. *Looking for Blackfellas' Point, an Australian History of Place*. Sydney: UNSW Press.

Martin, A. R. H. 1986. Late glacial and Holocene alpine pollen diagrams from the Kosciusko National Park, New South Wales, Australia. *Review of Palaeobotany and Palynology* 47, 367–409.

Martin, A. R. H. 1994. Kurnell Fen: an eastern Australian coastal wetland, its Holocene vegetation, relevant to sea-level change and aboriginal land use. *Review of Palaeobotany and Palynology* 80, 311–32.

Mathews, R. H. 1904. Ethnological notes on Aboriginal tribes of NSW and Victoria. *Journal of the Proceedings of the Royal Society of New South Wales* 38, 203–381.

Mulvaney, J. & J. Kaminga 1999. *Prehistory of Australia*. St Leonards: Allen & Unwin.

Stuiver, M. & T. F. Braziunas 1993. Dataset for radiocarbon calibration. *The Holocene* 3, 289–305.

Sullivan, H. 1983. Aboriginal usage of the forest environment, in D. Byrne (ed) The Five Forests, Volume 2, 1–102. Unpublished report to NSW National Parks and Wildlife Authority.

Thom, B. G. & P. S. Roy 1985. Relative sea levels and coastal sedimentation in southeast Australia in the Holocene. *Journal of Sedimentary Petrology* 55, 257–64.

Tibby, J. 1992. The mid to late Holocene history and development of the Tuross Lakes area, south coast, New South Wales. Unpublished BA dissertation, Department of Geography, Monash University.

Tindale, N. B. 1974. *Aboriginal Tribes of Australia*. Berkeley: University of California Press.

Chapter 22

Late Holocene Landscape Change Around Bwindi-Impenetrable Forest, Central Africa: Human Dimensions in Palaeoecological Research from Tropical Wetlands

Robert Marchant

INTRODUCTION

The reconstruction of vegetation community dynamics from pollen preserved within sediments is a technique that has been in a continual state of development for the past 100 years. This development has witnessed the growth of regional reconstructions of vegetation change, where the data from different sites are combined and compared within a temporal framework, usually provided by radiocarbon dating. This chapter will be based around two themes relating to this technique. Firstly, pollen data from six sites in the Rukiga Highlands of Uganda are presented as regional maps of vegetation composition and distribution, and the changes in vegetation are discussed in the context of possible forcing mechanisms, and secondly, data from palaeoecological sources and information on cultural changes are discussed in terms of interactions that may have taken place.

THE STUDY AREA

The Rukiga Highlands are an area characterised by a deeply incised and steeply undulating topography that straddles the equator in southwestern Uganda (Fig. 22.1). Within the valleys a range of swamps have been formed, in part due to tectonically induced changes in drainage patterns, and it is under these swamps that sediments and pollen accumulate. The natural vegetation of the Rukiga Highlands is montane forest, the largest remnant of which occurs within Bwindi-Impenetrable Forest National Park. This forested area has benefited from protected area status since the early 1930s, initially as a Forest Reserve and more recently as a National Park and a UNESCO World Heritage Site (Fig. 22.1). The protected status relates principally to the resident mountain gorilla population that comprises around 35% of the global population. The majority of the Rukiga Highlands is dominated by intensive subsistence agriculture that supports very high population densities of 500 km^2 (Taylor *et al.* 1999), and this is manifested by intensive hillslope terracing (Fig. 22.2), which becomes progressively less intensive towards the present boundary of the National Park.

    The Rukiga Highlands have been a focus of palaeoecological study in central Africa since the pioneering

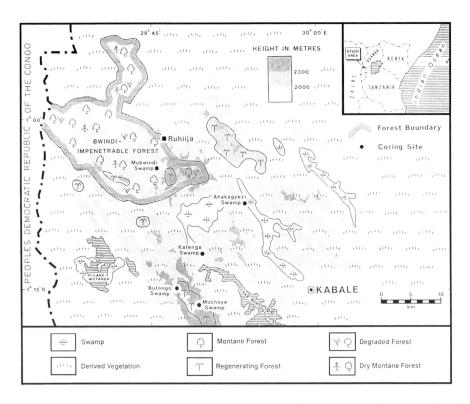

Fig. 22.1 The Rukiga Highlands showing regional relief and the location of swamps that have yielded pollen-based records of vegetation history. The present day vegetation of the Rukiga Highlands and boundary of Bwindi-Impenetrable Forest National Park are shown.

work of Morrison (1961) and Hamilton (1972). This was developed by (Taylor 1990, 1992, 1993) and more recently by Marchant *et al.* (1997). As a result of this early focus there are presently six pollen-based records (Fig. 22.1) that permit a regional reconstruction of vegetation dynamics. Crucially for this area of diverse topography, these sites span the altitudinal range of the Rukiga Highlands, and are from a range of different catchment types and sizes. Three of the sites (Ahakagyezi, Mubwindi and Muchoya Swamps) have a chronology provided by radiocarbon dating, which indicates that the sedimentary records extend beyond the last glacial period (Fig. 22.3) (Taylor 1990, 1992, 1993, Marchant *et al.* 1997), and importantly for the reconstruction presented here, these records are characterised by a high-resolution late Holocene component (Marchant and Taylor 1998). Of the sites so far studied, only the catchment for Mubwindi Swamp continues to support a dense cover of montane forest, being located within the present confines of the National Park. The combination of different locations, rapidly accumulating sediments that allow high-resolution analysis, and the application of radiocarbon dating allows a 3-dimensional reconstruction of the vegetation of the Rukiga Highlands to be produced for the late Holocene.

CULTURAL CONTEXT

From an archaeological viewpoint, the Rukiga Highlands is very poorly researched, with the majority of the

Fig. 22.2 Highly developed system of agricultural terraces from the Rukiga Highlands, southwest Uganda. Agricultural practice of inter-cropping in narrow strips has developed over a long period. The crops are harvested at different times and have different rooting depths preventing soil loss in an area of steep topography and high rainfall.

evidence for societal and cultural change being derived from locations outside the immediate study area. Particularly rich in archaeology is the area adjacent to the western shores of Lake Victoria, where sophisticated iron-smelting technology had developed by approximately 2500 to 2000 yr BP; this type of smelting has been closely associated with the Dimple Ware pottery style found throughout eastern and central Africa (Phillipson 1986). In Rwanda and Burundi, Dimple Ware has been found at twenty sites, seven of which are associated with slag heaps dated to approximately 1700 yr BP (Sutton 1971). Iron Age sites with Dimple Ware pottery *in situ* have also been described from Lolui and Ssese islands (Lake Victoria), where they have been dated to 1600 yr BP (Soper 1971). The areas experienced rapid cultural change around 1000 yr BP as communities became more settled and centralised states with systems of governance developed (Phillipson 1988).

The extent and legacy of this culture is exemplified in Mubende District, where a great line of earthworks is dated to between 650 and 500 yr BP (Robertshaw 1996). Within the Rukiga Highlands the evidence for cultural change comes mainly from anecdotal evidence. It is thought that iron working was a jealously guarded secret, only being carried out by certain clans – the BaSingola from Bwindi and the BaRunga and BaHeesi from Bukova, near Lake Bunyonyi (Turyahikayo-Rugyem 1942). Although there were plenty of iron ore deposits and forests for charcoal production, the smelting industry was restricted to these clans. Prominent iron workings close to the Rukiga Highlands are found at Kayonza (Baitwabobo 1972a), Nyabwinhenye in Bufumbia and Muko in Rubanda county (Turyahikayo-Rugyem 1942). Trade hinterlands are thought to cover Congo, Rwanda and western Tanzania (Kamuhangire 1972), with iron goods being traded over a large area using existing salt routes (Turyahikayo-Rugyem 1942).

Population influx and expansion of iron and agricultural technology led to the establishment of a widespread agricultural and pastoral community in the Rukiga Highlands (Rwandusya 1972). The BaTwa, BaTutsi and BaKiga are the three main cultural groups that live together, albeit in precarious harmony, being hunters, pastoralists and agriculturists respectively (Turyanhikayo-Rugyem 1942, Baitwabobo 1972b). However, the population has increased only relatively recently; by 150 yr BP the BaKiga were moving into Kigezi, driven by attacks from the BaTwa in the southwest and the BaTutsi in the south (Kagambirwe 1972), and the high population of today has been attributed to tribal wars in Rwanda and internal religious rising (Kagambirwe 1972). It has also been under pressure for a variety of reasons. For example, a rinderpest epidemic swept the area in the 1880s, a widespread famine occurred in 1897, and a series of

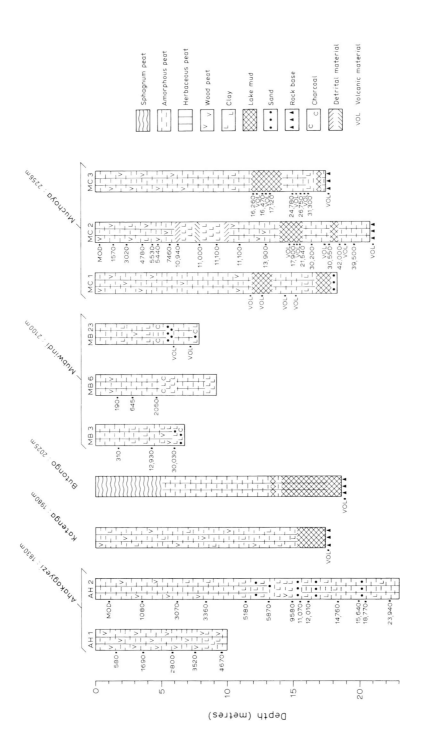

Fig. 22.3 Stratigraphic sequences from the Rukiga Highlands with associated radiocarbon chronology.

battles between the BaKiga and the BaTwa, particularly under the leader Basebya (who was killed in 1913), brought chaos to the southern parts of the area at the end of the nineteenth century (Turyahikayo-Rugyem 1942). A series of bitter wars with the BaTwa, quelled by the British administration in 1912, also caused animosity between different cultural groups.

METHODS

Sediments from the sites shown in Fig. 22.1 were abstracted using a combination of D-section and Hiller corers. The resulting stratigraphy and associated radiocarbon data are presented above in Fig. 22.3. Samples for pollen analysis from all locations were prepared using the standard preparatory technique (Faegri and Iversen 1989), and at least 500 pollen grains were counted for each sample, in line with the recommendation of Hamilton (1972). The pollen data were organised within six broad categories: degraded forest, derived vegetation, dry montane forest, montane forest, regenerating forest and swamp vegetation. The contribution of each of these was calculated as a percentage of pollen types originating from non-swamp inhabiting taxa (*i.e.* non-local species). These categories are depicted on a present-day vegetation map of the Rukiga Highlands (Fig. 22.1), which serves as a key for interpreting the vegetation distribution of earlier times. Maps have been produced for 2000, 800 and 0 yr BP, these periods depicting significant change in vegetation composition and distribution.

Derived vegetation described all land that is actively managed and includes farmed land and forest plantations. The main agricultural crops of the region are bananas, beans, white and sweet potatoes, finger millet (*Eleusine*), sorghum and wheat. Within the Rukiga Highlands, terracing is a common agricultural practice as it allows agricultural production on the steep slopes (Figs 22.2 and 22.4). Degraded forest includes significant amounts of taxa indicative of secondary forest, such as *Alchornea*, *Croton*, *Dombeya*, Ericaceae and *Hagenia*, in conjunction with large amounts of Poaceae pollen. Regenerating forest includes bamboo (dominated by *Synarundinaria alpina*) (Hamilton 1982), while broad-leaved species within this category are *Dombeya goetzenii*, *Macaranga kilimandscharica* and *Neoboutonia macrocalyx*. Other common taxa associated with regenerating communities are species of *Dodonaea*, *Plantago* and *Rumex*. Montane forest includes broadleaf, hardwood trees together with some conifers. This vegetation type is currently found in central Africa at altitudes between 1000 and 3000 m. Key taxa are *Celtis*, *Chrysophyllum*, *Croton*, *Drypetes*, *Faurea*, *Hagenia*, *Ilex*, *Nuxia*, *Olea*, *Podocarpus* and *Prunus*. It is the dominant type of vegetation found at present within Bwindi-Impenetrable Forest Swamp and varies considerably in composition, depending on altitude, climate and the nature of the catchment (Hamilton *et al.* 1986); at present, forest, sedges, grasses and even *Sphagnum* spp. can dominate uncultivated swamps.

Combining the symbols assigned to each category, it is possible to produce the mosaic of vegetation types represented by the pollen data. When there is within-category variation in vegetation composition that cannot be reconstructed with the limited number of categories, this will be described within the text. To place the reconstructions in context, the present-day composition and distribution of the vegetation within the Rukiga Highlands will be described first. It will then be used to aid in the reconstruction of past vegetation composition and distribution. The regional vegetation for the present day, using the classification outlined above, is shown in Fig. 22.1, and the vegetation distributions at 2000 and 800 yr BP are shown in Figs 22.5 and 22.6 respectively.

PRESENT DAY VEGETATION

The Rukiga Highlands are currently characterised by intensive agricultural activity (derived vegetation) with the steep hillsides being draped in a series of terraces on which crops of beans, Irish and sweet

Fig. 22.4 Under-developed system of agricultural terraces adjacent to Bwindi-Impenetrable forest National Park, the Rukiga Highlands, southwest Uganda. Agricultural activity abuts the National Park boundary.

potatoes, millet, sorghum and wheat are grown (Fig. 22.2). The valley bottoms often contain small banana and plantain plantations, and are also the main locations for settlement.

VEGETATION AT 2500 TO 1000 YR BP

Although there is an undated sedimentary hiatus at Butongo Swamp, sediments from all the swamps in the Rukiga Highlands are thought to record this time period (Fig. 22.3). On the strength of the similarly high amounts of pollen from degraded forest and derived vegetation, and the lack of stratigraphic change, it is suggested that the pollen records from Butongo and Katenga Swamps are contemporaneous with those radiocarbon dated to this period at Ahakagyezi, Mubwindi and Muchoya. The regional vegetation for 2000 yr BP is depicted in Fig. 22.5.

The pollen spectra from Ahakagyezi Swamp indicate that from 2000 yr BP until approximately 800 yr BP there was continued expansion of forest on the swamp surface; this was of a more mixed nature than previously recorded, but was still dominated by *Syzygium*. Within the catchment there was a dense covering of montane forest. The pollen record from the Mubwindi Swamp indicates that the montane forest was relatively closed, and a fringing band of relatively mesic forest dominated by *Syzygium* was present around the swamp margins (Marchant and Taylor 1998). The record from Butongo Swamp indicates forest disturbance, with increases in *Alchornea*, *Macaranga*, *Neoboutonia* and *Polyscias*, and concomitant decreases in *Ilex* and *Olea*. Soon after these fluctuations there is a rapid change in the vegetation on the swamp surface to one dominated by *Sphagnum* spp. The possible impetus for such a change in local vegetation is likely to result from a change in the nutrient status of the swamp, possibly due to deforestation within the catchment, which is thought to indicate a change to more ombrotrophic conditions (Hamilton *et al.* 1986). An anthropogenic cause has similarly been suggested for the initiation of peat formation in the British Isles (Moore 1973, 1975, Moore and Bellamy 1974) and Scandinavia (Hafsten and Solem 1976).

Within the Muchoya Swamp catchment there is a time-transgressive decrease in *Celtis*, Ericaceae, *Ilex*, *Olea* and *Podocarpus* stemming from approximately 2200 yr BP. This is mirrored by increases in taxa present within disturbed, open habitats such as *Dodonaea*, *Dombeya*, *Hagenia*, *Macaranga*, *Nuxia*, *Rumex* and *Vernonia*. These changes in the pollen spectra are linked with rising charcoal concentrations, which provide further evidence of possible human-induced clearance (Taylor 1992). This period appears to show

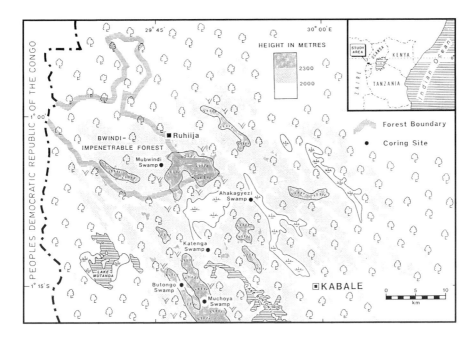

Fig. 22.5 Vegetation composition and distribution of the Rukiga Highlands at 2000 yr BP.

initial forest clearance within the Rukiga Highlands, the timing and location of which are contrary to previous evidence of clearances being detected from approximately 4000 yr BP (Hamilton *et al.* 1986). The period about 4000 yr BP marks a significant shift to a drier, more seasonal environment that has been detected throughout equatorial Africa (Burney 1993, Elenga *et al.* 1994, 1996, Marchant and Taylor 1998, Marchant and Hoogiemsra, 2004, Vincens *et al.* 1999). Therefore, the interpretation of clearance events within palaeoecological records must be carried out carefully – cause and effect relationships are complex and what may appear to have an anthropogenic origin can also stem from climate change.

VEGETATION AT 1000 TO 200 YR BP

The regional vegetation for 800 yr BP is shown in Fig. 22.6. From approximately 1000 yr BP there was an altitudinal change in the focus of vegetation degradation. Within the Ahakagyezi Swamp catchment there was a decline in components of established montane forest such as *Celtis*, *Olea* and *Podocarpus*, and within the Mubwindi Swamp catchment there was an increase in taxa from a more open habitat such as *Dodonaea*, Ericaceae, *Polyscias* and *Vernonia*. Although the forest around Mubwindi Swamp has remained intact, there is an indication of vegetation disturbance in close proximity, specifically recorded by increases in *Dodonaea* and *Vernonia* pollen and charcoal (Marchant and Taylor 1998). However, the dramatic changes in vegetation recorded at the other sites in the Rukiga Highlands are not characteristic of the Mubwindi Swamp catchment. Locations of this disturbance may have been the patches of bamboo forest within Bwindi-Impenetrable Forest. This vegetation association is currently found at similar altitudes to Muchoya Swamp and is thought to indicate an advanced stage of mature forest regeneration, supporting the suggestion of earlier forest clearance.

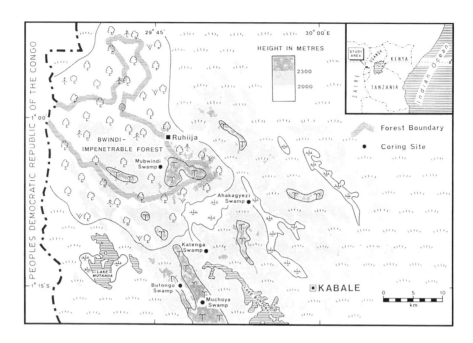

Fig. 22.6 Vegetation composition and distribution of the Rukiga Highlands at 800 yr BP.

The significant spread in forest clearance to lower altitudes appears concomitant with the abandonment of previously cleared high altitude sites, and is thought to be in response to a growing agricultural population; it appears that higher altitude sites were eventually abandoned, with land at lower altitudes increasingly being the focus of agricultural activity. It is not certain whether this process was concentrated within a specific time period, or occurred in a more diffuse manner. The shift of agricultural populations to lower altitudes has also been documented for the Rakai district to the west (MacLean 1995), which may have been caused by soil degradation at the highest altitudes, combined with a period of technological innovation.

A movement of settlement and farming from highland to lower altitudes may be why the Mubwindi Swamp catchment has managed to retain a virtually intact forest cover (Jolly *et al.* 1997). The decline in the *Syzygium* dominated swamp forest at Ahakagyezi Swamp may be significant in a transition of agricultural activity to lower altitudes. Schmidt (1980), notes that the wood of *Syzygium* is favoured as a source of charcoal for smelting in northwest Tanzania today. As such, the *Syzygium* decline may therefore correspond with a need to seek out new resources in order to maintain the developing iron industry (Taylor and Marchant 1995), particularly as up to 2000 kg of charcoal are required for a single smelt (Kamuhangire 1972).

As a caveat to such a picture of anthropogenically driven change, increases in aridity or seasonality of climate may also have played a role in changing the forest composition. Stager *et al.* (1997) report a sharp decrease in the precipitation:evaporation ratio affecting Lake Victoria from around 600 years ago, and it is possible that reduced levels of effective precipitation may have favoured certain taxa, for example *Podocarpus* and *Celtis*. The timing of this change corresponds with a peak in the amount of *Podocarpus* and taxa that favour drier soils, and a very noticeable peak in charcoal at Mubwindi Swamp (Marchant and

Taylor 1998). However, the very dramatic changes in vegetation are likely to result from human activity. The area around the Mubwindi Swamp catchment did not experience the wide-scale clearance recorded at the other sites within the Rukiga Highlands, and farther afield, although there has been a degree of disturbance.

VEGETATION AT 200 YR BP TO THE PRESENT DAY

Changes in montane forest composition within the Mubwindi Swamp catchment at approximately 200 yr BP may represent the onset of marginally more humid climates. Nicholson (1996) and Stager *et al.* (1997) suggest that levels of effective precipitation in central Africa increased from around 200 years ago. These possible climate-induced changes in the vegetation are not recorded elsewhere in the Rukiga Highlands as a result of forest clearance. However, in view of the increased abundance of taxa representing gappy and regenerating forest, together with the relatively recent radiocarbon date, perhaps the most likely cause of changes in pollen spectra was the selective felling of trees for timber. Significantly, this time period includes the period of the First World War, when Uganda Forest Department data record a heightened interest in *Podocarpus* from Bwindi-Impenetrable Forest as a source of wood for rifle stocks (Marchant and Taylor 1998).

Another possible factor could be the increased disturbance of the vegetation by elephants (*Loxodonta loxodonta africana*), which can be a significant factor in determining vegetation composition (Buechner and Dawkins 1961, Butynski 1986, Plumtree 1994). This effect may have been magnified due to the increased concentration of the animals into the Mubwindi Swamp catchment (where the animals are concentrated at present during the two dry seasons) by regional forest clearance. Alternatively, the floristic change could represent an increased input of pollen from a wide range of degraded habitats within the Rukiga Highlands, thus diluting the input of forest pollen from within the Mubwindi Swamp catchment. More recently, the driving force in determining forest composition within Bwindi-Impenetrable Forest was one of human activity, the main control being the activities of pit-sawyers that commenced around 100 years ago. Although there is now a ban on pit-sawing within the forest, the gaps originally opened up appear to be maintained by elephants and other, more vigorous, grazing animals.

Despite these relatively minor perturbations, the forest within the Mubwindi Swamp catchment has remained intact. There are a number of possible causes for the persistence of Bwindi-Impenetrable Forest when other areas have been a focus of forest clearance. These include: (a) the ground is more dissected and potentially less favourable to agriculture, (b) the area presently covered by the forest was furthest from the main influx of agriculturists and simply therefore the last area to be reached, or (c) there has been some degree of protection by the (until recently) resident indigenous BaTwa population. As regards the first possibility, there is no evidence for great differences in levels of soil fertility or erodibility. A highly developed system of terraced agriculture is present throughout the Rukiga Highlands (Fig. 22.2) on slope angles comparable to those retained within Bwindi Impenetrable Forest. Thus this possibility does not appear to have been likely in the case of late Holocene forest clearance. Unfortunately, there are no direct archaeological reports from the Rukiga Highlands, although there is a range of information from the regions surrounding the study area, indicating a well-developed agricultural population with iron-smelting technology by 2000 yr BP. The information also indicates that population movements have occurred from all geographical locations surrounding the present-day boarders of Bwindi-Impenetrable Forest, hence the second possibility may not have occurred.

As for the third possibility, a precursor of Bwindi-Impenetrable Forest may have corresponded to some form of disputed border between territories, and the associated economic and political instability would have placed severe limitations on the development of sedentary agriculture. Such a border may have related to the highly centralised societies in the Great Lakes Region between Lake Victoria and Western Rift Valley (Taylor *et al.* 1999, 2000). These commenced during the early part of the present millennium (Robertshaw 1996), and it is possible that any forests which were intact at the time of state formation and located towards the outer limits of the kingdoms' sphere of influence were 'protected' as a natural deterrent to potential invaders. There may also have been some degree of protection imparted by the BaTwa against the colonising agriculturists. Although the degree of this protection cannot be quantified by this research, one prominent BaTwa (Bandusya) often laments that his forefathers were once mortal enemies of the pioneering agriculturists, keeping all strangers out of the forest (Kingdon 1990).

Whatever the result of the cultural contacts between different groups in the Rukiga Highlands, we can only suggest what may have occurred, as modern analogues are unknown. What is quite obvious from the palaeoecological records, however, is that agricultural land use had spread rapidly since approximately 2000 yr BP within the Rukiga Highlands, with some protection being afforded to Bwindi-Impenetrable Forest. This protection has continued to present day, now under the guise of National Park legislation.

CULTURAL INTERCONNECTIONS WITH HOLOCENE CLIMATE CHANGE – A NECESSARILY COMPLEX MODEL

Investigating past societies in recent prehistory starts by examining patchy observations such as orally transmitted memories, written accounts and material artefacts, which are then contextualised, and making associations and inferences, so that they become more coherent. This is done according to the rules of the currently dominant paradigm of reductionism-empiricist science, although this is by no means the only way; many people did, and do, contextualize the world around them on the basis of different rules. A variety of theories have emerged to put the archaeological and historical findings and past environmental reconstructions into a more universal, abstract perspective. However, a complete and satisfactory theory of human-environment interactions has not yet been constructed, and it probably never will be due to the myriad connections between and within the cultural-environmental system, many of these being specific to a given locality or region. Nevertheless, it is worthwhile inspecting the building blocks linking cultural and environmental change, some of which come from the natural sciences, some from the social sciences. Some theories invoke an external agent driving social change, such as domination by another people and/or possible deterioration of essential environmental resources, while others have sought the explanation in some sort of endogenous change.

Hierarchies have emerged to specify human impact; for instance, increased deforestation is a 'natural' consequence of intensification of agriculture and increasing population density. In the latter case, correlation is easily taken for explanation – a possibly over-simplistic view. As the majority of historical and palaeoclimatic research has been performed by European and North American researchers, and ultimately set by their research agendas, these approaches often work under *a priori* assumptions that humans are controllers/modifiers of the environment, driven by a need to amass material wealth and resources, in short that human impact equates to deforestation. Such assumptions are explored within this section.

There are many open questions about the first steps in which humans attempted to exert (probably not pre-conceived) more control over their environment. Relevant for the present study are the following:

– which factors made certain environments feasible and attractive for human habitation
– how did agricultural cropping, animal herding and the use of trees for wood and fodder start
– was agriculture a response to deteriorating conditions for a gathering-hunting way of life
– were the first farmers clearing more or less pristine forests, for crops, wood and fodder, and in the process causing more widespread environmental change

– are there traces of environmental feedback as a consequence of human interference with natural processes?

Environmental zones such as river basins or high elevation areas where the natural vegetation may not be so dense are thought to have been favourable settlement sites (Taylor and Marchant 1995). In the first stages of agricultural development, the geography of a habitat must have been an important determinant. However, explanations of socio-political complexity from geographical features such as hills, rivers, soils or vegetation is a difficult paradigm to adopt. Resources such as good soils, pasture, food supply, sources of water and mineral deposits would have contributed to the earliest stages of the unfolding socio-political complexity and consequent conflicts. Good places to shelter or materials to build dwellings and make clothes are other necessary 'resources'. Organic resources such as reeds for dwellings, fibres such as cotton and hemp for textiles, and animal skins and bones for a variety of purposes are usually closely related to land-use activities. The combination of environmental diversity and fertile soils allowed multi-cropping, and subsistence prospects improved further due to the excellent grazing afforded by deforestation. A decisive factor in how these processes interact would have been the degree and character of social organisation, and superimposed on this would be the backdrop of environmental change; humans had to adapt in one way or another to their environment and to its slow, or sudden, changes.

A significant problem in palaeoecological research is attempting to distinguish between natural and human-induced change, particularly when there is little or fragmented (spatially or temporally) evidence for cultural change. However, if information is available from archaeological sources we are able to be more specific with our interpretations, and in turn our research questions. Unfortunately, the study region is not rich in archaeological or anthropological research, although there is information for other locations nearby indicating that changes detectable from the pollen data would have been driven by human impact. For example, there is evidence that fire has been widely used to facilitate hunting and clear vegetation for animal husbandry and agriculture (Marchant 1997). The exploitation of forests for wood and the processing of ore and salt have also left their impact on the landscape (Marchant 1997). The initial clearance at high altitude and subsequent migrations may be driven by the need to seek natural resources, but such an explanation must be tempered with the developing socio-political complexity.

Feedback between humans and the environment can take many forms. In some cases it has been direct and clear, for instance erosion as a consequence of forest clearance. Sometimes there has been a delayed effect, such as a change in regional climate, particularly moisture regime, following vegetation change. For example, one of the main drivers of recent environmental change in the Rukiga Highlands area is land-cover change at low altitudes, where large-scale drainage of swamps and subsequent conversion to agricultural land appears to have resulted in a reduction of local atmospheric moisture in the form of early morning mists that were characteristic of the steep topography. A consequent impact of this has been a steepening of the climatic lapse rate (the rate at which climate changes with altitude). Sometimes such feedbacks are even more complex and indirect, particularly when large investments are made to control the environment, such as the highly developed system of agricultural terraces (Fig. 22.2). Under such circumstances the population becomes more vulnerable to environmental changes such as weather extremes or climate change.

The concept of resilience has been introduced to ecology as a measure of the degree to which societies are able to withstand such environmental change. The way in which peoples perceived and interpreted environmental change, the appropriateness of these interpretations and their ability for innovative and practical responses are crucial elements in a system's resilience. However, more refined notions of stability and resilience are needed that can be adapted to local situations. By way of contrast to conventional, equilibrium-based analyses, complex systems approaches emphasise the fact that ecological and human systems are often in transient states, are inherently non-linear and are metastable. Within these stable domains, the system may fluctuate wildly, but as long as it remains within the boundaries of the domain, it is resilient, and is thus able to persist despite a high degree of disturbance.

CONCLUSIONS

The record of vegetation history supplied by wetland sediments in the Rukiga Highlands indicates that superimposed upon what appears to have been continuous variations are three relatively important changes in the composition of montane forest. The first, and indeed, the precursor for future changes, appears to centre on 2000 yr BP. This human-induced forest clearance was focused at the highest altitudes and was relatively restricted in its areal extent. At around 800 yr BP, there was a shift in the focus of human activity to lower altitudes, which was also accompanied by a shift to a less humid climate. The third period of change occurred approximately 200 yr BP when there appears to have been selective utilisation of trees within the present confines of Bwindi-Impenetrable Forest. More recently, beginning approximately 80 yr BP, some recovery of forest may have taken place as a result of a reduction in the activities of pit sawyers.

The vegetation record indicates that a precursor of the present Bwindi-Impenetrable Forest has been in existence throughout the late Holocene, during which it has been only moderately affected by human activity. According to pollen data from a number of sites in the Rukiga Highlands, other areas of lower montane forest have not fared so well. Taken together, the evidence indicates that the onset of forest clearance in western Uganda during the late Holocene was catchment specific. The evidence from the Rukiga Highlands thus serves as a warning against attempts to extrapolate evidence of the timing of forest clearance from one, or a limited number, of pollen based studies across broad expanses of central Africa, or indeed other locations.

The designation of Bwindi-Impenetrable Forest as a Forest Reserve in the early 1930s, and more recently as a National Park and World Heritage Site, can explain its persistence under considerable pressure from increasing land shortages during the present century. The reasons why Bwindi-Impenetrable Forest retained forest cover throughout the late Holocene are not apparent from this study, although as well as being a driver of vegetation change, it appears that human populations may be responsible for the preservation of the forest resource. For example, a distinguishing feature of the current forested area is a pygmy population (BaTwa) who until recently were resident within the forest, and it is suggested that this population may have been important in maintaining the coherence of the forest prior to legislation.

ACKNOWLEDGEMENTS

The work presented in this study was supported by the University of Hull, the Bill Bishop Memorial Fund, the British Institute in East Africa and the Natural Environment Research Council. Thanks are due to the Government of Uganda for research permission, to Myooba Godfy for his invaluable help and company, and to the staff and students of the Institute of Tropical Forest Conservation field station at Ruhiija. Keith Scurr is thanked for production of the diagrams.

REFERENCES

Baitwabobo, S. R. 1972a. Foundations of the Rujumbura society, in D. Denoon (ed.) *A History of Kigezi in South-west Uganda*, 35–61. Kampala: National Trust.
Baitwabobo, S. R. 1972b. Bashambo rule in Rujumbura, in D. Denoon (ed.) *A History of Kigezi in South-west Uganda*, 73–92 Kampala: National Trust.

Buechner, H. K. & H. C. Dawkins 1961. Vegetation change induced by elephants and fire in the Murchison Fall National Park, Uganda. *Journal of Ecology* 42, 752–66.

Burney, D. A. 1993. Late Holocene environmental change in arid south-west Madagascar. *Quaternary Research* 40, 98–106.

Butynski, T. M. 1986. Status of elephants in the Impenetrable Forest, Uganda. *African Journal of Ecology* 23, 189–93.

Elenga, H., D. Schwartz & A. Vincens 1994. Pollen evidence for Late Quaternary vegetation and inferred climate changes in Congo. *Palaeogeography, Palaeoclimatology, Palaeoecology* 109, 345–46.

Elenga, H., D. Schwartz, A. Vincens, J. Bertaux, C. Denamur, L. Martin, D. Wirrmann D. & M. Serrant 1996. Holocene pollen data from Kitina Lake (Congo): palaeoclimatic and palaeobotanical changes in the Mayombe Forest area. *Comptes Rendus dé l'Académia des Sciences Serie II Fasicule A – Sciences de la Terre et des Planètes* 323, 403–10.

Faegri, K. & J. Iversen 1989. *Textbook of Pollen Analysis*, Fourth Edition. Chichester: Wiley.

Hafsten, U. N. & T. Solem 1976. Age, origin and palaeoecological evidence of blanket bogs in Nord Trondelag, Norway. *Boreas* 5, 119–41.

Hamilton, A. C. 1972. The interpretation of pollen diagrams from highland Uganda. *Palaeoecology of Africa* 7, 45–149.

Hamilton, A. C. 1982. *Environmental History of East Africa*. London & New York: Academic Press.

Hamilton, A. C., D. Taylor & J. Vogel 1986. Early forest clearance and environmental degradation in southwest Uganda. *Nature* 320, 164–7.

Jolly, D., D. Taylor, R. Marchant, A. Hamilton, R. Bonnefille, G. Buchet & G. Riollet 1997. Vegetation dynamics in central Africa since 18,000 yr. BP: pollen records from the interlacustrine highlands of Burundi, Rwanda and western Uganda. *Journal of Biogeography* 24, 495–512.

Kamuhangire, E. R. 1972. *Causes and Consequences of Land Shortage in Kigezi*. Makerere: Department of Geography Occasional paper 23.

Kingdon, J. 1990. *Island Africa: the evolution of Africa's rare animals* and *plants*. London: Collins.

MacLean, M. R. 1995. Late Stone Age and Early Iron Age agriculture settlement patterns in Rakai District, southwestern Uganda, in J. Sutton (ed.) *The Growth of Farming Communities in Africa from the Equator Southwards*, 121–145 Nairobi: The British Institute in Eastern Africa.

Marchant, R. A. 1997. Late Pleistocene and Holocene history of Mubwindi Swamp, southwest Uganda Unpublished PhD thesis, University of Hull.

Marchant, R. A., D. M. Taylor & A. C. Hamilton 1997. Late Pleistocene and Holocene history of Mubwindi Swamp, southwest Uganda. *Quaternary Research* 47, 316–28.

Marchant, R. A. & D. M. Taylor. 1998. Dynamics of montane forest in central Africa during the late Holocene: a pollen-based record from southwest Uganda. *The Holocene* 8, 375–81.

Marchant, R. A. & H. Hooghiemstra. 2004. Rapid environmental change in tropical African and Latin American about 4000 years before present: a review. Earth Science Reviews 66, 217–260.

Moore, P. D. 1973. The influence of prehistoric cultures upon the initiation and spread of blanket bog in upland Wales. *Nature* 241, 350–4.

Moore, P. D. 1975. Origin of blanket mires. *Nature* 245, 267–9.

Moore, P. D. & D. J. Bellamy 1974. *Peatlands*. London: Elk Sciences.

Morrison, M. E. S. 1961. Pollen analysis in Uganda. *Nature* 190, 483–486

Nicholson, S. E. 1996. Environmental change within the historical period, in W. M. Adams, A. S. Goudie & A. Orme (eds) *The Physical Geography of Africa*, 41–53 Oxford: Oxford University Press.

Phillipson, D. W. 1986. Life in the Lake Victoria Basin. *Nature* 320, 110–11.

Phillipson, D. W. 1988. *African Archaeology*. Cambridge: Cambridge University Press.

Plumtree, A. J. 1994. The effects of trampling damage by herbivores on the vegetation of the Parc-National-Des-Volcans, Rwanda. *African Journal of Ecology* 32, 115–29.

Robertshaw, P. 1996. Archaeological survey, ceramic analysis, and state formation in western Uganda. *The African Archaeological Review* 12, 105–31.

Rwandusya, Z. 1972. The origin and settlement of the people of Bufumbria, in D. Denoon (ed.) *A History of Kigezi in South-west Uganda*, 85–102. Kampala: National Trust.

Schmidt, P. R. 1980. Early Iron Age settlements and industrial locales in West Lake. *Tanzania Notes and Records* 84, 77–94.

Schoenburn, D. L. 1993. We are what we eat. Ancient agriculture between the Great Lakes. *Journal of African History* 34, 1–31.

Soper, R. C. 1971. Iron Age archaeological sites of the Chabi Sector of Murchison Falls National Park, Uganda. *Azania* 6, 53–8.

Stager, J. C., B. Cumming & L. Meeker 1997. A high-resolution 11,400-yr diatom record from Lake Victoria, East Africa. *Quaternary Research* 47, 81–9.

Sutton, J. E. G. 1971. Temporal and spatial variability in African iron furnaces, in K. Haaland & P. L. Shinnie (eds) *African Iron Working, Ancient and Traditional*. Bergen: Norwegian University Press.

Taylor, D. M. 1990. Late Quaternary pollen records from two Ugandan mires: evidence for environmental change in the Rukiga Highlands of southwest Uganda. *Palaeogeography, Palaeoclimatology, Palaeoecology* 80, 283–300.

Taylor, D. 1992. Pollen evidence from Muchoya Swamp, Rukiga Highlands (Uganda), for abrupt changes in vegetation during the last *ca.* 21,000 years. *Bullétin de la Société Geologie de France* 163, 77–82.

Taylor, D. M. 1993. Environmental change in montane south-west Uganda: a pollen record for the Holocene from Ahakagyezi Swamp. *The Holocene* 3, 324–32.

Taylor, D. & R. Marchant 1995. Human-impact in south-west Uganda: long term records from the Rukiga Highlands, Kigezi. *Azania* 30, 283–95.

Taylor, D. M., R. Marchant & P. Robertshaw 1999. A sediment-based history of medium altitude forest in central Africa: a record from Kabata Swamp, Ndale volcanic field, Uganda. *Journal of Ecology* 87, 303–15.

Taylor, D. M., P. Robetshaw & R. A. Marchant 2000. Environmental change and political upheaval in precolonial western Uganda. *The Holocene* 10, 527–36.

Turyanhikayo-Rugyem, B. 1942. The History of the Bakiga in South-west Uganda and Northern Rwanda Between About 1500 and 1930. Unpublished PhD thesis, University of Michigan.

Vincens, A., D. Schwartz, H. Elenga, I. Ferrera, A. Alexandre, J. Bertaux, A. Mariotti, L. Martin, J. D. Meunier, N. Nguetsop, M. Servant, S. Servant-Vildary, and D. Wirrman 1999. Forest response to climate changes in Atlantic Equatorial Africa during the last 4000 years BP and inheritance on the modern landscapes. *Journal of Biogeography* 26: 879–885.

Chapter 23

Reflections on Wetlands and their Cultural and Historic Significance

Malcolm Lillie, Stephen Ellis, Helen Fenwick and Robert Smith

> world wetlands cover *c.* 4–6% of the earth's land surface, [and whilst] the rate of wetland loss is not really known...losses of *c.* 90% have occurred in New Zealand and 60% have occurred in China. (Mitsch and Gosslink 1993, 35–9)

At the start of this volume we emphasised that the significance of waterlogged deposits for the preservation of the organic part of the cultural record could not be overstressed. It was also noted that the integration of wetland context and the myriad permutations of that context were only recently being realised by both the archaeological community and by those involved in wetland preservation and/or conservation.

The potential effects of wetland losses are only recently beginning to be quantified to any real extent (*e.g.* Mitsch and Gosslink 1993), and the implications for the culture-historic portion of the record remain poorly understood. This has been vividly highlighted by recent discourse relating to the current state of existing approaches to the management of the historic environment. To date this debate has been related to contexts ranging from local, through pan-European (*e.g.* Coles and Olivier 2001, Coles 2004, Fischer *et al.* 2004, Olivier 2004), and up to the global scale (*e.g.* Ramsar 1971, Mitsch and Gosslink 1993, Bernick 1998, Gumbley *et al.* 2005), at differing levels of analysis.

Wetlands, and their context, necessitate investigation at the regional level hence the regional issues element of the title of this work, the agenda however is global in that the threats to wetlands whether through agriculture (Williams 1993, Richards 1993), drainage (Coles 1993) or other vectors of change, severely compromise world wetland environments. Whilst the heritage element of the wetlands record will always be at the forefront of archaeological investigations, this element is one aspect of a diverse environmental context which now ensures that wetlands are recognised as important, albeit severely threatened, landscapes. For clarity, the following discussion presents a brief resumé of the aims and themes of this volume, considering the topics covered and assessing how these relate to global perspectives on wetlands.

It is now generally recognised that the impacts of de-watering in relation to the heritage dimension of wetlands, and global wetland ecosystems in general, are synergetic (Coles 2004, 185, Olivier 2004, 156) and detrimental to the wetland and its contained archive. As a consequence, global management strategies now encompass entire systems, as both natural and human activities impact upon them (*e.g.* Coles 2001, Crisman *et al.* 2001).

This volume has presented papers that reinforce the importance of wetland ecosystems in terms of their heritage and environmental component (*e.g.* Pryor, Crone and Clarke, Bermingham, Panter and Pasley, *this volume*). It has also outlined a number of current approaches to this record in terms of understanding burial contexts and developing strategies for the *in situ* management of the archaeological component and holistic approaches to wetlands. (*e.g.* Corfield, Lillie and Nicholas, *this volume*).

From the archaeological perspective, recent investigations in the Åmose region of Denmark (Fisher *et al.* 2004, 199), citing Ahlström and Jensen (2003), have asserted that decomposition 'comes to an end' when the cultural layers being compromised are permanently saturated by groundwater. This assertion is apparently upheld by the work of Corfield (1996, *this volume*) and Chapman and Cheetham (2002), amongst others.

However, despite the above observation, the introduction to this volume and the papers that follow all demonstrate the complex nature of wetlands, whether they be a mire, bog, loch, river, or a more discrete context such as a ditch, well or pit. In each environment the specific waterlogged context differs; the sedimentological and environmental matrix forms at differing rates and with different internal and external influences. In addition, human modification or interaction varies (*e.g.* Nicholas *this volume* a and b), and within the wetland (or waterlogged context) the cultural remains themselves can result in the formation of discrete micro-environments directly associated with the structure or object contained therein. This complexity is inherent at the local through to the global scale within the wetland environment, and as such, only holistic approaches to each level within the environment will facilitate the development of a generic understanding of context.

The fact that wetland context is complicated has, perhaps, resulted in some inherent limitations in the development and advancement of scientific research into wetlands, even where apparently specific questions are being investigated. During recent research, Smith (2005, 189) artificially manipulated water levels in aerobic lysimeters containing both peat and waste leachate, which resulted in changes in the level of saturation of the sediments, along with the sediment redox potential and the dissolved oxygen content of the water. This study has led to a greater understanding of the complex interactions that occur between the parameters discussed, and the cumulative effect that they have upon the bacterial degradation of oak wood. This is a situation which previous researchers have strongly advocated (*e.g.* Brunning *et al.* 2000, Hogan *et al.* 2001, Powell *et al.* 2001, Chapman and Cheetham 2002, Cheetham 2004), and one which has provided important insights into the influence of dissolved oxygen and pH on oak wood degradation (Smith 2005, 189–90).

In particular, variation in the redox potential and pH of the burial environment can cause changes within the bacterial community (AFRC Institute of Food Research 1988, Barlaz *et al.* 1989, Barlaz 1997), and Smith (2005, 191) has shown that bacterial population shifts subsequently alter the decomposition of the organic fraction within the lysimeters containing waste material. In contrast to the above suggestion that saturation stops degradation, Smith (2005) has shown that a complex series of interactions can influence oak wood degradation, and that the reaction of the burial environment to various external influences can differ depending on the material contained within the waterlogged context.

In view of these findings, the preservation of biogenic material, often for millennia in wetland contexts, is all the more significant. The quality of such preservation can be exceptional, as encountered by Zhilin (1999, *this volume*) in the Volga and Oka regions. Here, finds of bone and antler, including fish hooks, net fragments with parts of pine bark float in association, net sinkers and conical fish traps, are significant in providing important insights into Mesolithic culture and technology. All of these finds add an important dimension to our understanding of Mesolithic resource procurement strategies in Europe (Bonsall *et al.* 1997, Lillie and Richards 2000, O'Connell *et al.* 2000, Eriksson *et al.* 2003, Lillie *et al.* 2003), and in particular the contributions of freshwater fish resources. Perhaps of even more significance in this context is the suggestion that a fragment of basket-like fish trap found in the upper Volga (at the site of Stanovoye 4 at the Podozerskoye peat bog) and dated to *c*. 9250–8450 cal BC (GIN-10125 II and GIN-10125 I) represents the earliest find of its type in Europe.

Moving from the artefactual record and site-specific contexts, Larsson's contribution (*this volume*) is important in highlighting the fact that wetland contexts often provide us with a long chronological perspective on human-landscape interactions. In Scandinavian prehistory, wetlands were important because of their prevalence in the landscape, and the greater temporal resolution afforded by waterlogged artefacts and ecofacts enables a more holistic understanding of societal trends, ritual deposition, and changes in articulation between society and landscape. When touching on one of the more enigmatic aspects of prehistoric societies, the role of ritual, Larsson provides an important observation in relation to the current discussion, in that wetland development is dependent on a wide range of factors, such that even in the same landblock, wetlands can exhibit diverging developmental histories. This is the case not only in Scandinavia but also in other regions of the world where this degree of preservation is in evidence.

Following Larsson's holistic approach to landscape development Halkon and Fenwick present differing approaches to landscape study with overviews of investigations of landscape developments in the Foulness valley and Lincolnshire Marsh regions of England. The former study provides an overview of landscape change and human-landscape interactions from the earlier Holocene through to the Roman period, investigated using archaeological and palaeoenvironmental data. In contrast, Fenwick uses documentary and cartographic evidence to reconstruct landscape reclamation in the Lincolnshire Marsh from the Medieval period onwards. The nature of the landscapes being investigated necessitates differing approaches to their study; further reinforcing the suggestion that a wide range of techniques and approaches are required to understand the past and the role of humans in the past.

The research presented by Bulbeck and colleagues (*this volume*) also highlights the global diversity in wetlands, in that, despite the assertion that archaeological remains rarely preserve well in humid equatorial regions, the site of Utti Batue is well-preserved in its particular context. The importance of this site lies in its identification as the probable palace centre of Luwu, which is associated with the oldest and most prestigious Bugis kingdom in Sulawesi, dated to the fifteenth to sixteenth century AD. Not only is the preservation of this site significant in the context of equatorial regions, but also its importance is further enhanced by the fact that it is perhaps the best preserved southeast Asian palace centre dating to the 'early modern period'.

The above discussion has considered generic approaches to the understanding of wetland contexts and presented an overview of the main archaeological methods for investigating human-landscape interactions. An additional, and extremely important, dimension to the investigation of the past has been the increasing role of palaeoenvironmental investigation and reconstruction. The papers by Davies, Griffiths, and Reed all provide detailed considerations of the methods employed in palaeoecological reconstructions using molluscan fauna, non-marine ostracods, and diatoms, and highlight the fact that these potentially powerful interpretative tools remain relatively underutilised by archaeological researchers despite their considerable value to reconstruction of the past.

Bunting and Schofield, and Blackford *et al.*, use more 'traditional' approaches to reconstructing the past via palynological based studies. The former study investigates topographically confined wetlands to produce detailed vegetational histories of the sites being considered and highlight limitations inherent in interpreting pollen assemblages and hydroseral changes over time. Blackford *et al.* investigate the palaeoecology of the peatlands of the North York Moors, diametrically opposite to the constrained sites studied by Bunting and Schofield, the Moors are mainly upland blanket peats that have been variously impacted upon in the prehistoric and historic past. Using microfossils and high-resolution studies to enhance the palynological record, Blackford *et al.* attempt to understand/identify human impacts on the uplands over time. The evidence is enhanced by the use of multi-proxy data, but limitations still exist in the availability of analogous sites with which to 'ground-proof' the evidence.

The final section of the volume presents a short series of papers that integrate various methods to study human-landscape interactions. In the second of his two papers George Nicholas considers why wetlands attract humans, and how different wetlands influence the exploitation patterns associated with them. Clearly, wetlands are very rich ecological niches and can be more productive than other landscape contexts. Understanding human-landscape use and interaction within wetlands is an important dimension of the archaeological record, and one that cannot be achieved by archaeological studies alone. Hope *et al.* address this issue to some degree in their study of the Bobundara swamp in New South Wales, Australia. A detailed consideration of vegetation change is integrated with archaeological and ethnographic data to provide an holistic overview of landscape change over the past 7500 years. This investigation is an important addition to the database of human-landscape interactions currently being developed throughout the world.

The final paper presented in this volume (Marchant) links relatively recent vegetation change to cultural change in the Rukiga Highlands of Uganda from *c.* 2500 yrs BP to the present. Various impacts and culture shifts are discussed throughout this paper and possible reasons for complex feedbacks from human impact on the environment that are still occurring up to the present are outlined. It is fitting that the volume reaches the present, in palaeoecological terms, and considers varying ecological responses to human-landscape interactions through the use of the palaeoenvironmental record from the highlands of Uganda. This final location is perhaps not one that immediately springs to mind when we consider wetlands, but it is one that highlights even further the rich diversity of the world's wetlands from the perspective of palaeoenvironmental reconstruction. It also emphasises the importance of palaeoenvironmental techniques when understanding very recent culture-historic change, as opposed to the 'traditional' emphasis on long-term change in the earlier to mid-Holocene archaeological record, at the global level of analysis.

Clearly then, the corpus of evidence from investigations of wetland archaeological sites and environments increasingly reinforces the importance of research in such contexts (*e.g.* Bernick 1998, Coles and Olivier 2001, Purdy 2001). That synergy exists between the aims of archaeologists and those of others working to preserve and/or conserve wetlands is well established, and now appears to be integral to future strategies at all levels, from the local to the global.

In light of the above observations and the papers presented the editors hope that this volume contributes further to raising awareness of the importance of global wetlands. In particular, the scope of the volume emphasises the fact that wetlands function as contexts that have the potential to preserve material of cultural and historic significance, this material can inform us about human socio-economic developments and human-landscape interaction over time, particularly their interaction with wetlands, and directly inform us about modern ecosystem response to human interference and modification.

REFERENCES

AFRC Institute of Food Research 1988. *A Basic Study of Landfill Microbiology and Biochemistry*. DTI Report: ETSUB1159. Northumberland: Department of Trade and Industry.

Barlaz, M. A. 1997. Microbial studies of landfills and anaerobic refuse decomposition, in: C. J. Hurst (ed.) *Manual of Environmental Microbiology*, 541–7. Washington DC: American Society for Microbiology.

Barlaz, M. A., D. M. Schaefer, & R. K. Ham 1989. Inhibition of methane formation from municipal refuse in laboratory scale lysimeters. *Applied Biochemistry and Biotechnology* 20, 181–211.

Bernick, K. (ed.) 1998. *Hidden Dimensions: the cultural significance of wetland archaeology*. Vancouver: University of British Columbia Press.

Bonsall, C., R. Lennon, K. McSweeney, C. Stewart, D. Harkness, V. Boroneant, L. Bartosiewicz, R. Payton & J. Chapman 1997. Mesolithic and Early Neolithic in the Iron Gates: a palaeodietary perspective. *Journal of European Archaeology* 5, 50–92.

Brunning, R., D. Hogan, J. Jones, E. Maltby, M. Robinson & V. Straker 2000. Saving the Sweet Track: the *in situ* preservation of a Neolithic wooden trackway, Somerset, UK. *Conservation and Management of Archaeological Sites* 4, 3–20.

Chapman, H. P. & J. L. Cheetham 2002. Monitoring and modelling saturation as a proxy indicator for in situ preservation in wetlands: a GIS-based approach. *Journal of Archaeological Science* 29, 277–89.

Cheetham, J. L. 2004. An assessment of the Potential for in situ Preservation of Buried Organic Archaeological Remains at Sutton Common, South Yorkshire. Unpublished PhD thesis, University of Hull.

Coles, B. 1993. Wetland Archaeology: A wealth of evidence, in M. Williams (ed.) *Wetlands: A Threatened Landscape*, 145–80. Oxford: Blackwell.

Coles, B. 2004. Steps towards the heritage management of wetlands in Europe. *Journal of Wetland Archaeology* 4, 183–98.

Coles, B. J. & A. Olivier (eds) 2001. *The Heritage Management of Wetlands in Europe*. Belgium: Europae Archaeologiae Consilium and WARP.

Coles, J. M. 2001. Wetlands, archaeology and conservation at AD 2001, in: B. Coles & A. Olivier (eds) *The Heritage Management of Wetlands in Europe*, 171–84. Belgium: Europae Archaeologiae Consilium and WARP.

Corfield, M. 1996. Preventive conservation for archaeological sites, in A. Roy & P. Smith (eds) *Archaeological conservation and its consequences*, 32–37. London: International Institute for Conservation of Historic and Artistic Works.

Crisman, T. L., U. A. M. Crisman & J. Prenger 2001. Wetlands and archaeology: the role of ecosystem structure and function, in: B. Purdy (ed) *Enduring Records: the environmental and cultural heritage of wetlands*, 254–61. Oxford: Oxbow Books.

Eriksson, G., L. Lõugas & I. Zagorska 2003. Stone Age hunter-fisher-gatherers at Zvejnieki, northern Latvia: radiocarbon, stable isotope and archaeozoology data. *Before Farming* 1, 1–25.

Fischer, A., H. Schlichtherle & P. Pétrequin 2004. Steps towards the heritage management of wetlands in Europe: response and reflection. *Journal of Wetland Archaeology* 4, 199–206.

Gumbley, W., D. Johns & G. Law 2005. Management of wetland archaeological sites in New Zealand. *Science for Conservation* 245.

Hogan, D. V., P. Simpson, A. M. Jones, & E. Maltby 2001. Development of a protocol for the reburial of organic archaeological remains, in: P. Hoffmann, J. M. Spriggs, T. Grant, C. Cook & A. Recht (eds) *Proceedings of the 8th ICOM Group on Wet Organic Archaeological Materials Conference*, 187–213. Bremerhaven: Druckerei Ditzen GmbH und Co. KG.

Lillie, M. C. & M. P. Richards 2000. Stable isotope analysis and dental evidence of diet at the Mesolithic-Neolithic transition in Ukraine. *Journal of Archaeological Science* 27, 965–72.

Lillie, M.C., M.P. Richards & K. Jacobs 2003. Stable isotope analysis of twenty-one individuals from the Epipalaeolithic cemetery of Vasilyevka III, Dnieper Rapids region, Ukraine. *Journal of Archaeological Science* 30, 743–52.

Mitsch, W. J. & J. G. Gosslink 1993. *Wetlands*, 2nd edition. New York: Van Nostrand Reinhold.

O'Connell, T. C., M. A. Levine & R. E. M. Hedges 2000. The importance of fish in the diet of central eurasian peoples from the Mesolithic to the early Iron Age, in conference volume: *Late Prehistoric Exploitation of the Eurasian Steppe: Volume II*, 303–327. Symposium held at the McDonald Institute for Archaeological Research, Cambridge 12–16 January 2000.

Olivier, A. 2004. Great Expectations: the English Heritage approach to the management of the historic environment in England's wetlands. *Journal of Wetland Archaeology* 4, 155–68.

Powell, K. L., S. Pedley, G. Daniel & M. Corfield 2001. Ultrastructural observations of microbial succession and decay of wood buried at a Bronze Age archaeological site. *International Biodeterioration and Biodegradation* 47, 165–73.

Purdy, B. (ed.) 2001. *Enduring Records: the environmental and cultural heritage of wetlands*. Oxford: Oxbow Books.

Ramsar Convention on Wetlands (Iran, 1971) [http://www.ramsar.org].

Richards, J. F. 1993. Agricultural Impacts in Tropical Wetlands: Rice paddies for mangroves in South and Southeast Asia, in M. Williams (ed) *Wetlands: A Threatened Landscape*, 217–33. Oxford: Blackwell.

Smith, R. J. 2005. The Preservation and Degradation of Wood in Wetland Archaeological and Landfill Sites. Unpublished PhD thesis, University of Hull.

Williams, M. 1993. Agricultural Impacts in Temperate Wetlands, in M. Williams (ed) *Wetlands: A Threatened Landscape*, 181–216. Oxford: Blackwell.

Zhilin M. G. 1999. New Mesolithic peat sites on the Upper Volga, in. S. Kozlowski, J. Gurba & L. L. Zaliznyak (eds) *Tanged Points Cultures in Europe*, 295–310. Lublin: Wydawnictwo UMCS.